Vector Bundles and Representation Theory

CONTEMPORARY MATHEMATICS

322

Vector Bundles and Representation Theory

Conference on Hilbert Schemes, Vector Bundles and Their Interplay with Representation Theory
April 5–7, 2002
University of Missouri, Columbia

S. Dale Cutkosky
Dan Edidin
Zhenbo Qin
Qi Zhang
Editors

This volume contains the proceedings of a conference on Hilbert Schemes, Vector Bundles and Their Interplay with Representation Theory, held at the University of Missouri, Columbia, on April 5–7, 2002.

2000 *Mathematics Subject Classification.* Primary 14C05, 14D20, 14F05, 14J28, 17B10, 20C05, 32L05, 57R99.

Library of Congress Cataloging-in-Publication Data

Conference on Hilbert Schemes, Vector Bundles and Their Interplay with Representation Theory : (2002 : University of Missouri–Columbia)

Vector bundles and representation theory : Conference on Hilbert Schemes, Vector Bundles and Their Interplay with Representation Theory, April 5–7, 2002, University of Missouri, Columbia / S. Dale Cutkosky . . . [et al.], editors.

p. cm. — (Contemporary mathematics, ISSN 0271-4132 ; 322)

Includes bibliographical references.

ISBN 0-8218-3264-6 (alk. paper)

1. Vector bundles—Congresses. 2. Hilbert schemes—Congresses. 3. Representations of algebras—Congresses. I. Cutkosky, Steven Dale. II. Title. III. Contemporary mathematics (American Mathematical Society) ; v. 322.

QA612.63.C66 2002
514′.224—dc21 2003045317

∞ The paper used in this book is acid-free and falls within the guidelines
established to ensure permanence and durability.
Visit the AMS home page at http://www.ams.org/

10 9 8 7 6 5 4 3 2 1   08 07 06 05 04 03

Contents

Preface

The study of Hilbert schemes and vector bundles is a fundamental problem in algebraic geometry. Their connections with physics and representation theory were pioneered in the work of Penrose and Atiyah during the 1970s. They played central roles in the Donaldson theory and the Seiberg-Witten theory from 1983 to 1995. From 1995, physicists working in string theory have speculated many surprising but deep results concerning Hilbert schemes, vector bundles and their interplay with representation theory. For instance, a mathematical version of the S-duality conjecture formulated by C. Vafa and E. Witten revealed a beautiful connection between stable vector bundles on algebraic surfaces and representations of certain infinite-dimensional Lie algebras. In the case of rank one, this leads to an elegant relation between the Hilbert schemes of points on algebraic surfaces and the representations of the infinite-dimensional Heisenberg algebras. Another example of recent interplay between vector bundles and representation theory motivated by physics is the relation among principal bundles over elliptic Calabi-Yau manifolds, representation of compact Lie groups and the physics F-theory. These recent physics discoveries have been leading to intensive studies of and rapid advances in the theory of Hilbert schemes, vector bundles and representation theory.

It is under such a background that in April 2002, the Department of Mathematics at the University of Missouri hosted a conference in Columbia, Missouri on Hilbert Schemes, Vector Bundles and Their Interplay with Representation Theory. The meeting brought together both senior and young researchers, including a number of graduate students, in algebraic geometry and representation theory.

Main speakers at the conference were W.-P. Li (HKUST), E. Izadi (University of Georgia), D. Morrison (Duke), K. Oguiso (University of Tokyo), J. Li (Stanford), W. Wang (MSRI & University of Virginia), R. Friedman (Columbia University), and H. Nakajima (Kyoto University). There were nineteen short communications presented by E. Gasparim (New Mexico State University), C.-H. Liu (Harvard), T. Nevins (MSRI), A. Iarrobino (Northeastern), G. J. Pearlstein (UC-Irvine), B. Kotzev (University of Missouri), Y. Kimiko (Kyoto University), D. Arcara (University of Georgia), A. Vitter (Tulane), W. Li (Oklahoma State University), C. Madonna (Univ. Roma 2), A. Caldararu (U. Mass), A. Mavlyutov (Indiana University), T. Luo (UT-Arlington), Y. Kachi (University of Tennessee), E. Markman (U. Mass), B. Purnaprajna (University of Kansas), P. Rao (University of Missouri-St. Louis), and X. Wu (University of South Carolina).

The meeting provided a forum f

opportunity to broaden their mathematical horizons, and promoted the research status of under-represented groups.

This proceedings volume contains 13 papers presented at the conference, and papers inspired by it. All of these papers were carefully refereed, and are in final form. They represent the latest developments in Hilbert schemes, vector bundles and their interplay with representation theory.

The conference was supported by the National Science Foundation under Grant No. 0118343, by the National Security Agency under Grant No. MSPF-02IC-002, and by the Math Department of the University of Missouri-Columbia. We thank them for their generous financial support. We also thank B. Frazier, D. W. Hodge, J. White, and K. Wilson for their help at various stages of the conference, and C. M. Thivierge for her assistance in preparing this proceedings volume.

S. Dale Cutkosky
Dan Edidin
Zhenbo Qin (Managing Editor)
Qi Zhang

Contemporary Mathematics
Volume **322**, 2003

Minuscule Representations, Invariant Polynomials, and Spectral Covers

Robert Friedman and John W. Morgan

Introduction

Let G be a simple and simply connected complex linear algebraic group, with Lie algebra $\mathfrak{g}$. Let $\rho\colon G \to \operatorname{Aut} V$ be an irreducible finite-dimensional representation of G, and let $\rho_*\colon \mathfrak{g} \to \operatorname{End} V$ be the induced representation of $\mathfrak{g}$. A goal of this paper is to study ρ_* and ρ, and in particular to give normal forms for the action of $\rho_*(X)$ and $\rho(g)$ for regular elements X of $\mathfrak{g}$ or regular elements $g \in G$. Of course, in the case of ρ_*, when X is semisimple, the action of $\rho_*(X)$ on V can be diagonalized, and its eigenvalues are given by evaluating the weights of ρ with respect to a Cartan subalgebra $\mathfrak{h}$ of $\mathfrak{g}$ on an element in $\mathfrak{h}$ conjugate to X. If on the other hand X is a principal nilpotent element of $\mathfrak{g}$, then X can be completed to an $\mathfrak{sl}_2$-triple (X, h_0, X_-), where h_0 is regular and semisimple. In this case, if $\mathfrak{h}$ is the Cartan subalgebra containing h_0, then the eigenvalues for $\rho_*(h_0)$ together with their multiplicities, which are given by evaluating the weights of ρ with respect to $\mathfrak{h}$ on h_0, completely determine V as an $\mathfrak{sl}_2$-module and hence determine the action of $\rho_*(X)$ on V. A minor modification of these ideas will then describe the action of $\rho_*(X)$ for every regular element X.

In this paper, we attempt to give a different algebraic model for the action of $\rho_*(X)$, where X is regular, and to glue these different models together over the set of all regular elements. Related methods also handle the case of $\rho(g)$, where g is a regular element of G. While we are only successful in case ρ is minuscule, and partially successful in case ρ is quasiminuscule, it seems likely that the techniques of this paper can be extended to give information about an arbitrary ρ. We believe that the techniques and results of this paper are also of independent interest. For example, they have been applied elsewhere to study sections of the adjoint quotient morphism [**7**].

2000 *Mathematics Subject Classification.* Primary: 14F05, 17B10; Secondary: 14H52, 20G05.

Key words and phrases. Minuscule representation, spectral cover, Weierstrass cubic, regular element, Kostant section.

The first author was partially supported by NSF grant DMS-99-70437. [illegible]
was partially supported by NSF [illegible]

To explain our results in more detail in the case of ρ_*, we begin by recalling some results of Kostant on the adjoint quotient of $\mathfrak{g}$. Let $\mathfrak{h}$ be the Lie algebra of a Cartan subgroup H of G and let W be the Weyl group of the pair (G, H). Kostant has shown [**11**] that the GIT quotient of $\mathfrak{g}$ by the adjoint action of G is isomorphic to $\mathfrak{h}/W$. Moreover, he has constructed an explicit cross-section $\overline{\Sigma}$ of the natural morphism $\mathfrak{g} \to \mathfrak{h}/W$, such that the image of $\overline{\Sigma}$ is contained in the dense open subset $\mathfrak{g}_{\mathrm{reg}}$ of regular elements of $\mathfrak{g}$.

Given the Cartan subgroup H, there are the corresponding weights of ρ. Fixing a choice of a set of positive roots for the pair (G, H), we shall let μ be the *lowest* weight of ρ. Let W_0 be the stabilizer of μ in W. Recall that the representation ρ is *minuscule* if all of its weights are conjugate under W, so that the set of all weights of ρ is identified with W/W_0. In this case, every weight has multiplicity one. The representation ρ is *quasiminuscule* if all of its nonzero weights are conjugate under W. In this case, every nonzero weight has multiplicity one.

Let $S = \mathrm{Sym}^* \mathfrak{h}^*$ be the affine coordinate ring of $\mathfrak{h}$, so that S is a polynomial ring. We have the invariant subrings S^W and S^{W_0}. The inclusions $S^W \to S^{W_0} \to S$ correspond to the finite morphisms $\mathfrak{h} \to \mathfrak{h}/W_0 \to \mathfrak{h}/W$. By Chevalley's theorem [**2**, p. 107, Thm. 3], S^W and S^{W_0} are polynomial algebras, i.e. $\mathfrak{h}/W$ and $\mathfrak{h}/W_0$ are isomorphic to affine space. Hence S^{W_0} is flat over S^W and is therefore a free S^W-module of rank equal to $\#(W/W_0)$. Note that μ is naturally an element of S^{W_0} and can be completed to a polynomial basis of S^{W_0}. The morphism $\overline{\Sigma}\colon \mathfrak{h}/W \to \mathfrak{g}$ induces a morphism $\rho_*\overline{\Sigma}\colon \mathfrak{h}/W = \mathrm{Spec}\, S^W \to \mathrm{End}\, V$, and is thus identified with an element of

$$(\mathrm{End}\, V) \otimes_{\mathbb{C}} S^W \cong \mathrm{Hom}_{S^W}(V \otimes_{\mathbb{C}} S^W, V \otimes_{\mathbb{C}} S^W),$$

which we also denote by $\rho_*\overline{\Sigma}$. With this said, we have the following:

THEOREM 1. *Let $\rho\colon G \to GL(V)$ be an irreducible finite-dimensional representation with lowest weight μ. Then there is a nonzero linear map $f\colon V \to S^{W_0}$, unique up to a nonzero scalar, with the following two properties:*

(i) *Let $\hat{f}\colon V \otimes_{\mathbb{C}} S^W \to S^{W_0}$ be the homomorphism defined by f via extension of scalars. Then the following diagram commutes:*

$$\begin{array}{ccc} V \otimes_{\mathbb{C}} S^W & \xrightarrow{\hat{f}} & S^{W_0} \\ {\scriptstyle \rho_*\overline{\Sigma}}\downarrow & & \downarrow{\scriptstyle \mu\cdot} \\ V \otimes_{\mathbb{C}} S^W & \xrightarrow{\hat{f}} & S^{W_0}. \end{array}$$

(ii) *If $g\colon V \to S^{W_0}$ is any other linear map such that the extension of g to $\hat{g}\colon V \otimes_{\mathbb{C}} S^W \to S^{W_0}$ makes the above diagram commute, then $g = sf$ for some $s \in S^{W_0}$.*

If ρ is minuscule, then $\hat{f}$ is an isomorphism. If ρ is quasiminuscule, then $\hat{f}$ is surjective, and $\mathrm{Ker}\, \hat{f} = \mathrm{Ker}\, \rho_*\overline{\Sigma}$.

In fact, we give an explicit description of the map f in Section 3.

Using very different methods, and based on his results in [**9**], Victor Ginzburg has sent us the sketch of a proof that $\hat{f}$ is surjective for every irreducible representation ρ (cf. also [**10**]).

There are two different ways we can interpret Theorem 1. In case ρ is minuscule, it describes, up to conjugation, the action of every regular element of $\mathfrak{g}$ on V, or equivalently the action of $\rho_*\overline{\Sigma}(x)$ for every $x \in \mathfrak{h}/W$, in terms of the algebra structures on S^W and S^{W_0}:

THEOREM 2. *Suppose that ρ is minuscule. Let $x \in \mathfrak{h}/W$ and let $\mathfrak{m}_x \subseteq S^W$ be the maximal ideal of x, so that the scheme-theoretic fiber over x in $\mathfrak{h}/W_0$ has coordinate ring equal to $S^{W_0}/\mathfrak{m}_x \cdot S^{W_0}$. Then the map $\hat{f}$ induces an isomorphism $\overline{f}_x \colon V \to S^{W_0}/\mathfrak{m}_x \cdot S^{W_0}$. Under this isomorphism, multiplication by μ on the right-hand side becomes the action of $\rho_*(\overline{\Sigma}(x))$. In particular, for the ideal $\mathfrak{m}_0 \cdot S^{W_0}$ generated by the homogeneous W-invariant functions of positive degree, there is an isomorphism $\overline{f}_0 \colon V \to S^{W_0}/\mathfrak{m}_0 \cdot S^{W_0}$ such that multiplication by $\mu \in S^{W_0}$ on $S^{W_0}/\mathfrak{m}_0 \cdot S^{W_0}$ corresponds to the action $\rho_*(X)$ of a principal nilpotent element X on V.*

Of course, the ring S^{W_0} is just a polynomial ring and μ can be chosen to be a coordinate function, so that the action of μ on S^{W_0} is easy to understand. However, the finite dimensional $\mathbb{C}$-algebras $S^{W_0}/\mathfrak{m}_x \cdot S^{W_0}$ and the action of μ on them are much more complicated. Thus for example if $x = 0$ the theorem tells us that the nilpotent element μ of $S^{W_0}/\mathfrak{m}_0 \cdot S^{W_0}$ has Jordan blocks which are the same as those given by the action of a principal nilpotent element X on V.

There is a partial analogue of Theorem 2 in the quasiminuscule, non-minuscule case. In this case, the action of $\rho_*\overline{\Sigma}$ on $V \otimes_{\mathbb{C}} S^W$ corresponds to the action of the matrix

$$\begin{pmatrix} 0 & * \\ 0 & \mu \end{pmatrix},$$

where we view μ as acting in block form on the free S^W-module S^{W_0}. Here, for any given $x \in \mathfrak{h}/W$, the off-diagonal term $*$ can in principle be determined. We discuss this off-diagonal term in Section 5 for the adjoint representation of a simply laced Lie algebra.

The second interpretation of the surjection $\hat{f} \colon V \otimes_{\mathbb{C}} S^W \to S^{W_0}$ in the minuscule or quasiminuscule case is that it describes relations between invariant polynomials for W and W_0. In particular, an explicit description of f leads to a set of generators for S^{W_0} as an S^W-module. We give examples of such relations in the fourth section.

One basic ingredient in the proof of Theorem 1 is a result concerning the Kostant section $\overline{\Sigma}$ which (in a slightly different form) was proved by Kostant in [**12**]. In general, the image of $\overline{\Sigma}$ is complicated. However, if we make the base change $\mathfrak{h} \to \mathfrak{h}/W$, then, up to the adjoint action of a morphism $\mathfrak{h} \to G$, the structure of $\overline{\Sigma}$ simplifies considerably:

THEOREM 3. *Let X be a principal nilpotent element of $\mathfrak{g}$ such that $\mathfrak{h}$ is contained in the unique Borel subalgebra containing X, and let Σ be the composition of $\overline{\Sigma}$ with the natural projection $\mathfrak{h} \to \mathfrak{h}/W$. Then there exists a morphism $\lambda \colon \mathfrak{h} \to G$ such that, for all $h \in \mathfrak{h}$,*

$$\mathrm{Ad}(\lambda(h))(\Sigma(h)) = h + X.$$

In other words, after making the base change $\mathfrak{h} \to \mathfrak{h}/W$, we can conjugate the section Σ via the morphism λ into the much simpler morphism from $\mathfrak{h}$ to $\mathfrak{g}$ defined by $h \mapsto h + X$, where X is a fixed principal nilpotent in a Borel subalgebra normalized by $\mathfrak{h}$. Using the explicit description of the Kostant section, we are able to give a very explicit construction following result concerning $\overline{\Phi}$, which is analogous to Theorem 3 but somewhat weaker:

THEOREM 4. *Let B be a Borel subgroup containing H with unipotent radical U, and let Φ be the composition of $\overline{\Phi}$ with the natural projection $H \to H/W$. Then there exists a morphism $\phi\colon H \to G$ and a morphism $u\colon H \to U$ such that, for all $h \in H$,*

$$\phi(h)\Phi(h)\phi(h)^{-1} = hu(h).$$

In contrast to the proof of Theorem 3, the argument does not use any special properties of Steinberg's construction and would apply to any section of the adjoint quotient morphism. In particular, the morphism u is not given explicitly. It is natural to conjecture that, for an appropriate choice of ϕ in the theorem, we can take u to be a constant, in other words that the pullback of the Steinberg section to H is conjugate via a morphism from H to G to the morphism $h \mapsto hu$ for a certain principal unipotent element in a Borel subgroup normalized by H. Using Theorem 4, the proofs of Theorem 1 and hence of Theorem 2 go over to the case of the Steinberg section, and to the action of $\rho(g)$ on V, for a regular element g in G and a minuscule or quasiminuscule representation ρ. In particular, if $\mathcal{S}$ is the affine coordinate ring of H, then there is a morphism $\hat{g}\colon V \otimes_{\mathbb{C}} \mathcal{S}^W \to \mathcal{S}^{W_0}$ which intertwines the action of $\rho(\overline{\Phi})$ and multiplication by μ, and $\hat{g}$ is an isomorphism in case ρ is minuscule and surjective in case ρ is quasiminuscule. It is natural to conjecture that $\hat{g}$ is always surjective; this does not seem to follow directly from the result of Ginzburg in the Lie algebra case.

Our strategy in the proof of Theorem 1 is to relate the problem to a question concerning holomorphic G-bundles on a cuspidal curve. More generally, let E be a Weierstrass cubic curve, i.e. a reduced irreducible curve of arithmetic genus one. Then E is either smooth, isomorphic to a nodal cubic curve in $\mathbb{P}^2$, or isomorphic to a cuspidal cubic curve in $\mathbb{P}^2$. Let E_{reg} be the Zariski open subset of smooth points. Fixing a base point $p_0 \in E_{\text{reg}}$ identifies E_{reg} with $\operatorname{Pic}^0 E$ and endows E_{reg} with the structure of a one-dimensional connected, commutative algebraic group, which is an elliptic curve (E smooth), or is isomorphic to the multiplicative group $\mathbb{C}^*$ (E nodal) or to the additive group $\mathbb{C}$ (E cuspidal). We assume now that E is singular and denote its normalization by $\widetilde{E} \cong \mathbb{P}^1$. A G-bundle on E which pulls back to the trivial G-bundle $\widetilde{E} \times G$ on $\widetilde{E}$ can be identified (after fixing a trivialization of the pullback) with an element $g \in G$, if E is nodal, or an element $X \in \mathfrak{g}$, if E is cuspidal. There is a related result for families. In this way, questions about G or $\mathfrak{g}$ can be translated into questions about G-bundles ξ on E. In case E is cuspidal, we can use the Kostant section to construct a universal G-bundle Ξ over $(\mathfrak{h}/W) \times E$ which pulls back to the trivial bundle over $(\mathfrak{h}/W) \times \widetilde{E}$. A similar construction works in the nodal case using the Steinberg section. Although the construction of Ξ seems to depend on a particular choice of a section of the adjoint quotient morphism, we show in [**7**], using many of the methods of this paper, that all such sections are conjugate in an appropriate sense. If E is smooth, modulo some minor technical difficulties, we can use the parabolic construction of [**5**]. Although the proof of Theorem 1 only needs the case where E is singular, minor modifications also handle the case of a smooth E. This case is interesting for the study of principal G-bundles on a smooth elliptic curve E as well as the vector bundles associated to them by a representation ρ of G. In fact, we were led to the results of this paper based on our previous work on G-bundles over smooth elliptic curves E.

To prove Theorem 1, we shall use the theory of spectral covers associated to semistable vector bundles over Weierstrass cubics. Given an irreducible representation $\rho\colon G \to \operatorname{Aut} V$ and a G-bundle ξ on E, we can form the associated vector bundle $\xi \times_G V$. If ξ pulls back to the trivial bundle on $\widetilde{E}$, then $\xi \times_G V$ is a semistable vector bundle on E of degree zero. It is well-known that, for every Weierstrass cubic E, there is an equivalence between the abelian category of semistable torsion free sheaves on E of degree zero and the abelian category of torsion sheaves on E [**13, 8**]. More precisely, given a semistable torsion free sheaf V on E, there is a zero-dimensional subscheme T_V in E, the *spectral cover*, which is an effective Cartier divisor if V is a vector bundle, and a sheaf $Q(V)$ supported on T_V which can be used to recover V from $Q(V)$ and the Poincaré sheaf on $E \times E$. The bundle V is *regular* if $Q(V)$ is a line bundle over T_V. More generally, given a family of semistable vector bundles $\mathcal{V}$ over $B \times E$, flat over B, the construction yields a relative effective Cartier divisor $T_{\mathcal{V}}$ in $B \times E$ and a sheaf $Q(\mathcal{V})$ over $T_{\mathcal{V}}$, from which we can recover $\mathcal{V}$. The bundle $\mathcal{V}$ restricts to a bundle which is regular and semistable on every fiber if and only if $Q(\mathcal{V})$ is a line bundle on $T_{\mathcal{V}}$. However, we shall need to consider more general situations, where $Q(\mathcal{V})$ is the pushforward of a line bundle on the normalization of $T_{\mathcal{V}}$.

In case E is cuspidal, we use the Kostant section $\overline{\Sigma}$ to construct a universal G-bundle Ξ over $(\mathfrak{h}/W) \times E$. Then we let $\mathcal{V}$ be the vector bundle over $(\mathfrak{h}/W) \times E$ associated to Ξ by the representation $\rho\colon G \to \operatorname{Aut} V$. Even though Ξ restricts to a regular G-bundle ξ_x on every fiber $\{x\} \times E$, the vector bundle $\mathcal{V}_x = \xi_x \times_G V$ is not in general regular. The spectral cover $T_{\mathcal{V}}$ and the sheaf $Q(\mathcal{V})$ are also quite complicated. For example, the irreducible components of $T_{\mathcal{V}}$ are indexed by the Weyl orbits of the weights of ρ, and each component has multiplicity equal to the multiplicity of the orbit. However, the Weyl orbit of the highest weight (or equivalently of the lowest weight) defines a reduced irreducible component T' of $T_{\mathcal{V}}$. Of course, ρ is minuscule if and only if $T' = T_{\mathcal{V}}$. We explicitly identify the normalization of T' with a finite normal cover of $\mathfrak{h}/W$. The surjection $Q(\mathcal{V}) \to Q(\mathcal{V})|T'$ corresponds to a surjection $\mathcal{V} \to \mathcal{V}'$, where $\mathcal{V}'$ is some coherent sheaf on $(\mathfrak{h}/W) \times E$. If ρ is minuscule, then $\mathcal{V} = \mathcal{V}'$. In the quasiminuscule case, we prove that $\mathcal{V}'$ is again a vector bundle. It is natural to conjecture that this is true for an arbitrary irreducible ρ. By analyzing how $\mathcal{V}'$ fails to be regular in codimension one, we are able to establish Theorem 1 but without explicitly determining $\hat{f}$. To determine $\hat{f}$ requires a more detailed analysis of the Kostant section which is given in Section 3. Similar but less explicit constructions using the Steinberg section work in case E is nodal.

The organization of this paper is as follows. In Section 1, we discuss bundles over singular curves and recall the theory of spectral covers. We give a sufficient criterion for a vector bundle to be given by pushing forward a line bundle on the normalization of the spectral cover. In Section 2, we study the spectral covers arising from the vector bundles associated to a universal G-bundle via an irreducible representation ρ, and in particular identify the normalization of the irreducible component T' associated to the Weyl orbit of the highest weight. These results are then applied to the case of a minuscule or quasiminuscule representation. In Section 3, we prove Theorem 3. We can then easily prove Theorem 1 by [illegible] the vector bundle results given in [illegible]

5, we study the quasiminuscule, non-minuscule case in more detail in the simply laced case, where the corresponding representation is the adjoint representation. It seems likely that similar methods will also describe the non-simply laced case. We conclude with some general conjectures concerning arbitrary representations.

It is a pleasure to thank Victor Ginzburg for explaining to us his work [**9, 10**] and for calling our attention to the reference [**12**].

1. Spectral covers of semistable vector bundles over Weierstrass cubics

Throughout this paper, E denotes a Weierstrass cubic, i.e. a reduced irreducible curve of arithmetic genus one, E_{reg} is the set of smooth points of E, and $p_0 \in E_{\mathrm{reg}}$ is a fixed point used as the origin of the group law on E_{reg}. If E is singular, then its unique singular point is either a node or a cusp, and we shall refer to E as either *nodal* or *cuspidal*. In this case, we denote the normalization by $\iota\colon \widetilde{E} \to E$. The point p_0 determines an isomorphism $E_{\mathrm{reg}} \to \mathrm{Pic}^0 E$. If E is nodal, $\mathrm{Pic}^0 E$ is canonically $\mathbb{C}^*$, and if E is cuspidal, the choice of a local coordinate z at $\iota^{-1}(E_{\mathrm{sing}})$ determines an isomorphism of $\mathrm{Pic}^0 E$ with $\mathbb{C}$. Let $\mathcal{P}$ be the Poincaré bundle over $E \times E$. (For the meaning of this sheaf and various properties of it in case E is singular, we refer to the proof of Lemma (0.3) in [**8**].) We shall use the symmetric form of $\mathcal{P}$: $\mathcal{P} = \mathcal{O}_{E\times E}(\Delta) \otimes \pi_1^*\mathcal{O}_E(-p_0) \otimes \pi_2^*\mathcal{O}_E(-p_0) \otimes H^0(E; K_E)$, where $\Delta \subseteq E \times E$ is the diagonal. In case E is smooth, the first factor is the parameter space $E = \mathrm{Pic}^0 E$, and the second factor is the original curve. Of course, the symmetry between the factors means that we can also view the first factor as the curve and the second as $\mathrm{Pic}^0 E$. In the singular case, for the purposes of this paper, it will suffice to consider the line bundle $\mathcal{P}|E_{\mathrm{reg}} \times E$. Our goal in this section is to describe basic results of [**13, 8, 16**] which describe semistable bundles of degree zero on E in terms of torsion sheaves on E. In [**8**], we mainly considered the case of bundles whose restriction to every fiber was regular. Here we shall need a slight generalization of those results. There are also analogous results relating semistable torsion free sheaves of degree zero on E to torsion sheaves on E, but we shall not need them.

1.1. Bundles over singular curves.

Assume in this section that E is singular and, in case E is cuspidal, that we have fixed once and for all a local coordinate z on $\widetilde{E}$ centered at the inverse image of the singular point, and in particular have determined an isomorphism $\mathrm{Pic}^0 E \cong \mathbb{C}$. We begin by describing vector bundles and G-bundles over E which become trivial when pulled back to $\widetilde{E}$. More generally, we shall consider parametrized versions. The arguments are standard, and left to the reader. (See [**6**] for the case of a single curve.)

THEOREM 1.1.1. *Suppose that E is cuspidal and that we have fixed the coordinate z on $\widetilde{E}$ near the inverse image of the singular point x. Let B be a scheme. Suppose that $\mathcal{V}$ is a vector bundle over $B\times E$ such that $(\mathrm{Id}\times\iota)^*\mathcal{V}$ restricts on every fiber to the trivial vector bundle of rank n. In this case, $\mathcal{W} = \pi_{1*}(\mathrm{Id}\times\iota)^*\mathcal{V}$ is a rank n vector bundle on B, and $(\mathrm{Id}\times\iota)^*\mathcal{V} \cong \pi_1^*\mathcal{W}$. Moreover, there is an equivalence of categories between the category of triples $(\mathcal{V}, \mathcal{W}, \Phi)$, where $\mathcal{V}$ is a rank n vector bundle on $B\times E$, $\mathcal{W}$ is a rank n vector bundle on B, and Φ is an isomorphism from $(\mathrm{Id}\times\iota)^*\mathcal{V}$ to $\pi_1^*\mathcal{W}$, and the category of pairs $(\mathcal{W}, \varphi)$, where $\mathcal{W}$ is a rank n vector bundle on B and φ is an element of* $\mathrm{End}(\mathcal{W})$. *Similar statements hold in the nodal case, where $\varphi \in \mathrm{Aut}(\mathcal{W})$.* □

REMARK 1.1.2. (1) For E cuspidal and $x = \iota^{-1}(E_{\text{sing}})$, the equivalence of categories is defined as follows: given $(\mathcal{W}, \varphi)$, $\mathcal{V}$ is the subsheaf of $\pi_1^*\mathcal{W}$ consisting of all local sections s such that, identically in b,

$$\frac{ds}{dz}(b, x) = \varphi(b)s(b, x).$$

If E is nodal, and $x, y \in \widetilde{E}$ are the two preimages of the singular point on E, then $\mathcal{V}$ is the subsheaf of $\pi_1^*\mathcal{W}$ consisting of all local sections s such that, for all b,

$$s(b, y) = \varphi(b)s(b, x).$$

In particular, it follows that if B is normal and $\mathcal{V}$ is a locally free sheaf as in the statement of the theorem, then $\mathcal{V}$ has the Hartogs property: if X is a subset of B of codimension at least 2 and $j\colon (B - X) \times E \to B \times E$ is the inclusion, then the natural map $\mathcal{V} \to j_*j^*\mathcal{V}$ is an isomorphism. In fact, a similar result is true for every locally free sheaf on $B \times E$, as long as B is normal.

(2) The construction is compatible with base change $B' \to B$.

(3) Suppose that $(\mathrm{Id} \times \iota)^*\mathcal{V}$ is trivialized. In this case, there is an equivalence of categories between the category of triples $(\mathcal{V}, V, \Phi)$, where $\mathcal{V}$ is a vector bundle on $B \times E$, V is a finite-dimensional vector space, and Φ is an isomorphism from $(\mathrm{Id} \times \iota)^*\mathcal{V}$ to $\mathcal{O}_{B\times\widetilde{E}} \otimes_{\mathbb{C}} V$, and the category of pairs (V, φ), where V is a finite-dimensional vector space and $\varphi\colon B \to \mathrm{End}\, V$ is a morphism. A morphism $B \to \mathrm{End}\, V$ is the same thing as a section of $\mathcal{O}_B \otimes_{\mathbb{C}} \mathrm{End}\, V$. In case $B = \mathrm{Spec}\, R$, this is in turn equivalent to an element of

$$\mathrm{End}\, V \otimes_{\mathbb{C}} R = \mathrm{End}_R(V \otimes_{\mathbb{C}} R, V \otimes_{\mathbb{C}} R).$$

After choosing a basis of V, this can further be identified with an $n \times n$ matrix with coefficients in R. Similar remarks hold in the nodal case. For example, in this case, φ is a morphism $B \to \mathrm{Aut}\, V$, and in case $B = \mathrm{Spec}\, R$, such a morphism is identified with an element of $\mathrm{Aut}_R(V \otimes_{\mathbb{C}} R)$.

(4) The construction is natural under direct image by a finite flat morphism. In other words, suppose that $\nu\colon B \to B'$ is a finite flat morphism. Thus $\nu_*\mathcal{O}_B$ is a locally free $\mathcal{O}_{B'}$-module. If $\mathcal{V}$ is a vector bundle on $B \times E$, then $(\nu \times \mathrm{Id})_*\mathcal{V}$ is a vector bundle on $B' \times E$. If in addition $(\mathrm{Id} \times \iota)^*\mathcal{V} \cong \pi_1^*\mathcal{W}$, then $(\mathrm{Id} \times \iota)^*(\nu \times \mathrm{Id})_*\mathcal{V} = \pi_1^*\nu_*\mathcal{W}$. In this case, the global section φ of $\mathit{End}(\mathcal{W})$ induces a global section of $\mathit{End}(\nu_*\mathcal{W})$, and this corresponds to the bundle $(\nu \times \mathrm{Id})_*\mathcal{V}$ on $B' \times E$.

In case B is a point and $\iota^*\mathcal{V}$ is trivialized, the rank n bundle $\mathcal{V}$ corresponds to an $n \times n$ matrix A (which is arbitrary, in the cuspidal case, and invertible, in the nodal case). For example, in the cuspidal case, if N is the nilpotent matrix defined by $N(e_i) = e_{i-1}$, $i \geq 1$, and $N(e_1) = 0$, where the e_i are the standard basis vectors, the corresponding vector bundle is I_n, in the notation of [**8**]. Here, for each $n > 0$, I_n is the unique indecomposable vector bundle over E of rank n up to isomorphism which has a filtration, all of whose successive quotients are $\mathcal{O}_E$. More generally, if $A = H + N$ is the Jordan decomposition of A, where H is semisimple, N is nilpotent, and $[H, N] = 0$, then the action of H decomposes V into a direct sum of eigenspaces $\bigoplus_i V_i$, with distinct eigenvalues

under N.

of degree zero, which we denote by λ_i. Then

$$\mathcal{V} \cong \bigoplus_i \left(\bigoplus_j I_{n_{ij}} \right) \otimes \lambda_i.$$

Of course, a similar statement holds for the nodal case.

An important example in this paper will be the Poincaré bundle $\mathcal{P}|E_{\mathrm{reg}} \times E$. It is easy to check the following:

PROPOSITION 1.1.3. *Suppose that E is cuspidal. The pullback of $\mathcal{P}|E_{\mathrm{reg}} \times E$ to $E_{\mathrm{reg}} \times \widetilde{E}$ is the trivial line bundle. Given a coordinate t on E_{reg} centered at p_0, under the correspondence of Theorem 1.1.1, the Poincaré bundle $\mathcal{P}|E_{\mathrm{reg}} \times E$ corresponds to the morphism $\mathbb{C} \to \operatorname{End}\mathbb{C}$ given by multiplication, i.e. corresponds to the function t.* □

A similar result handles the nodal case, where the coordinate t on $E_{\mathrm{reg}} \cong \mathbb{C}^*$ defines a isomorphism of algebraic groups from E_{reg} to $\operatorname{Aut}\mathbb{C}$, again by multiplication.

We shall also use Theorem 1.1.1 in the following situation. Let $\rho\colon G \to \operatorname{Aut} V$ be a representation of G and let $\rho_* : \mathfrak{g} \to \operatorname{End} V$ be the corresponding representation of the Lie algebra. Given $X \in \mathfrak{g}$, the element $\rho_*(X)$ defines a vector bundle over E. If $X = X_s + X_n$ is the Jordan decomposition of X, then $\rho_*(X) = \rho_*(X_s) + \rho_*(X_n)$ is the Jordan decomposition of $\rho_*(X)$. Let $\mathfrak{z}_{\mathfrak{g}}(X)$ be the Lie algebra centralizer of X in $\mathfrak{g}$. The element X is *regular* if and only if $\dim \mathfrak{z}_{\mathfrak{g}}(X)$ is equal to the rank of $\mathfrak{g}$ if and only if X_n is a principal nilpotent element in $\mathfrak{z}_{\mathfrak{g}}(X_s)$ Now suppose that we are given a morphism $f\colon B \to \mathfrak{g}$, for example the Kostant section $\mathfrak{h}/W \to \mathfrak{g}$ (which we shall describe later). Then $\rho_* \circ f\colon B \to \operatorname{End} V$ is a morphism, and hence defines a vector bundle $\mathcal{V}$ over $B \times E$ which becomes trivial over $B \times \widetilde{E}$. Of course, a similar picture holds in the nodal case.

1.2. Spectral covers.

DEFINITION 1.2.1. Let V be a semistable bundle of degree zero on E. Define $Q(V)$ as the cokernel of the natural global to local map $H^0(E; V \otimes \mathcal{O}_E(p_0)) \otimes_{\mathbb{C}} \mathcal{O}_E \to V \otimes_{\mathcal{O}_E} \mathcal{O}_E(p_0)$. Thus there is an exact sequence

$$0 \to H^0(E; V \otimes \mathcal{O}_E(p_0)) \otimes_{\mathbb{C}} \mathcal{O}_E \xrightarrow{A} V \otimes_{\mathcal{O}_E} \mathcal{O}_E(p_0) \to Q(V) \to 0.$$

We shall abbreviate $Q(V) = Q$ when V is clear from the context, and call Q the *Fourier-Mukai transform* of V.

We have the following from [**8**]:

LEMMA 1.2.2. *Suppose that V is semistable of degree zero and rank n. Then $h^0(E; V \otimes \mathcal{O}_E(p_0)) = n$, A is an injection from $\mathcal{O}_E^n$ to $V \otimes \mathcal{O}_E(p_0)$, and $Q(V)$ is a torsion sheaf of length n.* □

For example, if $L = \mathcal{O}_E(q - p_0)$ is a line bundle of degree zero, then $Q(L) = \mathcal{O}_E/\mathfrak{m}_q$.

Let T_V be the zero-dimensional subscheme of E defined by the vanishing of the section $\det A$ of $\bigwedge^n(V \otimes \mathcal{O}_E(p_0)) = \det V \otimes \mathcal{O}_E(np_0)$, in the above notation. By Cramer's rule, $Q = Q(V)$ is an $\mathcal{O}_{T_V}$-module. Clearly $Q(V_1 \oplus V_2) = Q(V_1) \oplus Q(V_2)$. The divisor T_V is additive over exact sequences: given $0 \to V' \to V \to V'' \to 0$, where V', V, V'' are semistable of degree zero, we have $T_V = T_{V'} + T_{V''}$ as divisors on E.

REMARK 1.2.3. (1) It follows from the defining exact sequence above that $H^0(E;Q) \cong H^0(E;V\otimes \mathcal{O}_E(p_0))\otimes H^1(E;\mathcal{O}_E)$ and hence that there is a non-canonical isomorphism $H^0(E;Q)\cong H^0(E;V\otimes\mathcal{O}_E(p_0))$.
(2) Mukai, in [**13**], defines a torsion sheaf Q' associated to V by:

$$Q' = R^1\pi_{1*}(\pi_2^*V\otimes\mathcal{P}^{-1}).$$

In fact, $Q' = Q\otimes\mathcal{O}_E(p_0)$, which we can see as follows: begin with the exact sequence

$$0\to\mathcal{P}^{-1}\to\pi_1^*\mathcal{O}_E(p_0)\otimes\pi_2^*\mathcal{O}_E(p_0)\to\mathcal{O}_\Delta\otimes\pi_1^*\mathcal{O}_E(p_0)\otimes\pi_2^*\mathcal{O}_E(p_0)\to 0,$$

and tensor with π_2^*V. If we apply $R^i\pi_{1*}$, we get an exact sequence

$$0\to H^0(E;V\otimes\mathcal{O}_E(p_0))\otimes\mathcal{O}_E(p_0)\to V\otimes\mathcal{O}_E(2p_0)\to R^1\pi_{1*}(\pi_2^*V\otimes\mathcal{P}^{-1})\to 0.$$

Hence $Q\otimes\mathcal{O}_E(p_0)\cong R^1\pi_{1*}(\pi_2^*V\otimes\mathcal{P}^{-1}) = Q'$.

The *support* of V is by definition the reduced support of T_V, or equivalently of Q. If $\operatorname{Supp} V = \{q_1,\dots,q_a\}$, where the q_i are distinct, then $V = \bigoplus_i V(q_i)$, where each $V(q_i)$ is supported at q_i. We shall only be concerned in this paper with bundles V such that the support of V is contained in E_{reg}. We can see the structure of Q concretely in this case: If λ is a line bundle of degree zero, let $I_k(\lambda) = I_k\otimes\lambda$. Given a point $q\in E_{\mathrm{reg}}$, we let $\lambda_q = \mathcal{O}_E(q-p)$ be the corresponding line bundle of degree zero. Then it is easy to check the following:

LEMMA 1.2.4. *Suppose that $V\cong\bigoplus_i I_{k_i}(\lambda_{q_i})$. Then $T_V = \sum_i k_iq_i$ and $Q\cong\bigoplus_i Q_i$, where each Q_i is supported at q_i and in fact $Q_i = \mathcal{O}_E/\mathfrak{m}_{q_i}^{k_i}$. Thus Q is a locally free $\mathcal{O}_{T_V}$-module of rank one if and only if $q_i\neq q_j$ for $i\neq j$, i.e. if and only if V is regular. Finally, T_V depends only on the S-equivalence class of V.* □

We have the following condition for a semistable vector bundle of degree zero over a singular curve E to have support on E_{reg}:

LEMMA 1.2.5. *Let E be singular. Suppose that V is a semistable vector bundle of rank n and degree zero over E. Then the following are equivalent:*

(i) $\operatorname{Supp} V\subseteq E_{\mathrm{reg}}$;
(ii) *V has a filtration whose successive quotients are line bundles of degree zero;*
(iii) *The pullback of E to the normalization $\widetilde{E}$ is the trivial rank n vector bundle.* □

LEMMA 1.2.6. *Suppose that E is cuspidal. Fixing a local coordinate at the preimage in $\widetilde{E}$ of the singular point to identify E_{reg} with $\mathbb{C}$, let t be the induced coordinate on E_{reg}. Let V be a vector bundle over E whose pullback $\widetilde{V}$ to $\widetilde{E}$ is trivial. Then, by Theorem 1.1.1, a trivialization of $\widetilde{V}$ determines an element $A\in$* $\operatorname{End} H^0(\widetilde{E};\widetilde{V})$. *The conjugacy class of A, viewed as an $n\times n$ matrix, is a complete invariant of the isomorphism class of V. Finally, T_V is the divisor in E_{reg} defined by the characteristic polynomial $p_A(t)$ in E_{reg}. A similar statement holds if E is nodal.* □

We return to the case of a general E and V. To recover V from Q, we have the following result of Mukai [**13**]:

PROPOSITION 1.2.7. *With $Q = Q(V)$ defined as above, ... isomorphism*

Proof. Let $H^0 = H^0(E; V \otimes \mathcal{O}_E(p_0))$. Using the exact sequence

$$0 \to H^0 \otimes \mathcal{O}_E(p_0) \to V \otimes \mathcal{O}_E(2p_0) \to Q \otimes \mathcal{O}_E(p_0) \to 0,$$

and the fact that $R^1\pi_{2*}(H^0 \otimes \pi_1^*\mathcal{O}_E(p_0)) = 0$, we see that it suffices to find a canonical isomorphism

$$\pi_{2*}(\pi_1^*(V \otimes \mathcal{O}_E(2p_0)) \otimes \mathcal{P})/\pi_{2*}(\pi_1^*(H^0 \otimes \mathcal{O}_E(p_0)) \otimes \mathcal{P}) \to V.$$

To analyze the quotient, we begin with the exact sequence

$$0 \to \mathcal{O}_{E\times E} \to \mathcal{O}_{E\times E}(\Delta) \to \mathcal{O}_\Delta \otimes K_E^{-1} \to 0.$$

Using $\pi_1^*(V\otimes\mathcal{O}_E(2p_0))\otimes\mathcal{P} = \pi_1^*(V\otimes\mathcal{O}_E(p_0))\otimes\pi_2^*\mathcal{O}_E(p_0)\otimes\mathcal{O}_{E\times E}(\Delta)\otimes H^0(E; K_E)$, we get an exact sequence

$$0 \to \pi_1^*(V \otimes \mathcal{O}_E(p_0)) \otimes \pi_2^*\mathcal{O}_E(-p_0) \to \pi_1^*(V \otimes \mathcal{O}_E(2p_0)) \otimes \mathcal{P} \to i_*(V) \to 0,$$

where $i_* \colon \Delta \to E \times E$ is the inclusion. Applying π_* then gives an exact sequence

$$0 \to H^0 \otimes \mathcal{O}_E(-p_0) \to \pi_{2*}(\pi_1^*(V \otimes \mathcal{O}_E(2p_0)) \otimes \mathcal{P}) \to V \to 0.$$

As for the term $\pi_1^*(H^0 \otimes \mathcal{O}_E(p_0)) \otimes \mathcal{P} = H^0 \otimes \pi_2^*\mathcal{O}_E(-p_0) \otimes \mathcal{O}_{E\times E}(\Delta)$, the fact that $\pi_{2*}\mathcal{O}_{E\times E} \cong \pi_{2*}\mathcal{O}_{E\times E}(\Delta)$ via the natural inclusion shows that

$$\pi_{2*}(\pi_1^*(H^0 \otimes \mathcal{O}_E(p_0)) \otimes \mathcal{P}) \cong H^0 \otimes \mathcal{O}_E(-p_0),$$

and it is easy to check that this isomorphism is compatible with the inclusion of $H^0 \otimes \mathcal{O}_E(-p_0)$ in $\pi_{2*}(\pi_1^*(V \otimes \mathcal{O}_E(2p_0)) \otimes \mathcal{P})$ given above. Thus we have identified $\pi_{2*}(\pi_1^*(V \otimes \mathcal{O}_E(2p_0)) \otimes \mathcal{P})/\pi_{2*}(\pi_1^*(H^0 \otimes \mathcal{O}_E(p_0)) \otimes \mathcal{P})$ with V. □

PROPOSITION 1.2.8. *Given a torsion sheaf Q on E, if we set*

$$V(Q) = \pi_{2*}(\pi_1^*(Q \otimes \mathcal{O}_E(p_0)) \otimes \mathcal{P}),$$

then $V(Q)$ is a semistable vector bundle of degree zero, and there is a canonical isomorphism from the cokernel of the natural map

$$H^0(E; V(Q) \otimes \mathcal{O}_E(p_0)) \otimes \mathcal{O}_E \to V(Q) \otimes \mathcal{O}_E(p_0)$$

to $Q \otimes \mathcal{O}_E(p_0)$.

Proof. Note that $V(Q) \otimes \mathcal{O}_E(p_0) \cong \pi_{2*}(\pi_1^*Q \otimes \mathcal{O}_{E\times E}(\Delta))$. Applying π_{2*} to the exact sequence

$$0 \to \pi_1^*Q \otimes \mathcal{O}_{E\times E}(\Delta - \pi_2^*p_0) \to \pi_1^*Q \otimes \mathcal{O}_{E\times E}(\Delta) \to \pi_1^*Q \otimes \pi_2^*\mathcal{O}_E(p_0)|\pi_2^*p_0 \to 0$$

then gives the result. □

In other words, $V(Q(V)) \cong V$, and $Q(V(Q)) \cong Q \otimes \mathcal{O}_E(p_0) \cong Q$. Thus we have:

COROLLARY 1.2.9. *The functor $V \mapsto Q(V)$ is an exact functor which defines an equivalence of categories from the abelian category of semistable torsion free sheaves of degree zero over E to the abelian category of torsion sheaves over E.* □

REMARK 1.2.10. Teodorescu [**16**] gives an alternate method for recovering V from Q. It suffices to recover the extension class of the extension

$$0 \to H^0 \otimes \mathcal{O}_E \to V \otimes \mathcal{O}_E(p_0) \to Q \to 0,$$

where $H^0 \otimes \mathcal{O}_E \cong H^0(E; Q) \otimes K_E$. But

$$\operatorname{Ext}^1(E; Q, H^0(E; Q)\otimes K_E) \cong \operatorname{Ext}^1(E; Q, K_E)\otimes H^0(E; Q) \cong H^0(E; Q)^*\otimes H^0(E; Q),$$

by Serre duality, and the extension class is $\mathrm{Id} \in H^0(E;Q)^* \otimes H^0(E;Q)$.

1.3. Families of vector bundles. We turn now to the relative version of the construction. Throughout this section, B denotes a scheme of finite type over $\mathbb{C}$ (or analytic space), and $\mathcal{V}$ is a rank n vector bundle over $B \times E$. Given $b \in B$, let $\mathcal{V}_b$ be the bundle on E induced by the restriction of $\mathcal{V}$ to the slice $\{b\} \times E$. We shall always assume that, for all $b \in B$, $\mathcal{V}_b$ is semistable of degree zero. We can then define $Q = Q(\mathcal{V})$ by the exact sequence

$$0 \to \pi_1^*\pi_{1*}(\mathcal{V} \otimes \pi_2^*\mathcal{O}_E(p_0)) \xrightarrow{A} \mathcal{V} \otimes \pi_2^*\mathcal{O}_E(p_0) \to Q \to 0,$$

and let $T_{\mathcal{V}}$ be the Cartier divisor in $B \times E$ defined by $\det A$. We call $T_{\mathcal{V}}$ the *spectral cover* of B defined by $\mathcal{V}$. Standard base change arguments show:

LEMMA 1.3.1. *The morphism $T_{\mathcal{V}} \to B$ is finite and flat of degree n. The coherent sheaf Q on $B \times E$ is an $\mathcal{O}_{T_{\mathcal{V}}}$-module, flat over B, and the rank of Q viewed as a vector bundle over B is also n. Finally, the above construction is compatible with base change: if $f \colon B' \to B$ is a morphism of schemes, then the spectral cover associated to $f^*\mathcal{V}$ is $f^*T_{\mathcal{V}}$ and the corresponding torsion sheaf in $B' \times E$ is f^*Q.* □

For example, suppose that L is a line bundle on $B \times E$ whose restriction to every slice $\{b\} \times E$ has degree zero. Then T_L is the section of $B \times \operatorname{Pic}^0 E = B \times E_{\mathrm{reg}}$ corresponding to L, and $Q(L) = \mathcal{O}_{T_L}$.

Let $p_1 = \pi_{1,2} \colon B \times E \times E \to B \times E$, let $q = \pi_{1,3} \colon B \times E \times E \to B \times E$, and let $r = \pi_{2,3} \colon B \times E \times E \to E \times E$. There is the relative version of Proposition 1.2.7:

LEMMA 1.3.2. *There is a natural isomorphism*

$$\mathcal{V} \cong q_*(p_1^*(Q \otimes \pi_2^*\mathcal{O}_E(p_0)) \otimes r^*\mathcal{P}). \quad \square$$

Similarly, the analogue of Proposition 1.2.8 holds: If Q is a sheaf on $B \times E$, finite and flat over B, and $\mathcal{V} \cong q_*(p_1^*(Q \otimes \pi_2^*\mathcal{O}_E(p_0)) \otimes r^*\mathcal{P})$, then

$$Q(\mathcal{V}) \cong Q \otimes \pi_2^*\mathcal{O}_E(p_0).$$

In other words, up isomorphism, $Q(\mathcal{V}) \cong Q$.

We can form the restriction of $r^*\mathcal{P}$ to $T_{\mathcal{V}} \times E$, which we shall denote $\mathcal{P}_{T_{\mathcal{V}}}$. Let $M = Q \otimes \pi_2^*\mathcal{O}_E(p_0)$, viewed as an $\mathcal{O}_{T_{\mathcal{V}}}$-module, and let $\nu \colon T_{\mathcal{V}} \to B$ be the projection. Then the lemma says that

$$\mathcal{V} \cong (\nu \times \mathrm{Id})_*(\pi_1^*M \otimes \mathcal{P}_{T_{\mathcal{V}}}).$$

There are natural analogues of the above lemmas for flat families $\pi \colon Z \to B$ of Weierstrass cubics with a section σ. Given a vector bundle $\mathcal{V}$ over Z whose restriction to every fiber of π is semistable of degree zero, we define Q and $T_{\mathcal{V}}$ by the exact sequence

$$0 \to \pi^*\pi_*(\mathcal{V} \otimes \mathcal{O}_Z(\sigma)) \xrightarrow{A} \mathcal{V} \otimes \mathcal{O}_Z(\sigma) \to Q \to 0,$$

and $T_{\mathcal{V}} = \det A$, and the obvious analogue of Lemma 1.3.1 holds. Similarly, there is a relative Poincaré sheaf $\mathcal{P} = \mathcal{O}_{Z \times_B Z}(\Delta) \otimes \pi_1^*\mathcal{O}_Z(-\sigma) \otimes \pi_2^*\mathcal{O}_Z(-\sigma) \otimes p^*L$ over $Z \times_B Z$, where L is the line bundle on B such that $\omega_{Z/B} = \pi^*L$, where $\omega_{Z/B}$ is the relative dualizing sheaf and $p \colon Z \times_B Z \to B$ is the projection, and a natural isomorphism

$$\mathcal{V} \cong \pi_{2*}(\pi_1^*(Q \otimes \mathcal{O}_Z(\sigma)) \otimes \mathcal{P})$$

LEMMA 1.3.3. *Suppose that $\mathcal{V}|\{b\} \times E$ has support in E_{reg} for every $b \in B$. The $\mathcal{O}_{T_\mathcal{V}}$-module Q is a line bundle if and only if the restriction of $\mathcal{V}$ to every fiber is regular.* □

As a consequence of the lemma, suppose that B is reduced and equidimensional and that $\mathcal{V}|\{b\} \times E$ has support in E_{reg} for every $b \in B$. Then T is generically reduced if and only if it is reduced, if and only if there is an open dense subset U of B such that $\mathcal{V}|\{b\} \times E$ is isomorphic to a direct sum of n distinct line bundles of degree zero for all $b \in U$, and in this case Q is a torsion free rank one $\mathcal{O}_{T_\mathcal{V}}$-module. (Here, Q is *torsion free* if there are no nonzero local sections of Q supported on a subscheme of $T_\mathcal{V}$ of strictly smaller dimension.) We shall need the following more precise version:

PROPOSITION 1.3.4. *Suppose that B is normal and that $T_\mathcal{V}$ is reduced and irreducible. Let $\hat{T}_\mathcal{V}$ be the normalization of $T_\mathcal{V}$, and let $\hat{\nu}\colon \hat{T}_\mathcal{V} \to B$ be the composition $\hat{T}_\mathcal{V} \to T_\mathcal{V} \xrightarrow{\nu} B$. For each $b \in B$, let $\mathcal{V}_b$ be the bundle corresponding to $\mathcal{V}|\{b\} \times E$. Suppose that there is a divisor $D \subseteq B$ and an open dense subset D' of D such that*

(i) *For $b \notin D$, $\mathcal{V}_b$ is regular and $T_\mathcal{V}$ is normal at all points of $(\nu)^{-1}(b)$.*
(ii) *Given $b \in D'$ and $q \in \operatorname{Supp} \mathcal{V}_b$, exactly one of the following holds:*
  (a) *$T_\mathcal{V}$ is normal at (b, q) and $\mathcal{V}_b(q)$ is regular;*
  (b) *$q \in E_{\text{reg}}$, there exists a neighborhood of (b, q) in $T_\mathcal{V}$ of the form $U_1 \cup U_2$, where each U_i is smooth, $(b, q) \in U_1 \cap U_2$, U_1 meets U_2 transversally, and $\mathcal{V}_b(q) \cong \mathcal{O}_E(q - p_0) \oplus \mathcal{O}_E(q - p_0)$.*

Then there exists a rank one reflexive sheaf M on $\hat{T}_\mathcal{V}$ such that

$$\mathcal{V} \cong (\hat{\nu} \times \mathrm{Id})_*(\pi_1^* M \otimes \mathcal{P}_{\hat{T}_\mathcal{V}}),$$

where $\mathcal{P}_{\hat{T}_\mathcal{V}}$ is the pullback of $\mathcal{P}_{T_\mathcal{V}}$ to $\hat{T}_\mathcal{V} \times E$. If $\hat{T}_\mathcal{V}$ is smooth, then M is a line bundle.

Proof. If $b \notin D$, then $T_\mathcal{V}$ is normal along $(\nu)^{-1}(b)$, so that $\hat{T}_\mathcal{V} = T_\mathcal{V}$ there, and Q is locally free of rank one in a neighborhood of $(\nu)^{-1}(b)$. Now suppose that $b \in D'$. If $T_\mathcal{V}$ is normal at (b, q), then again Q is locally free of rank one in a neighborhood of (b, q). Suppose that $T_\mathcal{V}$ is not normal at (b, q). We first find locally a rank two subbundle of $\mathcal{V}$ corresponding to the two branches of $T_\mathcal{V}$. There is a small analytic Stein neighborhood U of b in B such that $(\nu)^{-1}(U) = U' \cup U''$, where $(b, q) \in U'$, U' is connected, and $U' \cap U'' = \emptyset$. By hypothesis, if U is sufficiently small, we can write $U' = U_1 \cup U_2$, where U_1 and U_2 are smooth and meet transversally along some component D_0 of $(\nu)^{-1}(D)$. Since ν is finite and flat and the local degree of ν at (b, q) is 2, $\nu|U_i$ is an isomorphism of U_i onto U. Corresponding to $(\nu)^{-1}(U) = U' \cup U''$, we have $\mathcal{V}|U \times E = \mathcal{V}' \oplus \mathcal{V}''$, where the rank of $\mathcal{V}'$ is 2. The restriction of $\mathcal{P}_{T_\mathcal{V}}$ to U_i defines line bundles L_i on $U_i \times E$ and hence on $U \times E$. Here D_0 is the preimage in U_i of the set of points $u \in U$ where $L_1|\{u\} \times E \cong L_2|\{u\} \times E$. Since $(\mathcal{V}') \otimes L_1^{-1}|\{t\} \times E$ has a nonzero section for each $t \in U$, and has exactly one if $t \notin D$, it is easy to see that there is a nonzero map $L_1 \to \mathcal{V}'$. There is an effective divisor Y on $U \times E$ such that the map $L_1 \to \mathcal{V}'$ extends to a map $L_1 \otimes \mathcal{O}_{U \times E}(Y) \to \mathcal{V}'$ which does not vanish in codimension one. The quotient is thus of the form $L_2' \otimes I_Z$, where Z is a codimension two subscheme of $U \times E$ (or empty). Clearly $\mathcal{O}_{U \times E}(Y)$ restricts to the trivial line bundle on every fiber $\{u\} \times E$ and hence is pulled back from a line bundle on U. After shrinking U, we may assume that this line bundle is trivial. There is a homomorphism from L_2 to L_2'

which is an isomorphism away from D. Thus as before $L_2' \cong L_2 \otimes \mathcal{O}_{U\times E}(Y')$ and as before we may assume that the line bundle $\mathcal{O}_{U\times E}(Y')$ is trivial. By semistability, the scheme Z is the preimage of a subset Z' of U, which is either empty or of codimension 2. Thus in the complement of Z', there is an exact sequence

$$0 \to L_1 \to \mathcal{V}' \to L_2 \to 0.$$

For $u \in U - Z'$, $\mathcal{V}'|\{u\} \times E$ is a direct sum of two line bundles of degree zero. It follows that the extension

$$0 \to L_1|\{u\} \times E \to \mathcal{V}'|\{u\} \times E \to L_2|\{u\} \times E \to 0$$

is split for every $u \in U - Z'$. For each $u \in U - Z'$, let U_u be a Stein neighborhood of u. Now

$$\operatorname{Ext}^1(U_u \times E; L_2, L_1) = H^1(U_u \times E; (L_2)^{-1} \otimes L_1)) = H^0(U_u; R^1\pi_{2*}((L_2)^{-1} \otimes L_1)).$$

By the above discussion about $\mathcal{V}'|\{u\} \times E$, if s is the corresponding global section of $R^1\pi_{2*}((L_2)^{-1} \otimes L_1)$, then the image of s is zero in the fiber of $R^1\pi_{2*}((L_2)^{-1} \otimes L_1)$ over t for every $t \in U_u$. To see that s is actually zero, it suffices to prove that $R^1\pi_{2*}((L_2)^{-1} \otimes L_1)$ is supported along $D_0 = U_1 \cap U_2$ and that its length is one there. Each line bundle L_i defines a morphism g_i from U to $\operatorname{Pic}^0 E \cong E$, and the morphism $U_i \to T_{\mathcal{V}} \subseteq U \times E$ is given by (Id, g_i). Let $g = g_1 - g_2$, so that $D_0 = g^{-1}(p_0)$. Then $(L_2)^{-1} \otimes L_1$ is the pullback to $U \times E$ via $(g \times \mathrm{Id})^*$ of $\mathcal{P}$. The assumption that U_1 and U_2 meet transversally is the assumption that g is smooth at the origin and hence that $D_0 = g^*(p_0)$ as reduced divisors. By flat base change,

$$R^1\pi_{2*}((L_2)^{-1} \otimes L_1) = R^1\pi_{2*}(g \times \mathrm{Id})^*\mathcal{P} = g^*R^1\pi_{1*}\mathcal{P}.$$

It is a standard result that $R^1\pi_{1*}\mathcal{P}$ is supported at p_0 and has length one there. It follows that $R^1\pi_{2*}((L_2)^{-1} \otimes L_1)$ is supported along D_0 and has length one there. Hence $s = 0$.

Thus $\mathcal{V}'|U_u \cong (L_1 \oplus L_2)|U_u$. It follows that, at least over the complement of the set Z' of codimension at least two, $Q(\mathcal{V}')|(U - Z') \times E$, which is a direct summand of $Q(\mathcal{V})|(U - Z') \times E$, is $Q(L_1) \oplus Q(L_2)|(U - Z') \times E \cong \mathcal{O}_{U_1} \oplus \mathcal{O}_{U_2}|(U - Z') \times E$. Hence, in the complement of the codimension two subset $(\nu)^{-1}(Z')$ of $T_{\mathcal{V}}$, Q is a line bundle M_0 over the normalization $\hat{T}_{\mathcal{V}}$ of $T_{\mathcal{V}}$. Since $\hat{T}_{\mathcal{V}}$ is normal, M_0 extends uniquely to a rank one reflexive sheaf M on $\hat{T}_{\mathcal{V}}$. We have constructed M and an isomorphism from $\mathcal{V}$ to $(\hat{\nu} \times \mathrm{Id})_*(\pi_1^*M \otimes \mathcal{P}_{\hat{T}_{\mathcal{V}}})$ over $(B - Z') \times E$, where now Z' is some global subscheme of B of codimension at least two. Since the codimension of Z' in B is at least two, if $j\colon (B - Z') \times E \to B \times E$ is the inclusion, then it follows from (1) of Remark 1.1.2 that $\mathcal{V} \cong j_*j^*\mathcal{V}$. A similar result holds for $(\hat{\nu} \times \mathrm{Id})_*(\pi_1^*M \otimes \mathcal{P}_{\hat{T}_{\mathcal{V}}})$, since M is reflexive. Thus the isomorphism from $\mathcal{V}$ to $(\hat{\nu} \times \mathrm{Id})_*(\pi_1^*M \otimes \mathcal{P}_{\hat{T}_{\mathcal{V}}})$ over $(B - Z') \times E$ extends to give an isomorphism from $\mathcal{V}$ to $(\hat{\nu} \times \mathrm{Id})_*(\pi_1^*M \otimes \mathcal{P}_{\hat{T}_{\mathcal{V}}})$. The final statement follows since every rank one reflexive sheaf on a smooth scheme is a line bundle. □

1.4. Cohomology and extensions. For applications to the study of quasiminuscule, non-minuscule representations, we shall need a general result concerning extensions of a family of semistable bundles by the trivial bundle. We begin with the case of a single bundle over E:

PROPOSITION 1.4.1.

(i) $H^0(E;V) \cong \mathrm{Ker}\{\times t\colon Q \to Q\}$;
(ii) $H^1(E;V) \cong \mathrm{Coker}\{\times t\colon Q \to Q\} = Q/tQ$.

Proof. To prove (i), let $\mathbb{C}_{p_0}$ be the torsion module corresponding to $\mathcal{O}_E$, i.e. $\mathbb{C}_{p_0} = \mathcal{O}_E/\mathfrak{m}_{p_0} = \mathcal{O}_{E,p_0}/t\mathcal{O}_{E,p_0}$. Then, by the equivalence of categories given by the Fourier-Mukai correspondence.

$$H^0(E;V) = \mathrm{Hom}(\mathcal{O}_E, V) \cong \mathrm{Hom}(\mathbb{C}_{p_0}, Q) = \mathrm{Ker}\{\times t\colon Q \to Q\}.$$

To see (ii), the group $H^1(E;V)$ classifies the set of isomorphism classes of extensions $0 \to V \to \mathcal{E} \to \mathcal{O}_E \to 0$, and via the Fourier-Mukai correspondence this set is the same as the set of isomorphism classes of extensions $0 \to Q \to \tilde{Q} \to \mathbb{C}_{p_0} \to 0$, which is classified by $\mathrm{Ext}^1_{\mathcal{O}_E}(\mathbb{C}_{p_0}, Q)$. But $\mathrm{Ext}^1_{\mathcal{O}_E}(\mathbb{C}_{p_0}, Q) \cong Q/tQ$ via the resolution $0 \to \mathcal{O}_{E,p_0} \xrightarrow{\times t} \mathcal{O}_{E,p_0} \to \mathbb{C}_{p_0} \to 0$. □

A minor extension of the argument to the relative case gives:

COROLLARY 1.4.2. *Let B be a scheme, $\mathcal{V} \to B \times E$ a flat family of torsion free semistable sheaves of degree zero and $Q = Q(\mathcal{V})$ the corresponding sheaf on $B \times E$. Then, as sheaves on B,*

(i) $R^0\pi_{1*}\mathcal{V} \cong \mathrm{Ker}\{\times t\colon Q \to Q\}$;
(ii) $R^1\pi_{1*}\mathcal{V} \cong \mathrm{Coker}\{\times t\colon Q \to Q\} = Q/tQ$. □

As an application, we have:

COROLLARY 1.4.3. *Let B be a scheme, and let $\overline{\mathcal{V}} \to B \times E$ a family of torsion free semistable vector bundles of degree zero. Let $\mathcal{W}$ be a vector bundle on B. Then the relative extension sheaf $Ext^1_{\pi_1}(\overline{\mathcal{V}}, \pi_1^*\mathcal{W}) \cong R^1\pi_{1*}(\overline{\mathcal{V}}^\vee) \otimes \mathcal{W}$ is isomorphic to $(Q^\vee/tQ^\vee) \otimes \mathcal{W}$, where $Q^\vee = Q(\overline{\mathcal{V}}^\vee)$.* □

2. Covers of the moduli space of G-bundles

Throughout the remainder of this paper, the rank of G will be denoted by r, H is a maximal torus in G, $\mathfrak{h} = \mathrm{Lie}\, H$, and $\Lambda \subseteq \mathfrak{h}$ is the coroot lattice. The algebraic group $E_{\mathrm{reg}} \otimes_{\mathbb{Z}} \Lambda$, is abstractly isomorphic to the r-fold Cartesian product $(E_{\mathrm{reg}})^r$. If E is cuspidal, then $E_{\mathrm{reg}} \otimes_{\mathbb{Z}} \Lambda \cong \mathfrak{h}$, and if E is nodal then $E_{\mathrm{reg}} \otimes_{\mathbb{Z}} \Lambda \cong H$. The Weyl group $W = W(G,H)$ acts on Λ. We let $\mathcal{M} = (E_{\mathrm{reg}} \otimes_{\mathbb{Z}} \Lambda)/W$, and let $\widetilde{\nu}\colon E_{\mathrm{reg}} \otimes_{\mathbb{Z}} \Lambda \to \mathcal{M}$ be the natural projection. For E smooth, $\mathcal{M}$ is the moduli space of semistable holomorphic G-bundles on E, modulo S-equivalence. For E nodal, $\mathcal{M} = H/W$ can be identified with the adjoint quotient of G by the conjugation action of G, where quotient is to be taken in the GIT sense, i.e. again up to S-equivalence. By [**15**, §6], H/W is an affine space $\mathbb{A}^r$. Likewise, if E is cuspidal, then $\mathcal{M} = \mathfrak{h}/W$ is the (GIT) quotient of $\mathfrak{g}$ via the adjoint action of G. By Chevalley's theorem, $\mathfrak{h}/W$ is again an affine space $\mathbb{A}^r$. (In case E is nodal or E is cuspidal and G is not of type E_8, the parabolic construction gives a compactification of $\mathcal{M}$ [**6**], which however will not concern us here.)

2.1. Definition of certain covers. Fix a primitive weight μ for the pair (G,H), i.e. a surjective homomorphism $\Lambda \to \mathbb{Z}$ (minor modifications handle the case where μ is not primitive). Let $\Lambda_0 = \mathrm{Ker}\,\mu$ and let $W_0 = \mathrm{Stab}_W(\mu) \subseteq W$. Set $\hat{T}_0 = (E_{\mathrm{reg}} \otimes_{\mathbb{Z}} \Lambda)/W_0$. Thus there is a finite morphism $\hat{\nu}\colon \hat{T}_0 \to \mathcal{M}$. Since W_0 acts on Λ_0, it also acts on $E_{\mathrm{reg}} \otimes_{\mathbb{Z}} \Lambda_0$. We let $\mathcal{M}_0 = (E_{\mathrm{reg}} \otimes \Lambda_0)/W_0$. The homomorphism μ defines a W_0-invariant morphism $E_{\text{reg}} \otimes_{\mathbb{Z}} \Lambda \to E_{\text{reg}}$ and hence induces a morphism $r\colon \hat{T}_0 \to E_{\text{reg}}$.

LEMMA 2.1.1. *The space $\hat{T}_0$ is an étale fiber bundle over E_{reg} with fibers isomorphic to $\mathcal{M}_0$.*

Proof. There exists a primitive $v \in \Lambda$ which is W_0-invariant and whose image under μ is $d > 0$. Define the surjection $\Lambda \oplus \mathbb{Z} \to \mathbb{Z}$ by $(\lambda, k) \mapsto \mu(\lambda) - kd$. Clearly, the kernel is exactly $\Lambda_0 \oplus \mathbb{Z}$, included via the standard inclusion on $\Lambda_0 \oplus \{0\}$ and where $(0,1) \mapsto (v,1)$. Tensoring with E_{reg} gives a W_0-equivariant isomorphism $E_{\text{reg}} \otimes_{\mathbb{Z}} (\Lambda_0 \oplus \mathbb{Z}) = (E_{\text{reg}} \otimes_{\mathbb{Z}} \Lambda_0) \times E_{\text{reg}} \to (E_{\text{reg}} \otimes_{\mathbb{Z}} \Lambda) \times_{E_{\text{reg}}} E_{\text{reg}}$, where in the second factor the map $E_{\text{reg}} \to E_{\text{reg}}$ is multiplication by d. Taking the quotient by W_0 gives an isomorphism $\hat{T}_0 \times_{E_{\text{reg}}} E_{\text{reg}} \cong \mathcal{M}_0 \times E_{\text{reg}}$ as claimed. □

There is the following standard fact about groups generated by reflections:

LEMMA 2.1.2. *The group W_0 is generated by the reflections in W which fix μ, and hence by the reflections in the roots dual to the coroots lying in* $\operatorname{Ker} \mu = \Lambda_0$. □

In the main application, μ will be a minuscule or quasiminuscule weight, and in particular will be a fundamental weight ϖ_α for some simple root α unless G is of type A_n and μ is a root. Clearly, if $\mu = \varpi_\alpha$ for some α, then Λ_0 is spanned by the simple coroots β for $\beta \neq \alpha$. More generally, if Λ_0 is spanned by the coroots it contains, then Λ_0 is the coroot lattice for a subroot system R_0 of R of rank $n-1$ and W_0 contains the Weyl group of R_0. In fact, it follows from Lemma 2.1.2 that W_0 is exactly the Weyl group of R_0.

Next we turn to the possible singularities of the moduli space and their relation to the singularities of $\hat{T}_0$. Let $\mathcal{M}_{\text{sing}}$ be the singular locus of $\mathcal{M}$, and let $\mathcal{M}_{\text{reg}} = \mathcal{M} - \mathcal{M}_{\text{sing}}$. Note that $\mathcal{M}_{\text{reg}} = \mathcal{M}$ if E is singular, or if E is smooth and G is of type A_n or C_n, and in no other cases.

LEMMA 2.1.3. *Suppose that E is smooth. In the above notation, $(\hat{T}_0)_{\text{sing}} \subseteq \hat{\nu}^{-1}(\mathcal{M}_{\text{sing}})$.*

Proof. Let K be the compact form of G. Viewing a point $e \in E \otimes_{\mathbb{Z}} \Lambda$ as corresponding to a pair of elements $(x,y) \in T \times T$, where T is the maximal torus of K, let $K(e)$ be the centralizer of $\{x,y\}$ in K. By [**4**, Theorem 6.12], the image of e is a singular point of $\mathcal{M}$ if and only if $\pi_0(K(e)) \neq \{1\}$. Let R_e be the set of roots α such that e is in the kernel of the homomorphism $E \otimes_{\mathbb{Z}} \Lambda \to E$ defined by α, and let $W(R_e)$ be the subgroup of W generated by reflections in the elements of R_e. A formal argument (cf. [**1**, Lemma 7.1.4]) shows that there is an exact sequence

$$\{1\} \to W(R_e) \to \operatorname{Stab}_W(e) \to \pi_0(K(e)) \to \{1\}.$$

Then the image of e in $\mathcal{M}$ is a singular point of $\mathcal{M}$ if and only if $\operatorname{Stab}_W(e) \neq W(R_e)$.

Suppose that $e \in E \otimes_{\mathbb{Z}} \Lambda$ lies over a smooth point of $\mathcal{M}$. Then $\operatorname{Stab}_W(e) = W(R_e)$. Thus $\operatorname{Stab}_{W_0}(e)$ is the stabilizer in $W(R_e)$ of μ, and hence by Lemma 2.1.2 is generated by the reflections in $W(R_e)$ which fix μ. These generators fix $e \in E \otimes_{\mathbb{Z}} \Lambda$ and act as reflections on the tangent space to $E \otimes_{\mathbb{Z}} \Lambda$ at e. Hence by Chevalley's theorem $\hat{T}_0$ is smooth in a neighborhood of the image of e. □

2.2. Existence of universal conformal bundles. We now discuss the construction of universal G-bundles and universal vector bundles over $\mathcal{M} \times E$. Let E be a smooth elliptic curve. Recall the following from [**5**]: for each G, there is

- An integer $n > 0$ and an element $c \in Z(G)$ of order n;
- A G-bundle Ξ_0 over $\mathbb{A}^{r+1} \times E$ and an action of $\mathbb{C}^*$ on Ξ_0, lifting a linear action on $\mathbb{A}^{r+1}$,

with the following properties:

- $\Xi_0|\{x\} \times E$ is semistable for all $x \neq 0$;
- the induced $\mathbb{C}^*$-invariant morphism $\mathbb{A}^{r+1} - \{0\} \to \mathcal{M}$ induces an isomorphism from $\mathbb{A}^{r+1} - \{0\}/\mathbb{C}^*$ to $\mathcal{M}$.

Moreover, the primitive n^{th} roots of unity act trivially on $\mathbb{A}^{r+1}$, the quotient $\mathbb{C}^*/(\mathbb{Z}/n\mathbb{Z})$ acts effectively, and the action of $\mathbb{Z}/n\mathbb{Z}$ on Ξ_0 is via $\langle c\rangle$. It follows that, away from the singularities of $\mathcal{M}$, i.e. away from the non-free $\mathbb{C}^*$-orbits, there is an induced $G/\langle c\rangle$ bundle.

In practice, we shall modify this construction slightly. Let $\hat{G} = G \times_{\mathbb{Z}/n\mathbb{Z}} \mathbb{C}^*$, where $1 \in \mathbb{Z}/n\mathbb{Z}$ acts via $c \in G$ and as a primitive n^{th} root of unity ζ in $\mathbb{C}^*$. Given ζ, write $\zeta = \exp(2\pi\sqrt{-1}a/n)$, where $\gcd(a, n) = 1$. Of course, we could always fix $a = 1$ if desired. There is a $\mathbb{C}^*$-linearization of the trivial $\mathbb{C}^*$-bundle over $\mathbb{A}^{r+1}$, covering the given action on the base, as follows: $t(\lambda, x) = (t^{-a}\lambda, t \cdot x)$. Taking the product bundle $\mathbb{C}^* \times \Xi_0$ with its product linearization, we see that there is an induced $\hat{G}$-bundle $\widehat{\Xi}_a$ over $\mathcal{M}_{\text{reg}} \times E$.

LEMMA 2.2.1. *For each $x \in \mathcal{M}_{\text{reg}}$, the $\hat{G}$-bundle $\widehat{\Xi}_a|\{x\} \times E$ has a canonical lift to a G-bundle ξ_x, which is the regular G-bundle corresponding to the point x.*

Proof. Begin with the central extension

$$\{1\} \to G \to \hat{G} \to \mathbb{C}^* \to \{1\}.$$

By construction, the bundle $\hat{\xi}_x = \widehat{\Xi}_a|\{x\} \times E$ lifts to a G-bundle. Let $\det \hat{\xi}_x$ be the $\mathbb{C}^*$-bundle induced from the quotient map $\hat{G} \to \mathbb{C}^*$. The inclusion $\mathbb{C}^* \subseteq \hat{G}$ defines a surjective homomorphism $\operatorname{Aut} \hat{\xi}_x \to \operatorname{Aut} \det \hat{\xi}_x = \mathbb{C}^*$, and hence the lift of $\hat{\xi}_x$ is unique. □

For E nodal or cuspidal, we could use slight variations of this construction (with special care in case E is cuspidal and $G = E_8$). In fact, the moduli space $\mathcal{M}$ is isomorphic to an affine subspace $\mathbb{A}^r$ in $\mathbb{A}^{r+1} - \{0\}$ as we show in [**6**], and hence there is a universal G-bundle Ξ over $\mathcal{M} \times E$. On the other hand, the Steinberg and Kostant sections already give us universal G-bundles as described in [**6**]. As we shall show in [**7**], all of these sections are conjugate in the appropriate sense, so that the choice does not really matter. However, in the next section, for the cuspidal case, we shall use the explicit form of the Kostant section.

Now suppose that $\rho\colon G \to GL(V)$ is an irreducible representation. Then $\rho(c) = \zeta' \operatorname{Id}$, where ζ' is a root of unity whose order divides n. Thus there is $b \in \mathbb{Z}$ such that $\zeta^b = \zeta'$. The map $\chi_b(t) = t^b \operatorname{Id}$ is a homomorphism from $\mathbb{C}^*$ to $GL(V)$, and $\rho \otimes \chi_b$ factors to give a representation $\rho_b\colon \hat{G} \to GL(V)$. There is then the associated bundle $\mathcal{V}_{a,b} = \widehat{\Xi}_a \times_{\hat{G}} V$ over $\mathcal{M}_{\text{reg}} \times E$, where $\hat{G}$ acts on V via ρ_b. It is easy to see that the isomorphism type of the vector bundle $\mathcal{V}_{a,b}$ only depends on the product ab. For each $x \in \mathcal{M}_{\text{reg}}, \mathcal{V}_{a,b}|\{x\} \times E = \xi_x \times_G V = V_x$, and hence does not depend on a, b. Since the image of ρ is contained in $SL(V)$, the bundle V_x has trivial determinant. Thus $\det \mathcal{V}_{a,b}$ is pulled back from the factor $\mathcal{M}_{\text{reg}}$. We can take the associated spectral cover $T_{\mathcal{V}_{a,b}}$ in the sense of §1, and it is also independent of a, b. Finally, it is not hard to see that $Q(\mathcal{V}_{a,b})$ is independent of the choice of (a, b) up to twisting by a line bundle on $T_{\mathcal{V}_{a,b}}$. Fix some pair (a, b), and set $\mathcal{V} = \mathcal{V}_{a,b}$.

The fiber of $T_{\mathcal{V}}$ over x only depends on the S-equivalence class of ξ_x. Thus we have:

LEMMA 2.2.2. *Let $\mu_1, \dots, \mu_n$ be the weights of V, counted with multiplicity. Suppose that ξ_x is S-equivalent to the split bundle corresponding to the point $e \in E \otimes_{\mathbb{Z}} \Lambda$. Then the fiber over x of $T_{\mathcal{V}}$ is the divisor $\sum_{i=1}^{n} \mu_i(e)$.* □

COROLLARY 2.2.3. *Suppose that V is an irreducible representation, with lowest weight μ. Let $\hat{T}_0 = (E \otimes_{\mathbb{Z}} \Lambda)/W_0$ be as in §2.1. Then there is a component T_0 of $T_{\mathcal{V}}$, which has multiplicity one in $T_{\mathcal{V}}$, and a finite birational morphism $\varphi\colon \hat{T}_0 \to T_0$ covering the identity on $\mathcal{M}$, where the composite map $\hat{T}_0 \to T_0 \to \mathcal{M} \times E_{\text{reg}}$ is $(\hat{\nu}, r)$. Thus $\hat{T}_0$ is identified with the normalization of a component T_0 of $T_{\mathcal{V}}$.*

Proof. The reduction of the spectral cover $T_{\mathcal{V}} \subseteq \mathcal{M} \times E$ is the set of all points (x, p) such that there exists a lift of x to $e \in E_{\text{reg}} \otimes_{\mathbb{Z}} \Lambda$ and a weight μ_i such that $\mu_i(e) = p$. Define the map $\widetilde{\varphi}\colon E_{\text{reg}} \otimes_{\mathbb{Z}} \Lambda \to T_{\mathcal{V}}$ by $\widetilde{\varphi}(e) = (\widetilde{\nu}(e), \mu(e))$, where $\widetilde{\nu}\colon E_{\text{reg}} \otimes_{\mathbb{Z}} \Lambda \to \mathcal{M}$ is the quotient morphism. Clearly, $\widetilde{\varphi}$ descends to a morphism $\varphi\colon \hat{T}_0 \to T_{\mathcal{V}}$, covering the projection to $\mathcal{M}$, whose image is set-theoretically a component T_0 of $T_{\mathcal{V}}$. Since all weights in $W \cdot \mu$ have multiplicity one, T_0 is generically reduced and hence is reduced. It will suffice to show that φ has degree one. Since both $\hat{T}_0$ and T_0 map to $\mathcal{M}$ with degree n, this is clear. □

2.3. Some lemmas on weights. In this section, our goal is to analyze the morphism $\hat{T}_0 \to T_0$ of Corollary 2.2.3 more carefully. To do so, we shall have to analyze the weights in the W-orbit of μ. Given two weights μ_1, μ_2, define the divisor $D(\mu_1 - \mu_2) \subseteq E_{\text{reg}} \otimes_{\mathbb{Z}} \Lambda$ via

$$D(\mu_1 - \mu_2) = \{e \in E_{\text{reg}} \otimes_{\mathbb{Z}} \Lambda : \mu_1(e) = \mu_2(e)\}.$$

More generally, for any weight λ, let $D(\lambda) = \operatorname{Ker}\{\lambda\colon E_{\text{reg}} \otimes_{\mathbb{Z}} \Lambda \to E_{\text{reg}}\}$.

PROPOSITION 2.3.1. *Let V be an irreducible representation of G with lowest weight μ. Suppose that $\mu_1, \mu_2 \in W \cdot \mu$ are distinct weights of V in the W-orbit of μ. Then exactly one of the following holds:*

(i) *$\mu_1 - \mu_2 = k\alpha$ for some root α and $k \in \mathbb{Z}$. In this case, μ_2 is the unique element of $W \cdot \mu$ such that $\mu_1 - \mu_2$ is a nonzero real multiple of α.*

(ii) *There is an open dense subset of $D(\mu_1 - \mu_2)$ on which no root vanishes.*

Proof. First suppose that $\mu_1 - \mu_2 = k\alpha$ for some root α. Since the difference of two weights of an irreducible representation is a sum of roots, and the roots are primitive elements in the root lattice, $k \in \mathbb{Z}$. The affine line $\mu_1 + t\alpha$ meets the orbit $W \cdot \mu$ at least at the two points μ_1, μ_2. Since every element of $W \cdot \mu$ has the same length (with respect to any Weyl invariant inner product) and an affine line can meet a sphere in at most two points, we have proved (i).

every root α, $D(\alpha)$ meets $D(\mu_1 - \mu_2)$ in a proper subset. Since there are only finitely many roots, (ii) follows. □

Let us analyze these two possibilities separately. Before we do so, let us introduce the following notation:

DEFINITION 2.3.2. For $e \in E_{\mathrm{reg}} \otimes_{\mathbb{Z}} \Lambda$, let ξ_e be the regular G-bundle whose S-equivalence class is the image of e in $\mathcal{M}$. Let $\mathcal{V}_e$ be the vector bundle $\xi_e \times_G V$.

LEMMA 2.3.3. *In the above notation, suppose that $\mu_1 - \mu_2$ is not a multiple of a root. Then there is an open dense subset U of $D(\mu_1 - \mu_2)$ such that, for all $e \in U$, the following hold:*

(i) $\mathrm{Stab}_W(e) = \{1\}$.
(ii) *The multiplicity of every point of the divisor $\sum_{w\in W/W_0}(w\mu)(e)$ is at most two.*
(iii) *If ξ' is the H-bundle defined by e, then $\xi_e = \xi' \times_H G$, and the vector bundle $\mathcal{V}_e$ defined in Definition 2.3.2 is a direct sum of line bundles.*

Proof. An element $w \in W$ fixes a codimension one subset of $E_{\mathrm{reg}} \otimes_{\mathbb{Z}} \Lambda$ if and only if w fixes a codimension one subset of $\mathfrak{h}$ if and only if $w = r_\alpha$ is the reflection in a root α. In this case, since $r_\alpha(e) = e - \alpha(e)\alpha^\vee$ and $\alpha^\vee$ is primitive, the fixed set of r_α is $D(\alpha)$. Since by hypotheses $D(\alpha) \neq D(\mu_1 - \mu_2)$, (i) follows for an open dense subset of $D(\mu_1 - \mu_2)$.

To prove (ii), first suppose that $\mu', \mu'' \in W \cdot \mu$ and that $\mu' - \mu''$ vanishes along a component of $D(\mu_1 - \mu_2)$. Then, in $\mathfrak{h}$, $\mathrm{Ker}(\mu' - \mu'') = \mathrm{Ker}(\mu_1 - \mu_2)$. Hence $\mu' - \mu''$ is a multiple of $\mu_1 - \mu_2$. Thus, if μ', μ'', μ''' are three distinct elements of $W \cdot \mu$ and μ', μ'', μ''' all agree along a component of $D(\mu_1 - \mu_2)$, then $\mu' - \mu''$ and $\mu'' - \mu'''$ are both multiples of $\mu_1 - \mu_2$, and so μ', μ'', μ''' are all contained in an affine line in $\mathfrak{h}_{\mathbb{R}}$. But this is impossible since, as μ', μ'', μ''' are all conjugate via W, they all have the same length, and a sphere meets an affine line in at most two points.

To see (iii), it is clear that ξ_e is regular since the regular bundle corresponding to e fails to be split if and only if a root vanishes on e (by [**4**, 6.2], in case E is smooth, and by the direct descriptions of §1.1 in case E is singular). The vector bundle $\mathcal{V}_e$ is thus a direct sum of line bundles. □

The following handles the case where $\mu_1 - \mu_2$ is a multiple of a root:

LEMMA 2.3.4. *In the above notation, suppose that $\mu_1 - \mu_2 = k\alpha$ for some root α and some $k \in \mathbb{Z}$, $k \neq 0$. Then there is an open dense subset U of $D(\mu_1 - \mu_2) = D(k\alpha)$ such that, for all $e \in U$, the following hold:*

(i) *If $\mu', \mu'' \in W \cdot \mu$ and $\mu'(e) = \mu''(e)$, then $\mu' - \mu'' = \ell\alpha$ for some $\ell \in \mathbb{Z}$.*
(ii) *The multiplicity of every point of the divisor $\sum_{w\in W/W_0}(w\mu)(e)$ is at most two.*
(iii) $\mathrm{Stab}_W(e) = \{1, r_\alpha\}$ *if $e \in D(\alpha) \subseteq D(k\alpha)$ and* $\mathrm{Stab}_W(e) = \{1\}$ *if $e \in D(k\alpha) \setminus D(\alpha)$. Moreover, $r_\alpha(\mu') = \mu''$ for each pair of distinct weights $\mu', \mu'' \in W \cdot \mu$ such that $\mu'(e) = \mu''(e)$.*
(iv) *If $e \in D(k\alpha)\setminus D(\alpha)$, then ξ_e is split, and $\mathcal{V}_e$ is a direct sum of line bundles.*
(v) *Suppose that $e \in D(\alpha)$, and let G_α be the connected subgroup of G whose Lie algebra is $\mathfrak{g}^\alpha \oplus \mathfrak{g}^{-\alpha} \oplus \mathfrak{h}$. Then G_α is isomorphic to $SL_2 \times H'$ or to $SL_2 \times_{\mathbb{Z}/2\mathbb{Z}} H'$, where H' is the connected subtorus of H whose Lie algebra is* $\mathrm{Ker}\,\alpha$. *If ξ_e is the corresponding regular bundle, then $\xi_e = \xi' \times_{(SL_2 \times H')} G$,*

where $\xi' = \xi_1 \widehat{\boxtimes} \xi_2$ is the bundle over $SL_2 \times H'$ such that ξ_1 is the principal SL_2-bundle corresponding to I_2 and ξ_2 is the H'-bundle such that $\xi' \times_{H'} G$ is the split bundle whose image in $\mathcal{M}$ is e.

Proof. If $\mu', \mu'' \in W \cdot \mu$ and $\mu' - \mu''$ vanishes on a nonempty open subset of $D(\mu_1 - \mu_2)$, then $\mu' - \mu'' = t(\mu_1 - \mu_2) = \ell\alpha$ for some real number ℓ. By the proof of Proposition 2.3.1, $\ell \in \mathbb{Z}$, proving (i). Part (ii) and the first part of (iii) follow as in the proof of the previous lemma. The second part follows from the fact [**3**, p. 124, Prop. 3(i)] that $\mu' - j\alpha$ is a weight whose length is the same as that of μ' if and only if $j = \mu'(\alpha^\vee)$, and in this case by definition $\mu' - j\alpha = r_\alpha(\mu')$. Parts (iv) and (v) follow from the explicit description of the regular representative ξ_e given in [**4**] in the smooth case, and by the discussion of §1.1 in the singular case. □

We can now describe the morphism $\varphi\colon \hat{T}_0 \to T_0 \subseteq T_{\mathcal{V}}$ in more detail. Recall that φ is the morphism induced from $\widetilde{\varphi}\colon E_{\mathrm{reg}} \otimes_{\mathbb{Z}} \Lambda \to T_0 \subseteq \mathcal{M} \times E$ defined by $\widetilde{\varphi}(e) = (\widetilde{\nu}(e), \mu(e))$, where $\widetilde{\nu}\colon E_{\mathrm{reg}} \otimes_{\mathbb{Z}} \Lambda \to \mathcal{M}$ is the quotient map.

LEMMA 2.3.5. *In the above notation, suppose that $e \in E_{\mathrm{reg}} \otimes_{\mathbb{Z}} \Lambda$, and let $y \in \hat{T}_0$ be the image of e.*

(i) *If, for all $\mu' \in W \cdot \mu$ such that $\mu' \neq \mu$, $e \notin D(\mu - \mu')$, then $\varphi^{-1}(\varphi(y)) = \{y\}$, and the morphism $\varphi\colon \hat{T}_0 \to T_0$ is an isomorphism in a neighborhood of y.*

(ii) *If $\mu' \in W \cdot \mu$, $\mu' \neq \mu$, and e is a generic point of $D(\mu - \mu')$, where either $\mu - \mu'$ is not a multiple of a root, or $\mu - \mu' = k\alpha$ for some root α and $e \in D(k\alpha) \setminus D(\alpha)$, then there is an open neighborhood U in T_0 of $\varphi(y)$ such that $\varphi^{-1}(U) = U_1 \cup U_2$, $\varphi|U_i$ is an isomorphism onto a smooth divisor in $\mathcal{M} \times E$, and $\varphi(U_1)$ meets $\varphi(U_2)$ transversally.*

(iii) *If $\mu' \in W \cdot \mu$, $\mu' \neq \mu$, and e is a generic point of $D(\mu - \mu')$, where $\mu - \mu' = k\alpha$ for some root α and $e \in D(\alpha)$, then $\varphi\colon \hat{T}_0 \to T_0$ is an isomorphism in a neighborhood of y.*

Proof. Suppose that $e \notin D(\mu - \mu')$ for every $\mu' \in W \cdot \mu$ such that $\mu' \neq \mu$. Given $y' \in \hat{T}_0$ which is the image of $e' \in E_{\mathrm{reg}} \otimes_{\mathbb{Z}} \Lambda$, suppose that $\varphi(y) = \varphi(y')$. Then $e' = w(e)$ for some $w \in W$ and $\mu(e') = \mu(we) = \mu(e)$. It follows that $w^*\mu = \mu$, so that $w \in W_0$, i.e. e and e' have the same image $y \in \hat{T}_0$. Thus $\varphi^{-1}(\varphi(y)) = \{y\}$. If $w \in \mathrm{Stab}_W(e)$, then $(w\mu)(e) = \mu(e)$, and so $w\mu = \mu$, i.e. $w \in W_0$. Thus $\mathrm{Stab}_W(e) = \mathrm{Stab}_{W_0}(e)$, so that the morphism $\hat{\nu}\colon \hat{T}_0 \to \mathcal{M}$ is a local diffeomorphism near y. Since $\hat{\nu} = \pi_1 \circ \varphi$, the morphism φ must also be a local diffeomorphism onto its image near y. This proves (i).

Now suppose that the hypotheses of (ii) hold and let e' and y' be as above. By (ii) of Lemma 2.3.3 or (ii) of Lemma 2.3.4, if e is generic, then μ' is the unique element of $W \cdot \mu$ such that $\mu'(e) = \mu(e)$. Suppose that $\mu' = w\mu$. If $\varphi(y') = \varphi(y)$, then necessarily $e' = w^{-1}(e)$, and so $\varphi^{-1}(\varphi(y)) = \{y, y'\}$. Moreover, since $\mathrm{Stab}_W(e) = \{1\}$, $y \neq y'$ and the map $\hat{T}_0 \to \mathcal{M}$ is a local diffeomorphism onto its image at both y and y'. Hence the same is true for φ. Since $w\mu = \mu' \neq \mu$, it is easy to see that the two divisors meet transversally at $\varphi(y)$. This proves (ii).

Finally suppose that we are in Case (iii). In this case $\mathrm{Stab}_W(e) = \{1, r_\alpha\}$ and $\mu' = r_\alpha(\mu)$. Thus $\mathrm{Stab}_{W_0}(e) = \{1\}$. It follows as in the previous case that if map from $\hat{T}_0$ to $\mathcal{M}$ can be identified with $\mathbb{C}\cdot\alpha^\vee \subseteq \mathfrak{h}$. Since $\mu(\alpha^\vee)\neq 0$, it follows that the differential of φ at y is injective. This proves (iii). □

2.4. The case of a minuscule or quasiminuscule representation. We now apply the above results in case the representation ρ is minuscule or quasiminuscule. Given ρ, for each pair of integers (a,b), we have constructed a vector bundle $\mathcal{V}_{a,b}$ in §2.2, and the spectral cover $T_{\mathcal{V}_{a,b}}$ is independent of (a,b). As before, we fix some pair (a,b) and set $\mathcal{V}=\mathcal{V}_{a,b}$.

THEOREM 2.4.1. *Let E be singular, and suppose that ρ is minuscule. Then there exists a line bundle M on $\hat{T}_0$ such that $\mathcal{V}=(\hat{\nu}\times \mathrm{Id})_*(\pi_1^*M\otimes(r\times\mathrm{Id})^*\mathcal{P})$. A similar result holds if E is smooth, provided that we replace $\hat{\nu}\colon \hat{T}_0\to\mathcal{M}$ by $\hat{\nu}\colon(\hat{T}_0)_{\mathrm{reg}}\to\mathcal{M}_{\mathrm{reg}}$.*

Proof. In the minuscule case, $T_0=T_{\mathcal{V}}$. Note that the pullback of the Poincaré line bundle on $T_{\mathcal{V}}\times E$ to $\hat{T}_0$ is simply $(r\times\mathrm{Id})^*\mathcal{P}$. Thus, it suffices to show that the normalization morphism $\varphi\colon\hat{T}_0\to T_{\mathcal{V}}$ and the bundle $\mathcal{V}$ satisfy the hypotheses of Proposition 1.3.4. Let D_0 be the union in $\mathcal{M}$ of the images of the divisors $D(\mu-\mu')$, and let D_0' be the dense open subset of D_0 implicitly defined by Lemmas 2.3.3 and 2.3.4. Note that, since ρ is minuscule, if the difference $\mu-\mu'$ of two weights is a multiple of a root α, then in fact this multiple is ±1, so that Case (iv) of Lemma 2.3.4 does not arise.

If $x\in\mathcal{M}$ does not lie in D_0, then $T_{\mathcal{V}}$ is smooth over x and $\mathcal{V}_x$ is regular. If $x\in D_0'$ lies under a point of $D(\mu-\mu')$ satisfying the hypotheses of Lemma 2.3.3, then $\mathcal{V}_x$ satisfies (ii)(a) (with $\mathcal{V}_x(q)$ of rank one) or (ii)(b) of Proposition 1.3.4. Otherwise, we are in case (v) of Lemma 2.3.4, and $x\in\mathcal{M}$ lies under a generic point $e\in D(\alpha)$ for some root α. In this case, there is a connected subgroup G_α of G isomorphic to $SL_2\times H'$ or to $SL_2\times_{\mathbb{Z}/2\mathbb{Z}}H'$, the structure group of ξ_e reduces to G_α and lifts to $SL_2\times H'$, and on the SL_2 factor it is the principal bundle corresponding to I_2. Under the induced action of SL_2 on V, V decomposes into $\bigoplus_i W_i\oplus\bigoplus_j U_j$, where the W_i are irreducible, nontrivial representations of SL_2, and the U_i are the trivial representation. Since ρ is minuscule, $\dim W_i=2$ by e.g. [**3**, p. 128, Prop. 7]. Thus, for the action of $SL_2\times H'$ on V, V decomposes into $\bigoplus_i(W_i\otimes\chi_i)\oplus\bigoplus_j(U_j\otimes\chi_j)$, where the χ_i and χ_j are characters of H'. It then follows that the vector bundle $\mathcal{V}_e$ associated to ξ_e is of the form $\bigoplus_i(I_2\otimes\lambda_i)\oplus\bigoplus_j\lambda_j$. Here the λ_i in the first summand are the line bundles of degree zero corresponding to $\mu'(e)$ for $\mu'\in W\cdot\mu$ such that $\mu'-\mu''=\pm\alpha$ for some $\mu''\in W\cdot\mu$, and the λ_j in the second summand are the line bundles of degree zero corresponding to $\mu'(e)$ for $\mu'\in W\cdot\mu$ such that $\mu'(e)\neq\mu''(e)$ if $\mu''\in W\cdot\mu$, $\mu''\neq\mu'$. In particular, the line bundles corresponding to the various summands are all distinct. Thus, the vector bundle associated to ξ_e is regular. In this case, φ fulfills the hypotheses of (ii)(a) of Proposition 1.3.4, completing the proof of the theorem. □

REMARK 2.4.2. It is an interesting problem to identify the line bundle M of Theorem 2.4.1 in case E is smooth, where it will depend on the choice of (a,b). For groups of type A_n or C_n, the answer is essentially contained in [**8**, Corollary 3.4].

Now we turn to quasiminuscule representations.

THEOREM 2.4.3. *Let E be singular, and suppose that ρ is quasiminuscule but not minuscule. Define the sheaf $\overline{\mathcal{V}}$ by the exact sequence*

$$0 \to \pi_1^*\pi_{1*}\mathcal{V} \to \mathcal{V} \to \overline{\mathcal{V}} \to 0.$$

Then $\overline{\mathcal{V}}$ is a vector bundle, and its associated spectral cover is T_0, the image of $\hat{T}_0$. Moreover, there exists a line bundle M on $\hat{T}_0$ such that

$$\overline{\mathcal{V}} = (\hat{\nu} \times \mathrm{Id})_*(\pi_1^* M \otimes (r \times \mathrm{Id})^*\mathcal{P}).$$

A similar result holds if E is smooth, provided that we replace $\hat{\nu}\colon \hat{T}_0 \to \mathcal{M}$ by $\hat{\nu}\colon (\hat{T}_0)_{\mathrm{reg}} \to \mathcal{M}_{\mathrm{reg}}$.

Proof. Suppose that the multiplicity of the trivial weight in ρ is s. Let $e \in E_{\mathrm{reg}} \otimes_{\mathbb{Z}} \Lambda$ and let ξ_e be the regular vector bundle whose S-equivalence class corresponds to the image of e in $\mathcal{M}$. We first claim that $h^0(E; \xi_e \times_G V) = s$ for every e. Of course, if G is simply laced, then ρ is the adjoint representation, $s = r$, and the claim is essentially the definition of a regular bundle. In case G is not simply laced, the nonzero weights are the short roots. Let $e \in E_{\mathrm{reg}} \otimes_{\mathbb{Z}} \Lambda$ and let R_e be the set of roots which are trivial on e. Let X be a principal nilpotent element for the reductive Lie algebra $\mathfrak{z}_{\mathfrak{g}}(e)$ which is generated by $\mathfrak{h}$ and the root spaces $\mathfrak{g}^\alpha$, $\alpha \in R_e$. By the recipe for constructing ξ_e given in [**4**] in case E is smooth, and by the discussion in §1.1 in case E is singular, $H^0(E; \xi_e \times_G V)$ is given by the kernel of $\rho_*(X)$ acting on V_e, where V_e is the sum of the weight spaces (including those for the trivial weight) which annihilate e. The trivial weight contributes a subspace of V_e of dimension s and the remaining weight spaces annihilating e are one-dimensional subspaces of V_e corresponding to the short roots in R_e. We may assume that we have chosen X so that there is an $h_e \in \mathfrak{h} \subseteq \mathfrak{z}_{\mathfrak{g}}(e)$ such that (h_e, X) completes to an $\mathfrak{sl}_2$-triple in $\mathfrak{z}_{\mathfrak{g}}(e)$. Thus $(\rho_*(h_e), \rho_*(X))$ completes to some $\mathfrak{sl}_2$-triple in $\operatorname{End} V$. Since X is principal in $\mathfrak{z}_{\mathfrak{g}}(e)$, for all $\alpha \in R_e$, $\alpha(h_e)$ is an even nonzero integer. Thus the kernel of $\rho_*(h_e)$ is the weight zero subspace of V and so $\dim \operatorname{Ker}\{\rho_*(h_e)\colon V_e \to V_e\} = s$. Since all of the eigenvalues of $\rho_*(h_e)$ are even, it follows from the classification of representations of $\mathfrak{sl}_2$ that $\dim \operatorname{Ker}\{\rho_*(X)\colon V_e \to V_e\} = s$ as well.

Since $h^0(E; \xi_e \times_G V)$ is constant, the direct image $\pi_{1*}\mathcal{V}$ is locally free of rank s and the induced map $\pi_{1*}\mathcal{V}/\mathfrak{m}_x\pi_{1*}\mathcal{V} \to H^0(E; \mathcal{V}_x)$ is an isomorphism for every $x \in \mathcal{M}$. By semistability, the map $H^0(E; \mathcal{V}_x) \otimes \mathcal{O}_E \to \mathcal{V}_x$ is injective for every $x \in \mathcal{M}$. It follows that the induced homomorphism $\pi_1^*\pi_{1*}\mathcal{V} \to \mathcal{V}$ is injective on every fiber, and so the cokernel $\overline{\mathcal{V}}$ is a subbundle as claimed. Clearly the spectral cover corresponding to $\pi_1^*\pi_{1*}\mathcal{V}$ is just the trivial section of $\mathcal{M} \times E$ with multiplicity s, and hence $T_{\overline{\mathcal{V}}} = T_0$.

As in the proof of Theorem 2.4.1, we take D_0 to be the union in $\mathcal{M}$ of the images of the divisors $D(\mu - \mu')$, and D_0' to be the dense open subset of D_0 implicitly defined by Lemmas 2.3.3 and 2.3.4. To complete the proof, we must analyze the corresponding bundles $\overline{\mathcal{V}}_x$, where either $x \notin D_0$ or $x \in D_0'$. The arguments are very similar to those given in the proof of Theorem 2.4.1, and we shall be brief. The only essentially new case is when a weight, i.e. a short root α, vanishes. In this case, for a generic $x \in D(\alpha)$, there is a unique factor of $\mathcal{V}_x$ isomorphic to I_3, by an I_2 and the remaining $s-1$ copies of $\mathcal{O}_E$ are no longer present. Hence $\overline{\mathcal{V}}_x$ is regular. □

2.5. Some remarks on representations. Given a regular G-bundle ξ and an irreducible representation $\rho\colon G \to GL(V)$ of G, when is the associated vector bundle $\xi \times_G V$ regular? In this section, we state some results along these lines. The proofs are straightforward and will not be given.

PROPOSITION 2.5.1. *Let $\rho\colon G \to GL(V)$ be an irreducible representation. Then the following are equivalent:*

(i) *For every smooth elliptic curve E and every regular G-bundle ξ over E, the associated vector bundle $\xi \times_G V$ is a regular vector bundle.*
(ii) *For every regular element $g \in G$, $\rho(g)$ is a regular element of $GL(V)$.*
(iii) *The multiplicity of every weight of ρ is one, and for every pair μ_1, μ_2 of distinct weights of ρ, $\mu_1 - \mu_2$ is a root.*
(iv) *Either $G = SL_n$ and ρ is the standard representation or its dual, or $G = Sp(2n)$ and ρ is the standard representation.* □

A weaker but related question is the following: let ξ be the regular G-bundle S-equivalent to the trivial G-bundle. When is $\xi \times_G V$ regular? The answer is as follows:

PROPOSITION 2.5.2. *Let $\rho\colon G \to GL(V)$ be an irreducible representation. Then the following are equivalent:*

(i) *For every smooth elliptic curve E and every regular G-bundle ξ over E S-equivalent to the trivial bundle, the associated vector bundle $\xi \times_G V$ is a regular vector bundle.*
(ii) *For every regular element $X \in \mathfrak{g}$, $\rho_*(X)$ is a regular element of* $\operatorname{End} V$.
(iii) *If X is a principal nilpotent element in $\mathfrak{g}$, then $\rho_*(X)$ is a regular element of* $\operatorname{End} V$.
(iv) *The multiplicity of every weight of ρ is one, and for every pair μ_1, μ_2 of weights of ρ, $\mu_1 - \mu_2$ is a multiple of a root.*
(v) *The group G and the representation ρ are as in* (iv) *of Proposition* 2.5.1, *or $G = SL_2$ and ρ is any irreducible representation, or G is of type B_n or G_2 and ρ is the unique quasiminuscule, non-minuscule representation of G.* □

3. Kostant and Steinberg sections and the proof of the main theorem

3.1. Conjugation into the slice. Every $\operatorname{ad} G$-invariant function on $\mathfrak{g}$ may be identified with a W-invariant function on $\mathfrak{h}$. In this way, the adjoint quotient of $\mathfrak{g}$ is identified with $\mathfrak{h}/W$. Let $p\colon \mathfrak{g} \to \mathfrak{h}/W$ be the induced morphism, and let $\mathfrak{g}_{\mathrm{reg}}$ be the open dense subset of $\mathfrak{g}$ consisting of regular elements. In [**11**], Kostant constructed a section of p whose image lies in $\mathfrak{g}_{\mathrm{reg}}$ as follows. Let $\mathfrak{b}$ be a Borel subalgebra containing $\mathfrak{h}$, with nilpotent radical $\mathfrak{u}$. Thus, there is a unique set of positive roots R^+ for $(\mathfrak{g}, \mathfrak{h})$ such that $\mathfrak{u} = \bigoplus_{\alpha\in R^+} \mathfrak{g}^\alpha$. Fix a principal $\mathfrak{sl}_2$-triple (X, h_0, Y) adapted to $\mathfrak{b}$. By definition, this means that $X \in \mathfrak{u}$ is a principal nilpotent element of $\mathfrak{g}$, $[h_0, X] = 2X$, $[h_0, Y] = -2Y$, and $[X, Y] = h_0 \in \mathfrak{h}$. It follows automatically that $Y \in \mathfrak{u}_- = \bigoplus_{-\alpha\in R^+} \mathfrak{g}^\alpha$. Kostant has proved the following [**11**, Theorem 0.10]:

THEOREM 3.1.1. *Let Z be a vector subspace of $\mathfrak{g}$ which is complementary to* $\operatorname{Im}(\operatorname{ad} X)$ *and is invariant under* $\operatorname{ad} h_0$. *Then the affine space $Z+X$ is contained in $\mathfrak{g}_{\mathrm{reg}}$ and the morphism $Z+X \to \mathfrak{h}/W$ induced by the projection $\mathfrak{g} \to \mathfrak{h}/W$ is an isomorphism.* □

We call the inverse morphism $\overline{\Sigma}_Z \colon \mathfrak{h}/W \to Z+X \subseteq \mathfrak{g}_{\mathrm{reg}}$ the *Kostant section* defined by Z. Denote the composition $\mathfrak{h} \to \mathfrak{h}/W \to Z+X \to \mathfrak{g}_{\mathrm{reg}} \subseteq \mathfrak{g}$ by $\Sigma_Z \colon \mathfrak{h} \to \mathfrak{g}$. Clearly, Σ_Z is a W-invariant morphism from $\mathfrak{h}$ to $\mathfrak{g}$.

We would like to compare the $\overline{\Sigma}_Z$ for different choices of Z, and would also like to compare Σ_Z with the inclusion $i \colon \mathfrak{h} \to \mathfrak{g}$. Of course, Σ_Z can never be conjugate to the inclusion, not even pointwise, because not all elements of $\mathfrak{h}$ are regular in $\mathfrak{g}$. According to the next result, for every choice of Z, if we replace the inclusion morphism i by $i+X$, then the resulting morphism is indeed conjugate to Σ_Z via a morphism $\mathfrak{h} \to G$:

THEOREM 3.1.2. *Let (X, h_0, Y) be a principal $\mathfrak{sl}_2$-triple adapted to $\mathfrak{b}$ and let Z and Z' be linear complements to* $\operatorname{Im}(\operatorname{ad} X)$ *invariant under* $\operatorname{ad} h_0$. *Let $\overline{\Sigma}_Z$ and $\overline{\Sigma}_{Z'}$ be the Kostant sections defined by Z and Z' respectively. Let $i+X \colon \mathfrak{h} \to \mathfrak{g}$ be the map $h \mapsto h+X$. Then:*

(i) *$\overline{\Sigma}_Z$ and $\overline{\Sigma}_{Z'}$ are conjugate under the adjoint action of G on $\mathfrak{g}$, i.e. there exists a morphism $\psi \colon \mathfrak{h}/W \to G$ such that*

$$\operatorname{Ad}(\psi) \circ \overline{\Sigma}_Z = \overline{\Sigma}_{Z'}.$$

(ii) *The map $i+X$ is an embedding of $\mathfrak{h}$ into $\mathfrak{g}_{reg}$.*

(iii) *For each $h \in \mathfrak{h}$, $p(h+X) = p(h)$, so that $h+X$ and h are identified in the adjoint quotient. More precisely, there is a one-parameter subgroup $\beta \colon \mathbb{C}^* \to G$ such that, for all $h \in \mathfrak{h}$,*

$$\lim_{t \to 0} \operatorname{Ad} \beta(t)(h+X) = h.$$

(iv) *There exists a morphism $\Psi \colon \mathfrak{h} \to \mathfrak{u}_-$ such that, for all $h \in H$,*

$$\operatorname{Ad}(\exp(\Psi(h)))(h+X) = \Sigma_Z(h).$$

(v) *The morphism Ψ is the unique morphism from $\mathfrak{h}$ to $\mathfrak{u}_-$ with the property that, for all $h \in \mathfrak{h}$,*

$$\operatorname{Ad}(\exp(\Psi(h)))(h+X) \in Z+X.$$

By (i), all Kostant sections are equivalent under the adjoint action of G on $\mathfrak{g}$, a result which is subsumed in the main result of [**7**]. By (iv), after making the base change $\mathfrak{h} \to \mathfrak{h}/W$, we can conjugate any Kostant section until its image lies in a Borel subalgebra. This is of course not possible for the Kostant section itself, since otherwise we would be able to find a section to the morphism $\mathfrak{h} \to \mathfrak{h}/W$.

We will deduce Theorem 3.1.2 from a more general result, proved in §3.2.

3.2. More general $\mathfrak{sl}_2$-triples. For the moment, let (X, h_0, Y) denote an arbitrary $\mathfrak{sl}_2$-triple, not necessarily principal or adapted to $\mathfrak{b}$. The endomorphism $\operatorname{ad} h_0$ of $\mathfrak{g}$ is semisimple and its eigenvalues are integers. Let $\mathfrak{g}_\ell$ be the ℓ-eigenspace of $\operatorname{ad} h_0$. Thus $\mathfrak{g}$ is graded and the bracket is compatible with the grading. Moreover, in $\mathfrak{g}_\ell$ is nonzero. By the classification of finite-dimensional representations of $\mathfrak{sl}_2$, $\operatorname{ad} X\colon \mathfrak{g}_\ell \to \mathfrak{g}_{\ell+2}$ is injective if $\ell < 0$ and surjective if $\ell \geq -1$.

PROPOSITION 3.2.1. *Let (X, h_0, Y) be an $\mathfrak{sl}_2$-triple. Let $Z \subseteq \mathfrak{g}_{\leq 0}$ be a complementary linear subspace to the image of* $\operatorname{ad} X$ *invariant under h_0. Let $\tau \in \mathfrak{g}_{\leq 0}$. Then there is a unique $\psi \in \mathfrak{g}_-$ such that* $\operatorname{Ad}(\exp(\psi))(\tau + X) \in Z + X$.

Proof. Write $\tau = \tau_0 + \operatorname{ad} X(\tau_1)$, where $\tau_0 \in Z$ and we may assume that $\tau_1 \in \mathfrak{g}_-$, in fact that $d(\tau_1) \leq -2$. The proof of existence of ψ is by induction on $k = d(\tau_1) \leq -2$. If k is sufficiently negative, then $\tau_1 = 0$ and we can take $\psi = 0$. Suppose that we know the result for all τ such that $d(\tau_1) \leq k-1$, and let τ be an element such that $d(\tau_1) = k$. Let $\psi' = \tau_1$. Then

$$\begin{aligned}\operatorname{Ad}(\exp(\psi'))(\tau + X) &= \exp(\operatorname{ad}\psi')(\tau + X)\\ &= \tau + X - \operatorname{ad} X(\tau_1) + \sum_{n=1}^{\infty}\frac{1}{n!}\operatorname{ad}(\tau_1)^n(\tau) + \sum_{n=2}^{\infty}\frac{1}{n!}\operatorname{ad}(\tau_1)^n(X).\end{aligned}$$

We can write this as $\tau_0 + X + \alpha$, where $d(\alpha) \leq d(\tau_1)$. Let $\alpha = \tau_0' + \operatorname{ad} X(\tau_2)$, where $\tau_0' \in Z$ and $\tau_2 \in \mathfrak{g}_-$. Since Z is h_0-invariant, $d(\tau') \leq d(\alpha)$ and thus $d(\operatorname{ad} X(\tau_2)) \leq d(\alpha)$. We can thus assume that $d(\tau_2) \leq d(\alpha) - 2 \leq k - 2$. Thus by induction there exists a ψ_1 such that $\operatorname{Ad}(\exp(\psi_1))(\tau_0 + \alpha + X) \in Z + X$. Since $\exp(\psi_1)$ and $\exp(\psi')$ lie in the unipotent group G_-, $\psi = \log(\exp(\psi_1)\exp(\psi'))$ is defined and clearly $\operatorname{Ad}(\exp(\psi))(\tau + X) \in Z + X$. This completes the inductive step and proves the existence of ψ.

To see the uniqueness, suppose that $\operatorname{Ad}(\exp(\psi_i))(\tau + X) \in Z + X$ for $i = 1, 2$. Writing $\log(\exp(\psi_1)\exp(-\psi_2)) = \psi$, we see that it suffices to prove: if there exists $\zeta \in Z$ and $\psi \in \mathfrak{g}_-$ such that $\operatorname{Ad}(\exp(\psi))(\zeta + X) \in Z + X$, then $\psi = 0$. If $\psi \neq 0$, write $\psi = \psi_k + \psi'$, where ψ_k is homogeneous of degree $k < 0$ and $d(\psi') < k$. As before, we write

$$\operatorname{Ad}(\exp(\psi))(\zeta + X) = \zeta + X + [\psi_k, X] + \gamma,$$

where $d(\gamma) < d([\psi_k, X]) = k + 2$. Since $\zeta \in Z$ and Z is a vector subspace of $\mathfrak{g}$, it follows that $[\psi_k, X] + \gamma \in Z$. Since Z is invariant under the action of h_0, each homogeneous component of $[\psi_k, X] + \gamma$ lies in Z. But as $d(\gamma) < d([\psi_k, X])$, it follows that $[\psi_k, X] \in Z$. Since Z is a complement to the image of $\operatorname{ad} X$, $[\psi_k, X] = 0$, and hence $\psi_k = 0$ since $\psi_k \in \mathfrak{g}_k$ and $k < 0$. This is a contradiction. Hence $\psi = 0$. □

REMARK 3.2.2. Suppose that $\mathfrak{g}_{-1} = 0$, which is the case if (X, h_0, Y) is principal. Then, for all $g \in G_-$, $\operatorname{Ad}(g)(X) - X \in \mathfrak{g}_{\leq 0}$. It follows that, for all $g \in G_-$ and $z \in \mathfrak{g}_-$, $\operatorname{Ad}(g)(z + X) - X \in \mathfrak{g}_{\leq 0}$. Thus

$$(g, z) \mapsto \operatorname{Ad}(g)(z + X) - X$$

defines a morphism

$$G_- \times (Z + X) \to \mathfrak{g}_{\leq 0}.$$

It follows from Proposition 3.2.1 that this morphism defines an isomorphism from $G_- \times (Z + X)$ to $\mathfrak{g}_{\leq 0}$. In this essentially equivalent form, Proposition 3.2.1 is due to Kostant [**12**, Theorem 1.2].

3.3. A parametrized version. We continue to assume that (X, h_0, Y) is an arbitrary $\mathfrak{sl}_2$-triple. Let R be a finitely generated $\mathbb{C}$-algebra. The exponential map $\exp\colon \mathfrak{g}_- \to G$ extends to a map $\mathfrak{g}_- \otimes_{\mathbb{C}} R \to G(R)$, where $G(R)$ denotes the group of R-valued points of G, in other words the set of morphisms $\operatorname{Spec} R \to G$. Viewing $\mathfrak{g}_- \otimes_{\mathbb{C}} R$ as the set of morphisms $\operatorname{Spec} R \to \mathfrak{g}_-$, this map is the composition $\psi \mapsto \exp \circ \psi\colon \operatorname{Spec} R \to G$. The group $G(R)$ acts on $\mathfrak{g} \otimes_{\mathbb{C}} R$ via Ad. We still have the formula $\operatorname{Ad}(\exp(\psi)) = \exp(\operatorname{ad} \psi)$. From this, it is clear that the proof of Proposition 3.2.1 also proves:

PROPOSITION 3.3.1. *Let (X, h_0, Y) be an $\mathfrak{sl}_2$-triple. Let $Z \subseteq \mathfrak{g}_{\leq 0}$ be a complementary linear subspace to the image of* $\operatorname{ad} X$ *invariant under h_0. Let $\tau \in \mathfrak{g}_{\leq 0} \otimes_{\mathbb{C}} R$. Then there is a unique $\psi \in \mathfrak{g}_- \otimes_{\mathbb{C}} R$ such that* $\operatorname{Ad}(\exp(\psi))(\tau + X) \in Z \otimes_{\mathbb{C}} R + X \subseteq \mathfrak{g} \otimes_{\mathbb{C}} R$. □

REMARK 3.3.2. An examination of the inductive argument shows the following: suppose that $R = \mathbb{C}[s_1, \dots, s_n]$ is the coordinate ring of affine n-space with the usual grading, and that $\tau \in \mathfrak{g}_0 \otimes_{\mathbb{C}} R_1$, so that the components of τ are linear polynomials. Then $\psi = \sum_{k=1}^{N} \psi_{-k}$, where $\psi_{-k} \in \mathfrak{g}_{-2k} \otimes_{\mathbb{C}} R_{k+1}$ and so the components of ψ_{-k} are homogeneous polynomials of degree $k+1$.

We will also need the following lemma to prove Part (i) of Theorem 3.1.2:

LEMMA 3.3.3. *Let $Z, Z' \subseteq \mathfrak{g}_-$ be two linear subspaces, both complementary to* $\operatorname{Im}(\operatorname{ad} X)$ *and invariant under h_0. Then there exists a morphism $\gamma\colon Z \to \mathfrak{g}_-$ with* $\gamma(0) = 0$ *such that the morphism*

$$z \mapsto \operatorname{Ad}(\exp(\gamma(z)))(z + X) - X$$

is an isomorphism from Z to Z'.

Proof. Let R be the coordinate ring of Z. Applying Proposition 3.3.1 to the inclusion of Z in $\mathfrak{g}_-$, which corresponds to an element of $R \otimes_{\mathbb{C}} \mathfrak{g}_-$, we see that there is a unique morphism $\gamma\colon Z \to \mathfrak{g}_-$ with $\operatorname{Ad}(\exp(\gamma(z)))(z + X) \in Z' + X$ for all $z \in Z$. Clearly $\gamma(0) = 0$ by uniqueness. Define the morphism $\alpha\colon Z \to Z'$ by $\alpha(z) = \operatorname{Ad}(\exp(\gamma(z)))(z + X) - X$. Symmetrically, there is a unique morphism $\gamma'\colon Z' \to \mathfrak{g}_-$ such that, for all $z' \in Z'$, $\operatorname{Ad}(\exp(\gamma(z')))(z' + X) \in Z + X$. If we set $\beta(z') = \operatorname{Ad}(\exp(\gamma(z')))(z' + X) - X$, then β is a morphism from Z' to Z. By the uniqueness statements, it follows that

$$\operatorname{Ad}(\exp(\gamma'(\alpha(z)))) \circ \operatorname{Ad}(\exp(\gamma(z)))(z + X) = z + X$$

for all $z \in Z$, and hence that $\beta \circ \alpha = \operatorname{Id}$. Symmetrically, $\alpha \circ \beta = \operatorname{Id}$ as well, so that α and β are inverse to one another. □

3.4. Proof of Theorem 3.1.2. We now prove Theorem 3.1.2. Let (X, h_0, Y) be a principal $\mathfrak{sl}_2$-triple adapted to $\mathfrak{b}$. Since h_0 is regular, $\mathfrak{g}_0 = \mathfrak{h}$ and $\mathfrak{g}_- = \mathfrak{u}_-$. Let Z and Z' be complementary linear subspaces to $\operatorname{Im}(\operatorname{ad} X)$ and invariant under h_0. Both Z and Z' lie in $\mathfrak{g}_-$. By Lemma 3.3.3 there is a morphism $\gamma\colon Z \to \mathfrak{g}_-$ such that $z + X \mapsto \operatorname{Ad}(\exp(\gamma(z)))(z + X)$ is an isomorphism $I\colon Z + X$ to $Z' + X$. Since

The map $\mathbb{C} \to \mathfrak{h}$ defined by $a \mapsto ah_0$ exponentiates to a one-parameter subgroup $\beta\colon \mathbb{C}^* \to H$ such that $\operatorname{Ad}\beta(t)(X) = t^2X$. Thus, for all $h \in H$, $\operatorname{Ad}\beta(t)(h+X) = h+t^2X$, and hence $\lim_{t\to 0}\operatorname{Ad}\beta(t)(h+X) = h$. This proves Part (iii) of Theorem 3.1.2.

Viewing the inclusion $i\colon \mathfrak{h} \to \mathfrak{g}$ as a morphism from $\mathfrak{h}$ to $\mathfrak{g}_0$, it follows from Proposition 3.3.1 that there is a unique morphism $\Psi\colon \mathfrak{h} \to \mathfrak{g}_-$ such that

$$\operatorname{Ad}(\Psi(h))(h+X) \in Z + X$$

for all $h \in \mathfrak{h}$. This proves Part (v) of Theorem 3.1.2. Note that, by Remark 3.3.2, $\Psi = \sum_{k=1}^N \Psi_{-k}$, where $\Psi_{-k} \in \mathfrak{g}_{-2k} \otimes_{\mathbb{C}} R$ is a homogeneous polynomial of degree $k+1$.

Now define $\Sigma'_Z\colon \mathfrak{h} \to Z + X$ by

$$\Sigma'_Z(h) = \operatorname{Ad}\exp(\Psi(h))(h+X).$$

LEMMA 3.4.1. *The morphism Σ'_Z is invariant under W and the induced morphism from $\mathfrak{h}/W$ to $Z+X$ is the Kostant section. In other words, $\Sigma'_Z = \Sigma_Z$.*

Proof. Clearly $\Sigma'_Z(h)$ is conjugate to $h + X$, and hence by Part (iii) of Theorem 3.1.2 maps to the same point of $\mathfrak{h}/W$ as h. Since there is a unique such point in $Z + X$ with this property, namely $\Sigma_Z(h)$, it follows that $\Sigma'_Z = \Sigma_Z$. □

We have thus proved (iv) of Theorem 3.1.2. Since the image of the Kostant section lies in $\mathfrak{g}_{\mathrm{reg}}$, it follows that the image of $i + X$ has this property as well, proving (ii) of Theorem 3.1.2. This completes the proof of the theorem. □

3.5. Consequences for invariant polynomials. Let $\rho\colon G \to \operatorname{Aut} V$ be an irreducible representation and let $\rho_*\colon \mathfrak{g} \to \operatorname{End} V$ be the induced representation of Lie algebras. We can decompose V into a direct sum of weight spaces V_ν. Let μ be the lowest weight of ρ, with corresponding weight space V_μ, and let $\pi_\mu\colon V \to V_\mu$ be the induced projection, i.e. π_μ is the identity on V_μ and is zero on $V_\nu, \nu \neq \mu$. As in §2, let W_0 be the stabilizer in W of μ.

The endomorphism $\rho_*(h_0)$ of V is semisimple and its eigenvalues are integral. Thus $V = \bigoplus_k V_k$, where $\rho_*(h_0)$ acts on V_k with eigenvalue k, and each V_k is a sum of weight spaces. Clearly $\rho_*(X)(V_k) \subseteq V_{k+2}$. The eigenspace V_k corresponding to the minimal value of k is one dimensional, and in fact is exactly V_μ.

Let $S = \operatorname{Sym}^* \mathfrak{h}^*$ be the affine coordinate ring of $\mathfrak{h}$. The groups W and W_0 act on S, and we have the inclusions $S^W \subseteq S^{W_0} \subseteq S$, corresponding to the finite surjective morphisms $\mathfrak{h} \to \mathfrak{h}/W_0 \to \mathfrak{h}/W$. Note that the linear function μ is naturally an element of S and in fact lies in S^{W_0}. Of course, W acts on $\mathfrak{h}$ as a group generated by reflections. But, by Lemma 2.1.2, W_0 also acts on $\mathfrak{h}$ as a group generated by reflections. Thus, by Chevalley's theorem both S^W and S^{W_0} are polynomial algebras. We wish to compare these polynomial algebras.

LEMMA 3.5.1. *The ring S is a faithfully flat extension of S^{W_0}. Likewise, S^{W_0} is a faithfully flat extension of S^W. In particular, as an S^W-module, S^{W_0} is free of rank $\#(W/W_0)$.*

Proof. This is an immediate consequence of the fact that S, S^{W_0}, and S^W are regular and that the morphisms in question are finite and surjective. □

For the rest of this section we fix a linear complement $Z \subseteq \mathfrak{g}$ to $\operatorname{Im}(\operatorname{ad} X)$, invariant under h_0, and we let $\Sigma = \Sigma_Z$ and Ψ be as given in Theorem 3.1.2.

LEMMA 3.5.2. *Let $\lambda = \exp(-\Psi) \colon \mathfrak{h} \to U_-$, so that $\rho(\lambda)$ is a morphism from $\mathfrak{h}$ to* $\operatorname{Aut} V$. *Then, for all $h \in \mathfrak{h}$, we have the following equality in* $\operatorname{End} V$*:*

$$\rho(\lambda(h)) \circ \rho_*(\Sigma(h)) = \rho_*(h + X) \circ \rho(\lambda(h)).$$

Proof. This is immediate from Theorem 3.1.2. □

A morphism $\mathfrak{h} \to \operatorname{End} V$ is the same thing as an element of

$$(\operatorname{End}_{\mathbb{C}} V) \otimes_{\mathbb{C}} S \cong \operatorname{End}_S(V \otimes_{\mathbb{C}} S).$$

Thus, after choosing a basis of V, we can identify the morphism $\rho(\lambda)$ from $\mathfrak{h}$ to V with an $n \times n$ matrix with coefficients in S. The decomposition of V into the direct sum of its weight spaces means that we can write $\rho(\lambda)$ in block form as a sum of elements $\rho(\lambda)_{\nu_1,\nu_2} \in \operatorname{Hom}(V_{\nu_1}, V_{\nu_2})$, where ν_1, ν_2 are weights of ρ. It follows from Remark 3.3.2 that the entries of $\rho(\lambda)_{\nu_1,\nu_2}$ are homogeneous polynomials of the appropriate degree. Similarly, the morphism $\overline{\Sigma} \colon \mathfrak{h}/W \to \mathfrak{g}$ induces a morphism $\rho_*\overline{\Sigma} \colon \mathfrak{h}/W = \operatorname{Spec} S^W \to \operatorname{End} V$, and is thus identified with an element of $\operatorname{End} V \otimes_{\mathbb{C}} S^W$, which we continue to denote by $\rho_*\overline{\Sigma}$. The morphism $\rho_*\Sigma$ can likewise be viewed as an element of $\operatorname{End} V \otimes_{\mathbb{C}} S$. Since Σ is the pullback of $\overline{\Sigma}$, it follows that $\rho_*\Sigma$ is the image of $\rho_*\overline{\Sigma}$ under the inclusion $\operatorname{End} V \otimes_{\mathbb{C}} S^W \subseteq \operatorname{End} V \otimes_{\mathbb{C}} S$.

PROPOSITION 3.5.3. *Define the S-submodule M of* $\operatorname{Hom}_S(V \otimes_{\mathbb{C}} S, S)$ *by*

$$M = \{\phi \in \operatorname{Hom}_S(V \otimes_{\mathbb{C}} S, S) : \phi \circ \rho_*(\Sigma) = \mu \cdot \phi\}.$$

Then M is a free S-module of rank one, with a generator given by $(\pi_\mu \otimes \mathrm{Id}) \circ \rho(\lambda)$ (under any identification of V_μ with $\mathbb{C}$).

Proof. Since $\rho_*(\Sigma)$ and $\rho_*(i + X)$ are conjugate in $\operatorname{Aut}_S(V \otimes_{\mathbb{C}} S)$ (via $\rho(\lambda)$), it will suffice to prove the same statement with $\rho_*(\Sigma)$ replaced by $\rho_*(i + X)$ and M replaced by

$$M' = \{\phi \in \operatorname{Hom}_S(V \otimes_{\mathbb{C}} S, S) : \phi \circ \rho_*(i + X) = \mu \cdot \phi\}.$$

In this case, $\phi \in M'$ if and only if $\phi \circ \rho(\lambda) \in M$.

Fix a basis $\{v_i\}$ of V such that, for every $i, 1 \le i \le n$, $v_i \in V_{k_i}$. We may assume that, for $i \le j$, $k_i \le k_j$. In particular, $v_1 \in V_\mu$. In this basis, the element $\rho_*(i + X)$ is a lower triangular matrix whose diagonal entries are the weights of V, viewed as elements of S. Clearly, then, M' is identified with the set of all $\phi \colon V \otimes_{\mathbb{C}} S \to S$ which vanish on all weight spaces except for V_μ. This space is identified with $V_\mu^* \otimes_{\mathbb{C}} S$, where $V_\mu^* \subseteq V^*$ is the inclusion dual to the surjection $\pi_\mu \colon V \to V_\mu$. Thus, M' is a free rank one S-module with basis $\pi_\mu \otimes \mathrm{Id}$. It follows that $(\pi_\mu \otimes \mathrm{Id}) \circ \rho(\lambda)$ is a basis for M. □

Let W act on S in the usual way and trivially on V. There are induced actions of W on $V \otimes_{\mathbb{C}} S$ and on $\operatorname{Hom}_S(V \otimes_{\mathbb{C}} S, S)$, via $w \cdot \phi(v) = w \cdot \phi(w^{-1}v)$.

LEMMA 3.5.4. *The stabilizer W_0 of μ in W acts trivially on $(\pi_\mu \otimes \mathrm{Id}) \circ \rho(\lambda)$.*

Proof. By the previous proposition, $\tilde{f} = (\pi_\mu \otimes \mathrm{Id}) \circ \rho(\lambda)$ is an eigenvector for the action of $\rho_*(\Sigma)$ on $\operatorname{Hom}_S(V \otimes_{\mathbb{C}} S, S)$, with eigenvalue μ, and the corresponding eigenspace is free of rank one as an S-module. Since $\rho_*(\Sigma)$ is W-invariant, $w \cdot \tilde{f}$ is an eigenvector for $\rho_*(\Sigma)$ with eigenvalue $w\cdot\mu$. In particular, if $w \in W_0$, then $w\cdot\tilde{f} = c\tilde{f}$ for some $c \in S$. Since $\lambda \in G_-$, $\rho(\lambda)|V_\mu = \mathrm{Id}$. Thus, $(\pi_\mu \otimes \mathrm{Id}) \circ \rho(\lambda)|V_\mu \otimes_{\mathbb{C}} S$ is the identity. On the other hand, given $v \in V_\mu \otimes_{\mathbb{C}} S$ and $w \in W_0$, clearly $w^{-1}v \in V_\mu \otimes_{\mathbb{C}} S$ as well, so that $(w \cdot \tilde{f})(v) = w \cdot w^{-1}v = v$. It follows that $w \cdot \tilde{f}|V_\mu \otimes_{\mathbb{C}} S = \mathrm{Id}$, so that $c = 1$. Hence $\tilde{f}$ is W_0-invariant. □

COROLLARY 3.5.5. *Let M^{W_0} be the submodule of M fixed by W_0, so that*

$$M^{W_0} = \{\phi \in \mathrm{Hom}_{S^{W_0}}(V \otimes_{\mathbb{C}} S^{W_0}, S^{W_0}) : \phi \circ \rho_*(\Sigma) = \mu \cdot \phi\}.$$

Then M^{W_0} is a free rank one S^{W_0}-module, generated by $(\pi_\mu \otimes \mathrm{Id}) \circ \rho(\lambda)$.

Proof. We have a natural inclusion $M^{W_0} \otimes_{S^{W_0}} S \to M$. The element $(\pi_\mu \otimes \mathrm{Id}) \circ \rho(\lambda)$ lies in M^{W_0}, and so this inclusion is an isomorphism. Thus, $M^{W_0} \otimes_{S^{W_0}} S$ is free of rank one. Since S is faithfully flat over S^{W_0}, it follows that M^{W_0} is a free rank one S^{W_0}-module and that $(\pi_\mu \otimes \mathrm{Id}) \circ \rho(\lambda)$ is a generator. □

By Lemma 3.5.4 the map $(\pi_\mu \otimes \mathrm{Id}) \circ \rho(\lambda) \colon V \otimes_{\mathbb{C}} S \to S$ is invariant under W_0 and hence induces a map

$$(\pi_\mu \otimes \mathrm{Id}) \circ \rho(\lambda) \colon V \otimes_{\mathbb{C}} S^{W_0} \to S^{W_0}.$$

Let

$$f \colon V \to S^{W_0}$$

be the restriction of $(\pi_\mu \otimes \mathrm{Id}) \circ \rho(\lambda)$ to $V \otimes \{1\} \subset V \otimes_{\mathbb{C}} S^{W_0}$. Of course, the map f depends on λ, which depends on the choice of the slice Z.

COROLLARY 3.5.6. *Let $\mathfrak{m}_0 \subseteq S^W$ be the ideal generated by all W-invariant polynomials whose constant term is zero. Then the map*

$$\overline{f}_0 \colon V \to S^{W_0}/\mathfrak{m}_0 \cdot S^{W_0}$$

obtained by composing f with the natural quotient mapping $S^{W_0} \to S^{W_0}/\mathfrak{m}_0 \cdot S^{W_0}$ is independent of the choice of slice Z.

Proof. If Z' is another slice for $[X, \mathfrak{g}]$, then by Lemma 3.3.3 there is a map $\gamma \colon Z \to \mathfrak{u}_-$ with $\gamma(0) = 0$ such that

$$\mathrm{Ad}(\exp(\gamma(z)))(z + X) \in Z' + X$$

for all $z \in Z$. Let $\lambda = \exp(-\Psi)$ and $\lambda' = \exp(-\Psi')$, where Ψ, Ψ' are the maps given by Theorem 3.2.1 for the slices Z and Z', respectively. Then

$$\mathrm{Ad}(\exp(-\gamma(\Sigma_Z(h) - X))) \cdot (\lambda')^{-1} \cdot \mathrm{Ad}(\exp(\gamma(\Sigma_Z(h) - X))) \cdot \lambda \colon Z + X \to Z + X.$$

By uniqueness, it follows that

$$\exp(\gamma(\Sigma_Z(h) - X)) \cdot \lambda = \lambda' \cdot \exp(\gamma(\Sigma_Z(h) - X)).$$

Now $\exp(\gamma(\Sigma_Z(h) - X))$ is the exponential of a W-invariant map from $\mathfrak{h}$ to $\mathfrak{u}_-$, and thus $\exp(\gamma(\Sigma_Z(h) - X)) \equiv 1 \mod \mathfrak{m}_0$. Hence, $\lambda(h) \equiv \lambda'(h) \mod \mathfrak{m}_0 \cdot S^{W_0}$. □

A Kostant section can be viewed as parametrizing a universal family of regular, semistable G-bundles on the cuspidal Weierstrass cubic E whose support is contained in E_{reg}. We now apply Theorem 2.4.1 and Theorem 2.4.3 to the family of vector bundles on E obtained from this universal family by a minuscule or quasiminuscule representation ρ.

THEOREM 3.5.7. *Let $\rho\colon G \to GL(V)$ be an irreducible representation with lowest weight μ and let*

$$f = (\pi_\mu \otimes \mathrm{Id}) \circ \rho(\lambda)\colon V \to S^{W_0}$$

be the element given in Corollary 3.5.5. Then by extension of scalars f induces a homomorphism $\hat{f}\colon V \otimes_{\mathbb{C}} S^W \to S^{W_0}$ making the following diagram commute:

$$\begin{array}{ccc} V \otimes_{\mathbb{C}} S^W & \xrightarrow{\hat{f}} & S^{W_0} \\ {\scriptstyle \rho_*\overline{\Sigma}}\downarrow & & \downarrow{\scriptstyle \mu\cdot} \\ V \otimes_{\mathbb{C}} S^W & \xrightarrow{\hat{f}} & S^{W_0}. \end{array}$$

If ρ is minuscule, then $\hat{f}$ is an isomorphism. If ρ is quasiminuscule, then $\hat{f}$ is surjective.

Proof. By Corollary 3.5.5 the S^{W_0}-module map $f \otimes \mathrm{Id}\colon V \otimes_{\mathbb{C}} S^{W_0} \to S^{W_0}$ makes the analogous diagram commute, where S^W is replaced by S^{W_0} and $\rho_*\overline{\Sigma}$ by $\rho_*\Sigma$, and the set of all such maps $V \otimes_{\mathbb{C}} S^{W_0} \to S^{W_0}$ with this property is a free S^{W_0}-module with $f \otimes \mathrm{Id}$ as a generator. Let $\hat{f} = f \otimes \mathrm{Id}\,|V \otimes_{\mathbb{C}} S^W$. Since $\rho_*(\Sigma)$ is S^W-invariant we see that the diagram as defined in the theorem is commutative. It remains to show that $\hat{f}$ is surjective in the quasiminuscule case and an isomorphism in the minuscule case.

Let E be a cuspidal curve of arithmetic genus one. We consider the G-bundle over $(\mathfrak{h}/W) \times E$ which is trivial on $(\mathfrak{h}/W) \times \widetilde{E}$ and which is given by the Kostant section $\Sigma\colon \mathfrak{h}/W \to \mathfrak{g}_{\mathrm{reg}} \subset \mathfrak{g}$ under the correspondence of Theorem 1.1.1. Let $\mathcal{V} \to (\mathfrak{h}/W) \times E$ be the vector bundle induced from this G-bundle by ρ. This vector bundle is also trivialized on $(\mathfrak{h}/W) \times \widetilde{E}$ and is given by the map $\rho_*\Sigma\colon \mathfrak{h}/W \to \mathrm{End}(V)$.

Applying Theorem 2.4.1 and Theorem 2.4.3 and using the fact that every line bundle over $\mathfrak{h}/W_0$ is trivial since $\mathfrak{h}/W_0$ is isomorphic to an affine space, we see that there is a surjection from $\mathcal{V}$ to the vector bundle

$$(\hat{\nu} \times \mathrm{Id})_*(r \times \mathrm{Id})^*\mathcal{P},$$

where $\hat{\nu}\colon \mathfrak{h}/W_0 \to \mathfrak{h}/W$ is the covering projection and $r\colon \mathfrak{h}/W_0 \to \mathbb{C} \subseteq E$ is the map induced by the W_0-invariant weight $\mu\colon E_{\mathrm{reg}} \otimes_{\mathbb{Z}} \Lambda \to E_{\mathrm{reg}}$. By Proposition 1.1.3, $\mathcal{P} \to E_{\mathrm{reg}} \times E$ pulls back to the trivial line bundle on $E_{\mathrm{reg}} \times \widetilde{E}$ and hence is given by a map $E_{\mathrm{reg}} \to \mathrm{End}(\mathbb{C}) = \mathbb{C}$. This map is the fixed identification $E_{\mathrm{reg}} \cong \mathbb{C}$. Thus, when we use this identification to produce an isomorphism of the coordinate ring of E_{reg} with $\mathbb{C}[t]$, $\mathcal{P}$ becomes the element $t \in \mathbb{C}[t]$. Hence, $(r \times \mathrm{Id})^*\mathcal{P}$ is identified with the element $r^*(t)$ which is the element $\mu \in S^{W_0}$. In this way we identify $(r \times \mathrm{Id})^*\mathcal{P}$ with the element $\mu \in S^{W_0}$ viewed as an S^{W_0}-valued endomorphism of $\mathbb{C}$. Applying $(\hat{\nu} \times \mathrm{Id})_*$ produces the S^W-module S^{W_0} with the S^W-linear endomorphism given by multiplication by μ.

Theorem 2.4.1 now implies that there is a surjective S^W-linear homomorphism $f'\colon V\otimes_{\mathbb{C}} S^W \to S^{W_0}$ making the following diagram commute:

$$\begin{array}{ccc} V\otimes S^W & \xrightarrow{f'} & S^{W_0} \\ {\scriptstyle \rho_*\overline{\Sigma}}\downarrow & & \downarrow{\scriptstyle \mu\cdot} \\ V\otimes S^W & \xrightarrow{f'} & S^{W_0}. \end{array}$$

Its extension to a map $\hat{f}'\colon V\otimes_{\mathbb{C}} S^{W_0}\to S^{W_0}$ is an element of the free module M^{W_0} and hence $\hat{f}' = s(f\otimes \mathrm{Id})$ for some $s\in S^{W_0}$. Since $\hat{f}'$ is surjective, $s\in S^{W_0}$ is invertible. Restricting to $V\otimes_{\mathbb{C}} S^W$, we see that $f' = s\hat{f}$, and, since f' is surjective, so is $\hat{f}$. Finally, if ρ is minuscule, then f' is an isomorphism and hence $\hat{f}$ is an isomorphism as well. □

COROLLARY 3.5.8. *Suppose that ρ is minuscule. Let $x\in\mathfrak{h}/W$ and let $\mathfrak{m}_x\subseteq S^W$ be the maximal ideal of x. Then the scheme-theoretic fiber over x in $\mathfrak{h}/W_0$ has coordinate ring $S^{W_0}/\mathfrak{m}_x\cdot S^{W_0}$. The map $\hat{f}$ induces an isomorphism $\overline{f}_x\colon V\to S^{W_0}/\mathfrak{m}_x\cdot S^{W_0}$. Under this isomorphism, multiplication by μ on the right-hand side becomes the action of $\rho_*\overline{\Sigma}(x)$. In particular, under the isomorphism $\overline{f}_0\colon V\to S^{W_0}/\mathfrak{m}_0\cdot S^{W_0}$ given in Corollary 3.5.6, multiplication by $\mu\in S^{W_0}$ on $S^{W_0}/\mathfrak{m}_0\cdot S^{W_0}$ corresponds to the action $\rho_*(X)$ of the principal nilpotent element X on V.* □

In particular, the corollary describes the action of every regular element of $\mathfrak{g}$ on V up to conjugation.

Using his deep results on perverse sheaves on the loop group of the Langlands dual of $\mathfrak{g}$, Ginzburg has proved the following generalization of Theorem 3.5.7:

THEOREM 3.5.9. *Let ρ be an arbitrary irreducible representation. The map $\hat{f}\colon V\otimes_{\mathbb{C}} S^W\to S^{W_0}$ is always surjective.*

It is natural to ask if there is a more elementary and direct proof of Theorem 3.5.9. The image of $\hat{f}$ is the S^W-submodule of S^{W_0} generated by $(\pi_\mu\otimes\mathrm{Id})\circ\rho(\lambda)_{\nu,\mu}$, where $\rho(\lambda)_{\nu,\mu}\colon V_\nu\otimes_{\mathbb{C}} S\to V_\mu\otimes_{\mathbb{C}} S$ is the linear map induced by $\rho(\lambda)$. It follows the image of $\hat{f}$ is generated by homogeneous elements, when S^W and S^{W_0} are given their natural gradings as subrings of S. By the homogeneous Nakayama lemma, to prove Theorem 3.5.9, it would suffice to prove that the induced homomorphism $\overline{f}_0\colon V\to S^{W_0}/\mathfrak{m}_0\cdot S^{W_0}$ is surjective.

3.6. The Steinberg section. Let $\overline{\Phi}\colon H/W\to G_{\mathrm{reg}}$ be a section of Steinberg type. By this we mean that $\overline{\Phi}$ is any morphism from H/W to G_{reg}, the set of regular elements of G, which is a section of the adjoint quotient morphism $G\to H/W$. We begin by proving a weaker analogue of Theorem 3.1.2 for $\overline{\Phi}$:

THEOREM 3.6.1. *Let $\overline{\Phi}\colon H/W\to G_{\mathrm{reg}}$ be the Steinberg section, and let $\Phi\colon H\to G_{\mathrm{reg}}$ be the composition $H\to H/W\to G_{\mathrm{reg}}$. Then there exists a Borel subgroup B of G containing H, with unipotent radical U, and a morphism $\phi\colon H\to G$ such that, for all $h\in H$, $\phi(h)\Phi(h)\phi(h)^{-1}\in B$, and moreover $\phi(h)\Phi(h)\phi(h)^{-1} = hu(h)$ for some morphism $u\colon H\to U$.*

Proof. Suppose that we can show that there exists some Borel subgroup B and a morphism $\phi\colon H\to G$ such that, for all $h\in H$, $\phi(h)\Phi(h)\phi(h)^{-1}\in B$. Then after a further conjugation we may assume that B contains H. In this case, we can write

$\phi(h)\Phi(h)\phi(h)^{-1} = h'u(h)$ for some morphism $u\colon H \to U$, where h and h' lie in H and have the same image in H/W. It follows that there is a $w \in W$ such that, for an open dense set of $h \in H$, $h' = w(h)$. Hence we can assume that $h' = w(h)$ for all h. After further conjugating B by a representative for w in the normalizer of H in G, we can then assume that $\phi(h)\Phi(h)\phi(h)^{-1} = hu(h)$ for some morphism $u\colon H \to U$ as desired.

To find the morphism ϕ, we first claim that there exists a morphism $\overline{\phi}$ from H to the space $\mathcal{B} \cong G/B$ of Borel subgroups of G, such that, for all $h \in H$, $h \in \overline{\phi}(h)$. To see this, fix a Borel subgroup B and let $I \subseteq G \times (G/B)$ be the incidence variety:

$$I = \{(g, xB) : x^{-1}gx \in B\} = \{(g, xB) : g \in xBx^{-1}\}.$$

Identifying the set of all Borel subgroups of G with G/B, I is the set of pairs consisting of an element g of G and a Borel subgroup containing g. Clearly I is a closed subvariety of $G \times (G/B)$. The morphism $(g, xB) \mapsto (x, x^{-1}gx)$ defines an isomorphism from I to $G \times_B B$, where B operates on G by right multiplication and on itself by conjugation, with inverse $(g, b) \mapsto (gbg^{-1}, gB)$ (cf. [**14**, §4.3]). Let I_{reg} be the inverse image of G_{reg} under the projection $\pi_1\colon I \to G$. There is a morphism $\theta\colon I \to H$ defined by the composition

$$I \cong G \times_B B \to G \times_B H \cong (G/B) \times H \to H,$$

where the morphism $G \times_B B \to G \times_B H$ is induced by the homomorphism $B \to H$, and the morphism $(G/B) \times H \to H$ is projection onto the second factor. It is straightforward to verify that the following diagram is commutative:

$$\begin{array}{ccc} I & \xrightarrow{\pi_1} & G \\ {\scriptstyle\theta}\downarrow & & \downarrow \\ H & \longrightarrow & H/W. \end{array}$$

By a theorem of Grothendieck [**14**, §4.4], this diagram identifies $I \to H$ with a simultaneous resolution of the the morphism $G \to H/W$, in the terminology of [**14**]. Since the adjoint quotient morphism $G \to H/W$ is smooth along G_{reg}, the above diagram identifies I_{reg} with the fiber product $H \times_{H/W} G_{\mathrm{reg}}$. Using the morphisms $\overline{\Phi}\colon H/W \to G$ and $\pi_1\colon I \to G$, we can take the fiber product $(H/W) \times_G I$. By the above remarks (all products below are fiber products),

$$(H/W) \times_G I = (H/W) \times_{G_{\mathrm{reg}}} I_{\mathrm{reg}} \cong (H/W) \times_{G_{\mathrm{reg}}} (H \times_{H/W} G_{\mathrm{reg}}) \cong H.$$

The isomorphism $H \cong (H/W) \times_G I$ together with the projection $(H/W) \times_G I \to I$ identify an element h of H with a pair (x, B_h), where $x \in H/W$ is the image of h and B_h is a Borel subgroup containing $\overline{\Phi}(x) = \Phi(h)$. Here B_h is the image of h under the morphism $\overline{\phi}\colon H \to G/B$, which is the composition

$$H \cong (H/W) \times_G I \to I \to G/B,$$

and we identify G/B with the variety of all Borel subgroups of G.

Now suppose that we can lift the morphism $\overline{\phi}\colon H \to G/B$ to a morphism $\phi\colon H \to G$. It then follows that, for all $h \in H$, $\phi(h)\Phi(h)\phi(h)^{-1} \in B$ as claimed. To see that such a lift is possible, let $\underline{B}$ be the sheaf of morphisms from H to B, and similarly for $\underline{G}$ and $\underline{G/B}$. Then the following is an exact sequence of sheaves of sets in the étale topology:

$$\{1\} \to \underline{B} \to \underline{G} \to \underline{G/B} \to \{1\}.$$

In particular, we have the following long exact sequence (of pointed sets):

$$H^0(H;\underline{G}) \to H^0(H;\underline{G/B}) \to H^1(H;\underline{B}).$$

On the other hand, the solvable group B has a filtration by normal subgroups, such that the successive quotients are either $\mathbb{G}_m$ or $\mathbb{G}_a$. Since H is a torus, all line bundles on it are trivial, and since it is affine, all higher coherent sheaf cohomology vanishes. Thus $H^1(H;\underline{B})$ is trivial. It follows that every morphism from H to G/B lifts to a morphism from H to G, as claimed. This concludes the proof. $\square$

The method of proof above also proves a weak form of Theorem 3.1.2. However, it does not give the homogeneity statements.

Arguments very similar to those used in the proof of Theorem 3.5.7 then show the following:

THEOREM 3.6.2. *Let $\mathcal{S}$ be the affine coordinate ring of H. Let $\rho\colon G \to GL(V)$ be an irreducible representation with lowest weight μ and let*

$$g = (\pi_\mu \otimes \mathrm{Id}) \circ \rho(\phi)^{-1}\colon V \to \mathcal{S}^{W_0}.$$

Then by extension of scalars g induces a homomorphism $\hat{g}\colon V \otimes_{\mathbb{C}} \mathcal{S}^W \to \mathcal{S}^{W_0}$ making the following diagram commute:

$$\begin{array}{ccc} V \otimes_{\mathbb{C}} \mathcal{S}^W & \xrightarrow{\hat{g}} & \mathcal{S}^{W_0} \\ {\scriptstyle \rho_*\overline{\Phi}}\downarrow & & \downarrow{\scriptstyle \mu\cdot} \\ V \otimes_{\mathbb{C}} \mathcal{S}^W & \xrightarrow{\hat{g}} & \mathcal{S}^{W_0}. \end{array}$$

If ρ is minuscule, then $\hat{g}$ is an isomorphism. If ρ is quasiminuscule, then $\hat{g}$ is surjective. $\square$

In case ρ is minuscule, there is also a corollary to the above theorem, along the lines of Corollary 3.5.8, describing the action of $\overline{\Phi}(x)$ on V in terms of multiplication by μ on $\mathcal{S}^{W_0}/\mathfrak{m}_x \cdot \mathcal{S}^W$, which we leave to the reader to formulate.

Finally, it is natural to make the following conjecture, which does not seem to follow from the Lie algebra analogue:

CONJECTURE 3.6.3. *In the above notation, the homomorphism $\hat{g}$ is always surjective.*

4. Examples

The case of A_{n-1}. Let $\mathfrak{h}$ be the Cartan subalgebra of SL_n, so that $\mathfrak{h} \subset \mathbb{C}^n$ is identified with

$$\mathfrak{h} = \left\{ (x_1, \dots, x_n) | \sum_i x_i = 0 \right\}.$$

The coordinate ring of $\mathfrak{h}$ is $S = \mathbb{C}[x_1, \dots, x_n]/(\sum_i x_i)$. Let $\sigma_2, \dots, \sigma_n$ be the elementary symmetric functions in $x_1, \dots, x_n$ (note that $\sigma_1 = 0$ on $\mathfrak{h}$). Let $W = \mathfrak{S}_n$ be the symmetric group on n letters and let $W_0 = \mathfrak{S}_{n-1}$, which is embedded in W as the stabilizer of n. Thus W acts on S and $S^W = \mathbb{C}[\sigma_2, \dots, \sigma_n]$. Let $\sigma_i(k)$ be the ith elementary symmetric function in the variables $x_1, \dots, x_k$. Thus for example $\sigma_i(n) = \sigma_i$. Note that $x_n = -\sigma_1(n-1)$. More generally, we have the relation

$$(*) \qquad \sigma_i(n-1) + x_n\sigma_{i-1}(n-1) = \sigma_i.$$

Clearly

$$\begin{aligned} S^{W_0} &= \mathbb{C}[x_n, \sigma_1(n-1), \dots, \sigma_{n-1}(n-1)]/(x_n + \sigma_1(n-1) = 0) \\ &= \mathbb{C}[\sigma_1(n-1), \sigma_2(n-1), \dots, \sigma_{n-1}(n-1)]. \end{aligned}$$

By $(*)$, it follows that $S^{W_0} = S^W[x_n]$. Since x_n satisfies the monic equation $p(x) = \sum_{i=0}^{n}(-1)^i\sigma_i x^{n-i} = 0$, $S^{W_0} = S^W \cdot 1 \oplus \cdots \oplus S^W \cdot x_n^{n-1}$. Again by using $(*)$, we can write x_n^i as $\sigma_i(n-1)$ plus terms involving x_n^j for $j < i$ as well as elements of S^W. Thus

$$S^{W_0} = S^W \cdot 1 \oplus S^W \cdot \sigma_1(n-1) \oplus \cdots \oplus S^W \cdot \sigma_{n-1}(n-1)$$

as well, so that $1, \sigma_1(n-1), \dots, \sigma_{n-1}(n-1)$ is a basis for S^{W_0} as an S^W-module.

Let V be the defining n-dimensional representation of SL_n. The vector space V has the standard basis $e_1, \dots, e_n$. Define the $n \times n$ matrix $\overline{\Sigma}$ by

$$\overline{\Sigma} = \begin{pmatrix} 0 & 1 & 0 & 0 & \cdots & 0 & 0 \\ 0 & 0 & 1 & 0 & \cdots & 0 & 0 \\ 0 & 0 & 0 & 1 & \cdots & 0 & 0 \\ \vdots & \vdots & \vdots & \vdots & \ddots & \vdots & \vdots \\ (-1)^{n-1}\sigma_n & (-1)^{n-2}\sigma_{n-1} & (-1)^{n-3}\sigma_{n-2} & (-1)^{n-4}\sigma_{n-3} & \cdots & -\sigma_2 & 0 \end{pmatrix}.$$

As is well-known, $\overline{\Sigma}$ is a Kostant section for $\mathfrak{sl}_n$.

For each $k, 1 \le k \le n$, let $m_i(k)$ be the sum of all monomials of degree i in the variables $x_1, \dots, x_k$. Define the column vector $\mathbf{c}_k$ by

$$\mathbf{c}_k = \begin{pmatrix} 0 \\ 0 \\ \vdots \\ 1 \\ m_1(k) \\ \vdots \\ m_{n-k}(k) \end{pmatrix},$$

and let $A = (\mathbf{c}_1 \ \cdots \ \mathbf{c}_n)$.

It is not difficult to show that

$$A^{-1}\overline{\Sigma}A = \begin{pmatrix} x_1 & 1 & 0 & 0 & \cdots & 0 & 0 \\ 0 & x_2 & 1 & 0 & \cdots & 0 & 0 \\ 0 & 0 & x_3 & 1 & \cdots & 0 & 0 \\ \vdots & \vdots & \vdots & \vdots & \ddots & \vdots & \vdots \\ 0 & 0 & 0 & 0 & \cdots & 0 & x_n \end{pmatrix}$$

and that $A^{-1} =$

$$\begin{pmatrix} 1 & 0 & 0 & \cdots & 0 & 0 \\ -\sigma_1(1) & 1 & 0 & \cdots & 0 & 0 \\ \sigma_2(2) & -\sigma_1(2) & 1 & \cdots & 0 & 0 \\ \vdots & \vdots & \vdots & \ddots & \vdots & \vdots \\ (-1)^{n-1}\sigma_{n-1}(n-1) & (-1)^{n-2}\sigma_{n-2}(n-1) & (-1)^{n-3}\sigma_{n-3}(n-1) & \cdots & -\sigma_1(n-1) & 1 \end{pmatrix}.$$

Thus, $A^{-1} \in \mathrm{End}(V) \otimes S$ is the matrix realizing the endomorphism $\rho(\lambda)$ of Theorem 1 for the standard n-dimensional representation ρ of SL_n. The lowest weight of this representation is x_n and hence $\pi_\mu = x_n$. Thus, $(\pi_\mu \otimes \mathrm{Id}) \circ \rho(\lambda)$ is the row matrix given by the last row of A^{-1}. Note that the last row of A^{-1} is invariant under $W_0 = \mathfrak{S}_{n-1}$ and is in fact up to sign the basis for S^{W_0} over S^W described above.

We now consider the minuscule representation $\bigwedge^k V$. The lowest weight μ for this representation is $e_n + \ldots + e_{n-k+1}$ and hence the map $(\pi_\mu \otimes \mathrm{Id}) \circ \rho(\lambda)$ is the map given by the $k \times k$ minors of A^{-1} which involve the last k rows. The stabilizer of μ is $\mathfrak{S}_{n-k} \times \mathfrak{S}_k$, where the first factor acts on the first $n-k$ variables and the second on the last k variables. It is clear from the explicit description of the matrix for A^{-1} that each of the last k rows is invariant under the group $\mathfrak{S}_{n-k}$ and hence the determinants we are considering are also invariant under this group. The theorem implies that these determinants are also invariant under $\mathfrak{S}_k$ acting on the last k variables. In fact, this is easy to check directly.

Let $\tau_j(k)$ be the jth elementary symmetric function in the last k variables. We have the relations

$$\sigma_\ell = \sum_{i=0}^{\ell} \sigma_i(n-k)\tau_{\ell-i}(k), \ell = 1, \ldots, n.$$

The ring S^{W_0} is the quotient of the polynomial ring $\mathbb{C}[\sigma_i(n-k), \tau_j(k)]$ modulo the above relations. The theorem implies that the $k \times k$-minors of the $k \times n$ matrix obtained from A^{-1} by only considering the last k rows are a basis for S^{W_0} over S^W.

To complete the picture for SL_n, we consider the adjoint representation. In this case $\mu = e_n - e_1$ and the group W_0 is a copy of $\mathfrak{S}_{n-2}$, embedded in $\mathfrak{S}_n$ as the stabilizer of 1 and n. It is straightforward to check in this case that the image of $V \otimes_{\mathbb{C}} S^W$ in S^{W_0} is the S^W-submodule of S^{W_0} generated by $\sigma_{n-i}(n-1)x_1^{j-1}, 1 \leq i, j \leq n$, and that this submodule is all of S^{W_0}.

The case of C_n. The unique minuscule representation of $Sp(2n, \mathbb{C})$ is the defining $2n$-dimensional representation. Its weights are $\pm x_i, 1 \leq i \leq n$, with lowest weight $-x_1$. The ring S^W is a polynomial algebra $\mathbb{C}[\alpha_1, \ldots, \alpha_n]$, where α_i is the ith elementary symmetric function on $x_1^2, \ldots, x_n^2$. If W_0 is the stabilizer of the lowest weight $-x_1$, then S^{W_0} is a polynomial algebra with generators $x_1, \beta_1, \ldots, \beta_{n-1}$ where β_i is the ith elementary symmetric function on $x_2^2, \ldots, x_n^2$. As in the case of A_{n-1}, there is a set of relations

$$x_1^2\beta_{i-1} + \beta_i = \alpha_i, i = 1, \ldots, n.$$

Note that x_1 satisfies a monic polynomial over S^W of degree $2n$. It follows that $S^{W_0} = S^W[x_1] = S^W \cdot 1 \oplus S^W \cdot x_1 \oplus \cdots \oplus S^W \cdot x_1^{2n-1}$. In particular, $-x_1$ acts on $S^{W_0}/\mathfrak{m}_0 \cdot S^{W_0}$ with a Jordan block of length $2n$, corresponding to the fact that the image of a principal nilpotent element of $\mathfrak{sp}(2n)$ under the standard representation is a principal nilpotent matrix. In terms of bundles, this is equivalent to the statement that the bundle I_{2n} has a nondegenerate symplectic form.

The orthogonal representation of $Spin(2n)$. Choose coordinates $x_1, \ldots, x_n$ on the Cartan subalgebra of $\mathfrak{spin}(2n)$ so that the coroot lattice is the even integral lattice. The orthogonal representation of $Spin(2n)$ has weights $\pm x_1, \ldots, \pm x_n$, and $-x_1$ is the lowest weight. Let W_0 be the subgroup of the Weyl group W stabilizing $-x_1$. It is the Weyl group for $Spin(2n-2)$ in the variables $x_2, \ldots, x_n$. As before, we let α_i be the ith elementary symmetric function on $x_1^2, \dots, x_n^2$ and β_i the ith elementary symmetric function on $x_2^2, \dots, x_n^2$. The ring S^W is the ring generated by $\alpha_1, \dots, \alpha_n$ and the Pfaffian $\mathrm{Pfaff}_n = x_1 \cdots x_n$ subject to the one relation $\mathrm{Pfaff}_n^2 = \alpha_n$. Thus S^W is a polynomial algebra on $\alpha_i, 1 \le i < n$, and Pfaff_n. The algebra S^{W_0} is the algebra generated by x_1, $\beta_i, 1 \le i \le n-1$, and Pfaff_{n-1}, subject to the relation $\mathrm{Pfaff}_{n-1}^2 = \beta_{n-1}$, and thus S^{W_0} is a polynomial algebra generated by $x_1, \beta_i, 1 \le i < n-1$, and Pfaff_{n-1}. The ring $S^{W_0}/\mathfrak{m}_0 \cdot S^{W_0}$ is the quotient of S^{W_0} by the relations

$$x_1^2\beta_i + \beta_{i+1} = 0;$$
$$x_1 \mathrm{Pfaff}_{n-1} = 0.$$

Hence $S^{W_0}/\mathfrak{m}_0 \cdot S^{W_0}$ is generated by x_1 and Pfaff_{n-1} subject to the relations

$$x_1^{2n-2} = \mathrm{Pfaff}_{n-1}^2;$$
$$x_1 \mathrm{Pfaff}_{n-1} = 0.$$

It follows that

$$x_1^{2n-1} = x_1 \mathrm{Pfaff}_{n-1}^2 = 0.$$

A $\mathbb{C}$-basis for $S^{W_0}/\mathfrak{m}_0 \cdot S^{W_0}$ is $1, x_1, \dots, x_1^{2n-2}, \mathrm{Pfaff}_{n-1}$. As a module over $\mathbb{C}[x_1]$, $S^{W_0}/\mathfrak{m}_0 \cdot S^{W_0} \cong \mathbb{C}[x_1]/(x_1^{2n-2}) \oplus \mathbb{C}[x_1]/(x_1)$. This reflects the fact, which is easy to check directly, that a principal nilpotent element of $\mathfrak{spin}(2n)$ acts on V with two Jordan blocks, one of dimension $2n-1$ and one of dimension 1. This implies that the orthogonal bundle of rank $2n$ which is regular as a $Spin(2n)$-bundle and S-equivalent to the trivial bundle is $I_{2n-1} \oplus \mathcal{O}_E$. In particular, it is not regular as a vector bundle.

The case of E_6. Direct computation shows that the action of the principal nilpotent element X in the Lie algebra of type of E_6 on the 27-dimensional minuscule representation V has three Jordan blocks of dimensions 17, 9, and 1. The bottom of the Jordan block of rank 17 is the lowest weight space V_μ. Thus, Theorem 1 implies that multiplication by μ in $S^{W_0}/\mathfrak{m}_0 \cdot S^{W_0}$ is nilpotent of order 17. One can show that $S^{W_0} = S^W[\mu, a, b]$, where μ, a, b are homogeneous in degrees 1, 4, 8 respectively. Note that S^W is a graded polynomial algebra with generators in degrees $2, 5, 6, 8, 9, 12$. On the other hand, since W_0 is the Weyl group of D_5, and its action on $\mathfrak{h}$ decomposes as a copy of the standard representation of $W(D_5)$ plus a trivial factor, it follows that S^{W_0} has generators in degrees $1, 2, 4, 5, 6, 8$. Given these degrees and the fact that S^{W_0} has rank 27 over S^W, it is easy to check directly that S^{W_0} is generated as an S^W-algebra by elements in degrees $1, 4, 8$.

5. Quasiminuscule representations

Let ρ be a quasiminuscule, nonminuscule representation, and let $\mathcal{V}$ be the corresponding vector bundle over $\mathcal{M} \times E$. In this section, we describe the extension class corresponding to the extension

$$(*) \qquad 0 \to \pi_1^*\pi_{1*}\mathcal{V} \to \mathcal{V} \to \overline{\mathcal{V}} \to 0$$

of Theorem 2.4.3, in the case where G is simply laced and hence ρ is the adjoint representation. It seems likely that similar geometric methods can also describe the quasiminuscule, nonminuscule representation in the non-simply laced case. We shall just write out the case of $\mathfrak{g}$, i.e. where E is cuspidal, although minor modifications handle the nodal and smooth cases. Recall the notation of §2.1: we have the finite cover $\hat{\nu}\colon \hat{T}_0 \to \mathcal{M}$, as well as the normalization map $\varphi\colon \hat{T}_0 \to T_0 \subseteq \mathcal{M} \times E_{\mathrm{reg}}$, where T_0 is the image of $(\hat{\nu}, r)\colon \hat{T}_0 \to \mathcal{M} \times E_{\mathrm{reg}}$. Fixing a coordinate t for E at p_0, the pullback r^*t in the coordinate ring of $\hat{T}_0$, i.e. as an element of S^{W_0}, is just the weight μ. In the above notation, $\overline{\mathcal{V}} = (\hat{\nu} \times \mathrm{Id})_*(r \times \mathrm{Id})^*\mathcal{P}$.

5.1. The extension group. We begin by identifying the module of extensions for exact sequences whose first and third terms are above.

PROPOSITION 5.1.1. *Let $\mathcal{W}$ be a vector bundle over $\mathcal{M}$ and let M be the projective and hence free S^W-module corresponding to $\mathcal{W}$. Then $\mathrm{Ext}^1_{\mathcal{M}\times E}(\overline{\mathcal{V}}, \pi_1^*\mathcal{W})$, which is a module over S^W, is isomorphic to the S^W-module $(S^{W_0}/\mu S^{W_0}) \otimes_{S^W} M$.*

Proof. By Corollary 1.4.3, the module of such extensions is isomorphic to $(Q^\vee/tQ^\vee) \otimes_{S^W} M$, where $Q^\vee$ is the Fourier-Mukai transform of $(\overline{\mathcal{V}})^\vee$. Thus it suffices to show that $Q^\vee \cong S^{W_0}$. By Theorem 2.4.3, $\overline{\mathcal{V}}$ is the direct image under $\hat{\nu} \times \mathrm{Id}$ of the pullback of a line bundle on $\hat{T}_0$, necessarily trivial, tensored with $(r \times \mathrm{Id})^*\mathcal{P}$. Applying duality for the finite flat morphism $\hat{\nu} \times \mathrm{Id}$, and using the fact that the relative dualizing sheaf $\omega_{\hat{T}_0/\mathcal{M}}$ is a line bundle on $\hat{T}_0$, and hence is trivial as well, it follows that $(\overline{\mathcal{V}})^\vee$ is the direct image under $\hat{\nu} \times \mathrm{Id}$ of $(r \times \mathrm{Id})^*\mathcal{P}^{-1}$. Note that reflection in the root μ defines an element of W of order two normalizing W_0, and hence an involution τ of $\hat{T}_0$ such that $\hat{\nu} \circ \tau = \hat{\nu}$ and $r \circ \tau = -\mathrm{Id} \circ r$. Thus

$$\begin{aligned}(\hat{\nu} \times \mathrm{Id})_*(r \times \mathrm{Id})^*\mathcal{P}^{-1} &= (\hat{\nu} \times \mathrm{Id})_*(\tau \times \mathrm{Id})^*(r \times \mathrm{Id})^*\mathcal{P}^{-1} \\ &= (\hat{\nu} \times \mathrm{Id})_*(r \times \mathrm{Id})^*(-\mathrm{Id} \times \mathrm{Id})^*\mathcal{P}^{-1}.\end{aligned}$$

On the other hand, from the symmetric definition of $\mathcal{P}$, it follows that

$$(-\mathrm{Id} \times \mathrm{Id})^*\mathcal{P}^{-1} \cong \mathcal{P}.$$

Thus $(\overline{\mathcal{V}})^\vee \cong (\hat{\nu} \times \mathrm{Id})_*(r \times \mathrm{Id})^*\mathcal{P} \cong \overline{\mathcal{V}}$, and in particular $Q^\vee \cong S^{W_0}$. □

Let $D = T_0 \cap (\mathcal{M} \times \{p_0\})$ and let $\hat{D}$ be the pullback of D to $\hat{T}_0$. Since the affine coordinate ring of $\hat{D}$ is equal to $S^{W_0}/\mu S^{W_0}$, we can rephrase the proposition as follows:

COROLLARY 5.1.2. *In the above notation, the S^W-module $\mathrm{Ext}^1_{\mathcal{M}\times E}(\overline{\mathcal{V}}, \pi_1^*\mathcal{W})$ is isomorphic to the S^W-module of global sections of $\hat{\nu}^*\mathcal{W}|\hat{D}$.* □

Clearly, $\hat{D} = r^{-1}(0)$, and hence $\hat{D}$ is isomorphic to $\mathcal{M}_0$, in the terminology of §2.1. In particular, $\hat{D}$ is smooth. It follows from Lemma 2.3.5 that the induced morphism $\hat{D} \to D$ is birational. We shall say more about the structure of $\hat{D} \to D$ below.

5.2. Identification of the extension class. Assume now that ρ is the adjoint representation. The nonzero weights of ρ are the roots of $\mathfrak{g}$, and $\mathcal{V} = \mathrm{ad}\,\Xi$. First, we identify the vector bundle $\mathcal{W} = \pi_{1*}\mathcal{V}$:

PROPOSITION 5.2.1. *The vector bundle $\pi_{1*}\mathcal{V}$ is isomorphic to $\Omega^1_{\mathcal{M}}$.*

Proof. The bundle Ξ is a universal G-bundle over $\mathcal{M} \times E$. It follows that the Kodaira-Spencer homomorphism $T_{\mathcal{M}} \to R^1\pi_{1*}\,\mathrm{ad}\,\Xi$ is an isomorphism. (Of course, this is just a restatement of the result of Kostant [**11**] that the adjoint quotient morphism is smooth at a regular element.) Using relative duality, the isomorphism $(\mathrm{ad}\,\Xi)^\vee \cong \mathrm{ad}\,\Xi$ given by the Killing form, and the fact that the relative dualizing sheaf $\omega_{\mathcal{M}\times E/\mathcal{M}}$ is trivial, it follows that $(R^1\pi_{1*}\operatorname{ad}\Xi)^\vee \cong R^0\pi_{1*}\operatorname{ad}\Xi$. Thus $\Omega^1_{\mathcal{M}} \cong \pi_{1*}\operatorname{ad}\Xi = \pi_{1*}\mathcal{V}$, as claimed. □

There is another description of $\pi_{1*}\mathcal{V} \cong \Omega^1_{\mathcal{M}}$ as follows:

LEMMA 5.2.2. *The sheaf $\pi_{1*}\mathcal{V} \cong \Omega^1_{\mathcal{M}}$ corresponds to the S^W-module $(S\otimes_{\mathbb{C}}\mathfrak{h})^W$, where W acts on both factors of the tensor product by the natural actions.*

Proof. The sheaf $\Omega^1_{\mathcal{M}}$ is the coherent sheaf associated to the S^W-module $\Omega^1_{S^W/\mathbb{C}}$. By a theorem of Solomon (see for example [**2**, p. 135, ex. 3]), $\Omega^1_{S^W/\mathbb{C}} = (\Omega^1_{S/\mathbb{C}})^W$. But $\Omega^1_{S/\mathbb{C}} \cong S\otimes_{\mathbb{C}}\mathfrak{h}^*$, and the invariant bilinear form defines a W-equivariant isomorphism from $\mathfrak{h}^*$ to $\mathfrak{h}$. Thus $\Omega^1_{S^W/\mathbb{C}} \cong (S\otimes_{\mathbb{C}}\mathfrak{h})^W$. □

The extension class corresponding to $\mathcal{V}$ is thus a global section of $\hat{\nu}^*\Omega^1_{\mathcal{M}}|\hat{D}$. Our goal now will be to describe this section. Viewing D as a divisor on $\mathcal{M}\times\{p_0\}\cong\mathcal{M}$, there is the conormal morphism $I_D/I_D^2 \to \Omega^1_{\mathcal{M}}|D$. The image of this morphism is the conormal sheaf. At a smooth point of D, this subsheaf is a line subbundle which is the dual of the normal bundle to D. Since $\mathcal{M}$ is an affine space, $\operatorname{Pic}\mathcal{M}$ is trivial, and hence there is an $\bar{s}\in S^W$, unique up to a nonzero scalar, such that $I_D = (\bar{s})$. In fact, we can take $\bar{s} = \prod_{w\in W/W_0} w\mu$. Thus $d\bar{s}$ defines a section of $\Omega^1_{\mathcal{M}}|D$. There is then the induced section $\hat{\nu}^*d\bar{s}$ of $\hat{\nu}^*\Omega^1_{\mathcal{M}}|\hat{D}$. Since $\hat{D}\cong\mathcal{M}_0$, which is again an affine space, there exists an indivisible section s of $\hat{\nu}^*\Omega^1_{\mathcal{M}}|\hat{D}$, unique up to a constant, and an element $a\in\Gamma(\hat{D},\mathcal{O}_{\hat{D}})\cong S^{W_0}/\mu S^{W_0}$ such that $\hat{\nu}^*d\bar{s} = as$. Concretely, choosing a basis of the trivial vector bundle $\Omega^1_{\mathcal{M}}$ and hence of $\hat{\nu}^*\Omega^1_{\mathcal{M}}|\hat{D}\cong\mathcal{O}^r_{\hat{D}}$, we can write $s = (f_1,\dots,f_r)$, where the f_i have no common factor, and $as = (af_1,\dots,af_r)$ for functions f_i on $\hat{D}$.

THEOREM 5.2.3. *The above section s defines the extension class corresponding to $\mathcal{V}$. Moreover, s can be completed to a basis for $\hat{\nu}^*\Omega^1_{\mathcal{M}}|\hat{D}$. In other words, the sheaf $(\hat{\nu}^*\Omega^1_{\mathcal{M}}|\hat{D})/\mathcal{O}_{\hat{D}}\cdot s$ is a locally free sheaf of $\mathcal{O}_{\hat{D}}$-modules.*

Proof. We begin by verifying that, at a generic point x of D, the corresponding extension class lies in the image of the conormal line $N^{-1}_{D/\mathcal{M},x}$ in $\Omega^1_{\mathcal{M},x}$. For such an $x\in D$, we may assume that D and T_0 are smooth at x. Let ξ be the G-bundle over E corresponding to $x\in\mathcal{M}$. By the proof of Theorem 2.4.1 and 2.4.3, ξ is described as follows: let H' be the subtorus of H given by $\operatorname{Ker}\mu$ and let SL_2 be the subgroup of G corresponding to the Lie algebra spanned by $\mathfrak{g}^\mu$, $\mathfrak{g}^{-\mu}$, and $\mu^\vee$. Then there is the natural embedding $SL_2\times_{\mathbb{Z}/2\mathbb{Z}}H'\to G$. Let ξ_1 be the SL_2-bundle corresponding to I_2. Then there exists a generic H'-bundle ξ_2 such that, if ξ' is the bundle on $SL_2\times_{\mathbb{Z}/2\mathbb{Z}}H'$ induced by $\xi_1\boxtimes\xi_2$ on $SL_2\times H'$, then ξ is the induced G-bundle. Note that $\operatorname{ad}_{SL_2}\xi_1 = I_3$ and that $\operatorname{ad}_{H'}\xi_2 = \mathcal{O}_E^{r-1}$. There is the induced injection $\operatorname{ad}_{SL_2}\xi_1\oplus\operatorname{ad}_{H'}\xi_2\to\operatorname{ad}_G\xi$. The arguments of Theorem 2.4.3 imply that the corresponding homomorphism on H^1 is an isomorphism. In particular,

$$T_{\mathcal{M},x}\cong H^1(E;\operatorname{ad}_G\xi) = H^1(E;\operatorname{ad}_{SL_2}\xi_1)\oplus H^1(E;\operatorname{ad}_{H'}\xi_2).$$

Clearly, $H^1(E;\operatorname{ad}_{H'}\xi_2)$ is the image of the tangent space $T_{D,x}$ to D at x, and hence the normal bundle sequence is split at x. Another way to give the same splitting on the normal bundle sequence is as follows. The Killing form induces a quadratic form on $H^1(E;\operatorname{ad}\xi)$, which is easily seen degenerate with radical equal to $H^1(E; \mathrm{ad}_{SL_2}\xi_1)$, and this splits the sequence. Dualizing, there is a direct sum decomposition

$$\Omega^1_{\mathcal{M},x} \cong H^0(E; \mathrm{ad}\,\xi) = H^0(E; \mathrm{ad}_{SL_2}\xi_1) \oplus H^0(E; \mathrm{ad}_{H'}\xi_2).$$

Under the identification of $\Omega^1_{D,x}$ with $H^0(E; \mathrm{ad}_{H'}\xi_2)$, the above is a splitting of the conormal sequence such that the natural morphism $\Omega^1_{\mathcal{M},x} \to \Omega^1_{D,x}$ is the projection to $H^0(E; \mathrm{ad}_{H'}\xi_2)$.

The extension $(*)$ restricted to the slice $\{x\} \times E$ gives an extension

$$0 \to H^0(E; \mathrm{ad}\,\xi) \otimes_{\mathbb{C}} \mathcal{O}_E \to \mathrm{ad}\,\xi \to \overline{V} \to 0,$$

and hence defines an extension class $\varepsilon \in H^1(E; \overline{V}^\vee) \otimes H^0(E; \mathrm{ad}\,\xi)$. By construction, the part of $\overline{V}$ supported at $p_0 \in E$ is isomorphic to I_2 and so $H^1(E; \overline{V}^\vee) \cong \mathbb{C}$. Thus up to a scalar we can view ε as an element of $H^0(E; \mathrm{ad}\,\xi)$. We claim that, if V' is defined via the commutative diagram

$$\begin{array}{ccccccccc}
0 & \longrightarrow & H^0(E; \mathrm{ad}\,\xi) \otimes_{\mathbb{C}} \mathcal{O}_E & \longrightarrow & \mathrm{ad}\,\xi & \longrightarrow & \overline{V} & \longrightarrow & 0 \\
 & & \downarrow & & \downarrow & & \| & & \\
0 & \longrightarrow & H^0(E; \mathrm{ad}_{H'}\xi_2) \otimes_{\mathbb{C}} \mathcal{O}_E & \longrightarrow & V' & \longrightarrow & \overline{V} & \longrightarrow & 0
\end{array}$$

then the corresponding extension V' is split. There are many ways to see this. For example, it follows from the explicit knowledge of $\mathrm{ad}\,\xi$ given in the proof of Theorem 2.4.3. Another way is to use that fact that orthogonal projection from $\mathfrak{g}$ to $\mathfrak{h}'$ is equivariant with respect to the action of the subgroup $SL_2 \times_{\mathbb{Z}/2\mathbb{Z}} H'$. Thus there is an induced homomorphism of vector bundles from $\mathrm{ad}\,\xi$ to $H^0(E; \mathrm{ad}_{H'}\xi_2) \otimes_{\mathbb{C}} \mathcal{O}_E$. In turn, this homomorphism induces a surjection from V' to $H^0(E; \mathrm{ad}_{H'}\xi_2) \otimes_{\mathbb{C}} \mathcal{O}_E$ which splits the extension. In any case, we see that $\varepsilon \in H^0(E; \mathrm{ad}\,\xi) \cong \Omega^1_{\mathcal{M},x}$ lies in the kernel of the projection $\Omega^1_{\mathcal{M},x} \to \Omega^1_{D,x}$, and hence ε lies in the image of the conormal line $N^{-1}_{D/\mathcal{M},x}$. Thus, at a generic point of D, the extension class ε is equal to a nonzero multiple of $\hat{\nu}^* d\bar{s}$ at x and hence is a nonzero multiple of s at x.

Suppose that s' is the section of $\hat{\nu}^*\Omega^1_{\mathcal{M}}|\hat{D}$ corresponding to the extension $(*)$. The above shows that there exists a regular function c on $\hat{D}$ such that $s' = cs$. Fixing a basis for $\Omega^1_{\mathcal{M}}$, we can write $s' = cs = (cf_1, \dots, cf_r)$. To complete the proof of Theorem 5.2.3, it suffices to show the following: for all $y \in \hat{D}$, there exists an i such that $cf_i \notin \mathfrak{m}_y$. This will show that c is a nowhere vanishing function on $\hat{D}$ and hence a constant, and that s is a nowhere vanishing section of $\hat{\nu}^*\Omega^1_{\mathcal{M}}|\hat{D}$. To see this, let $x \in \mathcal{M}$ lie under y, let ξ be the corresponding G-bundle, and let $0 \to \mathcal{O}_E^r \to V \to \overline{V} \to 0$ be the corresponding extension, where we use the given basis of $\Omega^1_{\mathcal{M}}$ to trivialize the subbundle $H^0(E; V) \otimes \mathcal{O}_E$ of V. The fact about V that we shall need is that $h^0(E; V) = r$ and hence that the coboundary map $\delta\colon H^0(E; \overline{V}) \to H^1(E; \mathcal{O}_E^r)$ is injective. Write $\delta = (\delta_1, \dots, \delta_r)$, where $\delta_i \in \mathrm{Ext}^1(\overline{V}, \mathcal{O}_E) \cong \mathrm{Hom}(H^0(E; \overline{V}), H^1(E; \mathcal{O}_E))$. By Proposition 5.1.1, we have identified $Q = Q(V)$ with $\widetilde{R} = S^{W_0}/\mathfrak{m}_x S^{W_0}$ and $\mathrm{Ext}^1_{\mathcal{O}_E}(\overline{V}, \mathcal{O}_E^r)$ with $(\widetilde{R}/\mu\widetilde{R})^r$. Under this identification, the ith component δ_i of δ corresponds to the ith component of s', namely cf_i. Thus, if $cf_i \in \mathfrak{m}_y$ for every i, then cf_i lies in the subspace $\mathfrak{m}_y/\mu\widetilde{R}$ of $\widetilde{R}/\mu\widetilde{R}$, which is a proper subspace since $y \in \hat{D}$, i.e. $\mathfrak{m}_y$ is not the unit ideal. This would imply that all of the δ_i lie in a proper subspace of $\mathrm{Hom}(H^0(E; \overline{V}), H^1(E; \mathcal{O}_E))$, and hence that they have a common kernel. This contradicts the injectivity of δ. Thus $cf_i \notin \mathfrak{m}_y$ for some i, proving that both c and f_i are units at y. □

We can say more about the function a such that $\hat{\nu}^* d\bar{s} = as$ in the discussion preceding Theorem 5.2.3. To determine a up to a nonzero constant, it suffices to describe the divisor of zeroes of a in $\hat{D}$. We begin by describing the divisor D in more detail. Let $y \in \hat{D}$, so that by definition y is the image of a point $e \in E_{\text{reg}} \otimes \Lambda$ such that $\mu(e) = 0$. If $\pm\mu$ are the only roots which vanish at e, then by Lemma 2.3.5, $\hat{T}_0 \to T_0$ is a local isomorphism near y, and hence $\hat{D} \to D$ is also an isomorphism. Thus we may assume that there exists a root $\mu' \neq \pm\mu$ such that $\mu'(e) = 0$ as well. The possibilities are as follows (we leave the calculations to the reader):

LEMMA 5.2.4. *Let $e \in E_{\text{reg}} \otimes \Lambda$ lie over $y \in \hat{D}$. Suppose that there exists a root $\mu' \neq \pm\mu$ such that $\mu'(e) = 0$, and that e is a generic point of $D(\mu) \cap D(\mu')$.*

(i) *If μ and μ' are orthogonal, then $\varphi^{-1}(\varphi(y)) = \{y, y'\}$ consists of two distinct points, both T_0 and D have two local branches at $\varphi(y)$, meeting transversally, and $\varphi\colon \hat{T}_0 \to T_0$ is a local diffeomorphism at y onto one of the two branches of T_0 passing through $\varphi(y)$. There exist local coordinates $x_1, \dots, x_{r-2}$, s_1, t_2 for $\hat{T}_0$ at y and local coordinates $x_1, \dots, x_{r-2}$, t_1, t_2 for $\mathcal{M}$ at $\hat{\nu}(y)$ such that*
$$\hat{\nu}(x, s_1, t_2) = (x, t_1, t_2),$$
where $t_1 = s_1^2$. A local form for the equation of D is $t_1 t_2 = 0$. Finally, $\mu = r^* t = s_1$.

(ii) *If μ and μ' are not orthogonal, then they span a root system of type A_2. The preimage $\varphi^{-1}(\varphi(y)) = \{y\}$. There exist local coordinates $x_1, \dots, x_{r-2}$, s_1, s_2 for $\hat{T}_0$ at y and local coordinates $x_1, \dots, x_{r-2}$, σ_2, σ_3 for $\mathcal{M}$ at $\hat{\nu}(y)$ such that*
$$\hat{\nu}(x, s_1, s_2) = (x, \sigma_2, \sigma_3),$$
where $\sigma_2 = -(s_1^2 + s_2^2 + s_1 s_2)$ and $\sigma_3 = -(s_1^2 s_2 + s_1 s_2^2)$. A local form for the equation of D is $4\sigma_2^3 + 27\sigma_3^2 = 0$. Finally, $\mu = r^* t = s_1 - s_2$. □

COROLLARY 5.2.5. *Let $\hat{D}_{1,1}$ the the hypersurface in $\hat{D}$ which is the image of the set of $e \in E_{\text{reg}} \otimes \Lambda$ such that there exists a root $\mu' \neq \pm\mu$, orthogonal to μ, with $\mu'(e) = 0$. Let $\hat{D}_2$ the the hypersurface in $\hat{D}$ which is the image of the set of $e \in E_{\text{reg}} \otimes \Lambda$ such that there exists a root $\mu' \neq \pm\mu$, not orthogonal to μ with $\mu'(e) = 0$. Then the function a vanishes to order 1 along every component of $\hat{D}_{1,1}$ and to order 3 along every component of $\hat{D}_2$.* □

6. Further conjectures

In this final section, we speculate on how some of the previous results and conjectures might be generalized to an arbitrary irreducible representation ρ. For simplicity, we shall only discuss the case of $\mathfrak{g}$, where these questions might be accessible to the methods of Ginzburg. The general idea is to find a filtration of the corresponding vector bundle $\mathcal{V}$, or equivalently of the module $V \otimes_{\mathbb{C}} S^W$, so that the the associated graded module is free and action of $\rho_* \overline{\Sigma}$ on the associated graded can be described explicitly. For simplicity, we shall shift our point of view and work with highest weights instead of lowest weights in what follows.

6.1. Notation. Given an irreducible representation $\rho\colon G \to GL(V)$, let λ be a weight of ρ and let m_λ be the multiplicity of λ, so that $V = \bigoplus_\lambda V_\lambda$, where H acts on V_λ by the character corresponding to λ, and $\dim V_\lambda = m_\lambda$. Recall that, if λ is a weight, then so is $w\lambda$ for every $w \in W$. In particular, each W-orbit of weights contains a unique dominant weight. We define a partial ordering on the set of all dominant weights as follows: If λ_1 and λ_2 are two dominant weights, then $\lambda_1 \geq \lambda_2$ if the difference $\lambda_1 - \lambda_2$ is a rational linear combination of simple roots with nonnegative coefficients, and $\lambda_1 > \lambda_2$ to mean $\lambda_1 \geq \lambda_2$ and $\lambda_1 \neq \lambda_2$.

6.2. Statement of the conjecture. For each dominant weight λ of ρ, define the operator $P_\lambda \in \operatorname{End}(V \otimes_{\mathbb{C}} S^W)$ by

$$P_\lambda = \prod_{\lambda' \in W\cdot\lambda} (\rho_*\overline{\Sigma} - \lambda' \cdot \mathrm{Id}).$$

Note that P_λ does in fact lie in $\operatorname{End}(V \otimes_{\mathbb{C}} S^W)$, although the individual factors only lie in $\operatorname{End}(V \otimes_{\mathbb{C}} S)$, and that P_λ commutes with $\rho_*\overline{\Sigma}$. Moreover, $\operatorname{Ker} P_\lambda$ is a saturated submodule of $V \otimes_{\mathbb{C}} S^W$ which corresponds to the component of the spectral cover whose reduction is given by the Weyl orbit of λ. It seems natural, however, that the ordering of the weights must somehow be taken into account. To do so, make the following construction: For λ dominant, define

$$Q_\lambda = \prod_{\lambda' \leq \lambda} P_\lambda,$$

where the product is over all dominant weights $\lambda' \leq \lambda$, define and

$$Q_\lambda^0 = \prod_{\lambda' < \lambda} P_\lambda.$$

Then $F_\lambda = \operatorname{Ker} Q_\lambda \subseteq V \otimes_{\mathbb{C}} S^W$ is invariant under $\rho_*\overline{\Sigma}$, as is $F_\lambda^0 = \operatorname{Ker} Q_\lambda^0$, and $F_\lambda^0 \subseteq F_\lambda$. Note that, since the dominant weights of ρ might only be partially ordered, $\{F_\lambda\}$ need not be a filtration of $V \otimes_{\mathbb{C}} S^W$.

CONJECTURE 6.2.1. *Let λ be a dominant weight. Let W_λ be the stabilizer in W of λ and m_λ the multiplicity of λ in ρ. Then, in the above notation, there is an isomorphism of S^W-modules*

$$\hat{h}\colon F_\lambda/F_\lambda^0 \cong (S^{W_\lambda})^{m_\lambda},$$

and $\hat{h} \circ \rho_*\overline{\Sigma} = \lambda\hat{h}$.

In case λ is the highest weight of ρ, the conjecture is Ginzburg's result. At the other extreme, if $\lambda = 0$, the conjecture is contained in the following:

PROPOSITION 6.2.2. $\operatorname{Ker} \rho_*\overline{\Sigma}$ *is a free submodule of* $V \otimes_{\mathbb{C}} S^W$ *of rank* m_0.

Proof. If 0 is not a weight of ρ, then $\operatorname{Ker} \rho_*\overline{\Sigma}(x) = 0$ for each x such that $\overline{\Sigma}(x)$ is semisimple. Thus $\operatorname{Ker} \rho_*\overline{\Sigma} = 0$, and hence it is free of rank $m_0 = 0$. Thus we may assume that 0 is a weight of ρ. In this case, the proof uses the following result of Kostant [**11**, (5.1.1)]: Suppose that m_0, the multiplicity of the weight 0 in V, is positive. Let X be a regular element of $\mathfrak{g}$, let $\mathfrak{g}^X$ denote the centralizer of X, and define

$$V^{\mathfrak{g}^X} = \{v \in V : \rho_*(Y)(v) = 0 \text{ for all } Y \in \mathfrak{g}^X\}.$$

Then $\dim V^{\mathfrak{g}^X} = m_0$, and in particular it is independent of the choice of X.

For each $x \in \mathfrak{h}/W$, we just write $\mathfrak{g}^x$ for $\mathfrak{g}^{\overline{\Sigma}(x)}$. Then $\bigcup_{x\in\mathfrak{h}/W}\mathfrak{g}^x$ is an r-dimensional subbundle $\mathfrak{z}_{\overline{\Sigma}}$ of the trivial bundle $(\mathfrak{h}/W)\times\mathfrak{g}$. Since S^W is a polynomial algebra, the vector bundle $\mathfrak{z}_{\overline{\Sigma}}$ corresponds to a free S^W-submodule of $\mathfrak{g}\otimes_{\mathbb{C}} S^W$, with basis $X_1,\dots,X_r$ (although the main point is rather that the submodule is projective). Let $M=\bigcap_{i=1}^r \operatorname{Ker}\rho_*(X_i)\subseteq V\otimes_{\mathbb{C}} S^W$. Since $\dim V^{\mathfrak{g}^x}$ is independent of x, it follows that M is a projective, and hence free, submodule of $V\otimes_{\mathbb{C}} S^W$, clearly saturated. As $\overline{\Sigma}(x)\in\mathfrak{g}^x$ for every x, $M\subseteq \operatorname{Ker}\rho_*\overline{\Sigma}$. Finally, if x is the image of a regular semisimple element, then it is easy to see that $V^{\mathfrak{g}^x}=\operatorname{Ker}\overline{\Sigma}(x)$, and since M is saturated, we must have equality everywhere. Thus $\operatorname{Ker}\rho_*\overline{\Sigma}=M$, and in particular $\operatorname{Ker}\rho_*\overline{\Sigma}$ is free. □

References

[1] A. Borel, R. Friedman, and J.W. Morgan, *Almost commuting elements in compact Lie groups*, math.GR/9907007.

[2] N. Bourbaki, *Groupes et Algèbres de Lie*, Chap. 4, 5, et 6, Masson, Paris, 1981.

[3] N. Bourbaki, *Groupes et Algèbres de Lie*, Chap. 7 et 8, Masson, Paris, 1990.

[4] R. Friedman and J.W. Morgan, *Holomorphic principal bundles over elliptic curves*, math.AG/9811130.

[5] R. Friedman and J.W. Morgan, *Holomorphic principal bundles over elliptic curves II: The parabolic construction*, math.AG/0006174.

[6] R. Friedman and J.W. Morgan, *Holomorphic principal bundles over elliptic curves III: Singular curves and fibrations*, math.AG/0108104.

[7] R. Friedman and J.W. Morgan, *Automorphism sheaves, spectral covers, and the Kostant and Steinberg sections*, math.AG/0209053.

[8] R. Friedman, J.W. Morgan, and E. Witten, *Vector bundles over elliptic fibrations*, J. Algebraic Geometry **8** (1999), 279–401.

[9] V. Ginzburg. *Perverse sheaves on a loop group and Langlands' duality*, math.AG/9511007.

[10] V. Ginzburg. *Loop Grassmannian cohomology, the principal nilpotent and Kostant theorem*, math.AG/9803141.

[11] B. Kostant, *Lie group representations on polynomial rings*, Am. J. Math. **85** (1963), 327–404.

[12] B. Kostant, *On Whittaker vectors and representation theory*, Invent. Math. **48** (1978), 101–184.

[13] S. Mukai, *Duality between $D(X)$ and $D(\hat{X})$ with its application to Picard sheaves*, Nagoya Math. J. **81** (1981), 101–116.

[14] P. Slodowy, *Simple Singularities and Simple Algebraic Groups*, Lecture Notes in Mathematics **815**, Berlin New York, Springer, 1980.

[15] R. Steinberg, *Regular elements of semisimple algebraic groups*, Publ. Math. Inst. Hautes Études Sci. **25** (1965), 49–80.

[16] T. Teodorescu, *Semistable Torsion-Free Sheaves over Curves of Arithmetic Genus One*, Columbia University thesis (1999).

Department of Mathematics, Columbia University, New York, NY 10027

E-mail address: rf@math.columbia.edu, jm@math.columbia.edu

(2) [Og1] *Let X be a K3 surface with $NS(X) = \mathbf{Z}H$ and $(H^2) = 2n$. Then*

$$|FM(X)| = 2^{\tau(n)-1},$$

where $\tau(n)$ is the number of prime factors of n, e.g. $\tau(12) = \tau(6) = 2$, $\tau(8) = \tau(2) = 1$.

(3) [HLOY] *Let X be a K3 surface such that $\rho(X) = 2$ and $\det NS(X) = -p$, where p is a prime number. Let $h(p)$ be the class number of $\mathbf{Q}(\sqrt{p})$. Then*

$$|FM(X)| = \frac{h(p)+1}{2}.$$

§2. $FM(X)$ for K3 surfaces of Picard number 1

In this section, we prove the following theorem which describes the set of the FM partners of X geometrically:

Theorem 2.1. *Let X be a K3 surface such that $NS(X) = \mathbf{Z}H$ and $(H^2) = 2n$. Then*

$$FM(X) = \{M_H((r, H, s)) \mid rs = n,\ (r, s) = 1,\ r \geq s\}.$$

Proof. Note that

$$|\{(r, s)|rs = n, (r, s) = 1, r \geq s > 0\}| = 2^{\tau(n)-1} = |FM(X)|.$$

Here the first equality is elementary and the second equality is due to Corollary 1.6(2). Then, Theorem 2.1 follows from the next:

Proposition 2.2. *Let r (and respectively, r') be a positive integer satisfying $r|n$, $(r, \frac{n}{r}) = 1$ (respectively $r'|n$, $(r', \frac{n}{r'}) = 1$). Set $s = \frac{n}{r}$ and $s' = \frac{n}{r'}$. Then*

(1) $M_H((r, H, s)) \in FM(X)$.

(2) If $M_H((r', H, s')) \cong M_H((r, H, s))$ as abstract variety, then $r' = r$ or $r' = s$.

In what follows, we shall show this proposition.

Proof of (1). This is a special case of Theorem 1.3. □

Proof of (2). In what follows, we set $v(r) := (r, H, s)$. Let us consider the universal sheaf $\mathcal{E} \in \mathrm{Coh}(M_H(v(r)) \times X)$, and the associated FM transform:

$$\Phi^{\mathcal{E}}_{M_H(v(r)) \to X} : D(M_H(v(r))) \to D(X).$$

Then corresponding to this functor we have a Hodge isometry:

$$f_r := f^{\mathcal{E}}_{M_H(v(r)) \to X} : (\tilde{H}(M_H(v(r)), \mathbf{Z}), \mathbf{C}\omega_{M_H(v(r))}) \xrightarrow{\sim} (\tilde{H}(X, \mathbf{Z}), \mathbf{C}\omega_X) \quad (2.1)$$

with $f_{r'}((0,0,1)) = v(r')$.

Let $g : M_H(v(r')) \tilde{\to} M_H(v(r))$ be an isomorphism. This g induces a Hodge isometry

$$g^* : (H^2(M_H(v(r)), \mathbf{Z}), \mathbf{C}\omega_{M_H(v(r))}) \tilde{\to} (H^2(M_H(v(r')), \mathbf{Z}), \mathbf{C}\omega_{M_H(v(r'))}) .$$

Identifying the zero-th and four-th cohomology of the two varieties with $\mathbf{Z}$, we have a Hodge isometry, $\tilde{g} := (\mathrm{id}, g^*, \mathrm{id})$,

$$\tilde{g} : (\tilde{H}(M_H(v(r)), \mathbf{Z}), \mathbf{C}\omega_{M_H(v(r))}) \tilde{\to} (\tilde{H}(M_H(v(r')), \mathbf{Z}), \mathbf{C}\omega_{M_H(v(r'))}) .$$

Define

$$f_r((-1,0,0)) =: u(r) \ , \ f_{r'}((-1,0,0)) =: u(r')$$

for the Hodge isometries f_r and $f_{r'}$ above. Then

$$\begin{aligned} \langle v(r)^2 \rangle = \langle u(r)^2 \rangle = 0 \ , \ \langle u(r), v(r) \rangle = 1 \ , \\ \langle v(r')^2 \rangle = \langle u(r')^2 \rangle = 0 \ , \ \langle u(r'), v(r') \rangle = 1 \ . \end{aligned} \tag{2.3}$$

Now consider the Hodge isometry,

$$\tilde{\varphi} := f_{r'} \circ \tilde{g} \circ f_r^{-1} : (\tilde{H}(X, \mathbf{Z}), \mathbf{C}\omega_X) \tilde{\to} (\tilde{H}(X, \mathbf{Z}), \mathbf{C}\omega_X).$$

We have $\tilde{\varphi}(v(r)) = v(r')$ and $\tilde{\varphi}(u(r)) = u(r')$. Hereafter we consider the following restrictions:

$$\begin{aligned} \varphi := \tilde{\varphi}|_{\widetilde{NS}(X)} : \widetilde{NS}(X) \tilde{\to} \widetilde{NS}(X) \ , \\ \varphi_T := \tilde{\varphi}|_{T(X)} : (T(X), \mathbf{C}\omega_X) \tilde{\to} (T(X), \mathbf{C}\omega_X) \ . \end{aligned}$$

We also fix a basis for $\widetilde{NS}(X)$;

$$\widetilde{NS}(X) = \mathbf{Z}e \oplus \mathbf{Z}H \oplus \mathbf{Z}f \cong U \oplus \langle 2n \rangle \ ,$$

where e and f represent the (fundamental) classes of H^0 and H^4 respectively. Also we set

$$\begin{aligned} v(r) = re + H + sf \ , \ u(r) = le + kH + mf \ ; \\ v(r') = r'e + H + s'f \ , \ u(r') = l'e + k'H + m'f \ . \end{aligned}$$

Lemma 2.3. *Define $U_r := \mathbf{Z}u(r) \oplus \mathbf{Z}v(r)$, and denote by $\pi(r)$ a generator of $U_r^{\perp}$ in $\widetilde{NS}(X)$. Then up to sign, we have*

$$\pi(r) = 2n(-l + rk)e + 2n(m - sk)f + (rm - ls)H \ ,$$

and $\widetilde{NS}(X) = U_r \perp \mathbf{Z}\pi(r)$ (orthogonal direct sum). Similarly for $U_{r'} := \mathbf{Z}u(r') \oplus \mathbf{Z}v(r')$, we have $\widetilde{NS}(X) = U_{r'} \perp \mathbf{Z}\pi(r')$ and, up to sign,

$$\pi(r') = 2n(-l' + r'k')e + 2n(m' - s'k')f + (r'm' - l's')H \ .$$

Proof. We have $\mathrm{rk}\, U_r^{\perp} = 1$, which implies that the generator of $U_r^{\perp}$ is unique up to sign. Moreover, since $\det \widetilde{NS}(X) = -2n$ and $U_r \cong U$ is unimodular, one concludes that $\widetilde{NS}(X) = U_r \perp \mathbf{Z}\Pi(r)$ for some $\Pi(r)$ such that $\langle \Pi(r)^2 \rangle = 2n$. By using the equations (2.3), one can verify directly that $\pi(r)$ has the required properties for $\Pi(r)$,

$$\langle \Pi(r), u(r) \rangle = \langle \Pi(r), v(r) \rangle = 0 \ , \ \langle \Pi(r)^2 \rangle = 2n \ ,$$

from which we may conclude $\Pi(r) = \pm \pi(r)$. The same argument applies to $U_{r'}$ and $\pi(r')$. □

Lemma 2.4.

$$r'm' - l's' \equiv \pm(rm - ls) \ \ mod\, 2n \ \ .$$

Proof. Since $\mathrm{rk}\, T(X) = 19$, we see that $O_{Hodge}(T(X), \mathbf{C}\omega_X) = \langle \pm \mathrm{id} \rangle$, which implies $\varphi_{T(X)^*/T(X)} = \pm\mathrm{id}$. Since $T(X)$ and $\widetilde{NS}(X)$ are primitive and orthogonal to each other in the unimodular lattice $\tilde{H}(X, \mathbf{Z})$, we have $\varphi|_{\widetilde{NS}(X)^*/\widetilde{NS}(X)} = \pm\mathrm{id}$. Now, using the explicit form of $\pi(r)$, we have

$$\widetilde{NS}(X)^*/\widetilde{NS}(X) = \langle \frac{\pi(r)}{2n} \bmod \widetilde{NS}(X) \rangle = \langle (rm - ls)\frac{H}{2n} \bmod \widetilde{NS}(X) \rangle,$$

and similarly for $\pi(r')$,

$$\widetilde{NS}(X)^*/\widetilde{NS}(X) = \langle \frac{\pi(r')}{2n} \bmod \widetilde{NS}(X) \rangle = \langle (r'm' - l's')\frac{H}{2n} \bmod \widetilde{NS}(X) \rangle.$$

Moreover, since $\varphi : \widetilde{NS}(X) \xrightarrow{\sim} \widetilde{NS}(X)$ is an isometry and $\varphi(U_r) = U_{r'}$, we have

$$\varphi(\pi(r)) = \pm\pi(r') \ \ .$$

In particular, in $\widetilde{NS}(X)^*/\widetilde{NS}(X)$, we have

$$\varphi((rm - ls)\frac{H}{2n}) = \pm(r'm' - l's')\frac{H}{2n} \ \ .$$

On the other hand, by $\varphi|_{\widetilde{NS}(X)^*/\widetilde{NS}(X)} = \pm\mathrm{id}$, we have

$$\varphi((rm - ls)\frac{H}{2n}) = \pm(rm - ls)\frac{H}{2n} \ \ .$$

Since $\frac{H}{2n}$ is a generator of $\widetilde{NS}(X)^*/\widetilde{NS}(X) \simeq \mathbf{Z}/2n$, we finally obtain

$$r'm' - l's' \equiv \pm(rm - ls) \bmod 2n \ \ . \quad \square$$

Here for convenience we list the explicit equations which follow from the relations in (2.3);

$$rs = n \ \ , \ \ lm = nk^2 \ \ , \ \ -mr - ls + 2nk = 1 \tag{2.4}$$

$$r's' = n \ \ , \ \ l'm' = n(k')^2 \ \ , \ \ -m'r' - l's' + 2nk' = 1 \tag{2.5}$$

Now our proof of (2) in Proposition 2.2 will be completed by the following purely arithmetical lemma:

Lemma 2.5. *Let $r, s, l, m, k, r', s', l', m', k'$ be natural numbers satisfying (2.4) and (2.5). Then the following hold:*

Proof of (1). From the equations (2.4), (2.5), we have

$$mr + ls \equiv -1 \bmod 2n \ , \quad m'r' + l's' \equiv -1 \bmod 2n \ . \tag{2.6}$$

Then adding these equations with the given one, $m'r' - l's' \equiv mr - ls \bmod 2n$, we may conclude

$$mr \equiv m'r' \bmod n \ , \quad ls \equiv l's' \bmod n \ . \tag{2.7}$$

Consider the prime decomposition of n, and write

$$rs = r's' = n = p_1^{e_1} p_2^{e_2} \cdots p_k^{e_k} \ .$$

Note that from (2.6), we have $(r, s) = 1$. Therefore, if $p_i | r$ then $p_i^{e_i} | r$, and similarly if $p_i | r'$, then $p_i^{e_i} | r'$. Therefore if we show $\{p : \text{prime s.t. } p|r\} = \{p : \text{prime s.t. } p|r'\}$, then the desired relation $r = r'$ follows.

To show that the set of primes divisors of r and r' are equal, let us assume that there is a prime number p such that $p|r$ and $p \nmid r'$ (or vice versa, if necessary). Then, since p divides $n = rs = r's'$, we have $p|s'$. Since p divides s' and n, we see $p|ls$ from the second equation of (2.7). Now note that $p \nmid s$ by $(r, s) = 1$ and $p|r$. Therefore we have $p|l$. By $p|r$ and $p|l$, we have $p|mr + ls$, whence $p|(-1)$ by $p|n$ and the formula (2.6), a contradiction. □

Proof of (2). From the given equation, $m'r' - l's' \equiv -mr + ls \bmod 2n$, and the third equations in (2.4) and (2.5), we have $-2mr \equiv 1 + m'r' - l's' \equiv -2l's' \bmod 2n$. Similarly, we have $-2m'r' \equiv 1 + mr - ls \equiv -2ls \bmod 2n$. To summarize, we have

$$mr \equiv l's' \bmod n \ , \quad ls \equiv m'r' \bmod n \ .$$

Now assume $r' \neq s$, there is a prime p such that $p|s$ but $p \nmid r'$ (or vice versa if necessary). Then the same argument as in (1) leads us to a contradiction $-1 \equiv m'r' + l's' \equiv 0 \bmod p$. □

Now our proof of Proposition 2.2 is completed. □

§3. First Application

In this section, we describe the set of FM partners of a polarized K3 surface of Picard number 1 and of degree $2n$ in terms of a certain covering of the moduli of polarized K3 surfaces of degree $2n$. Note that if X is a K3 surface of Picard number 1 and of degree $2n$, then so are the FM partners Y of X by Lemma 1.5.

By a quasi-polarized K3 surface of degree $2n$, we mean a pair (X, H) where X is a K3 surface and H is a primitive nef and big line bundle on X with $(H^2) = 2n$. Choose a primitive embedding $\langle 2n \rangle \to \Lambda_{K3} = U^{\oplus 3} \oplus E_8(-1)^{\oplus 2}$ so that the image, denoted by $\mathbf{Z}h$, lies in the first copy of U. An isomorphism $\mu : H^2(X, \mathbf{Z}) \simeq \Lambda_{K3}$ with $\mu(H) = h$ is called an admissible marking, and the triple (X, H, μ) is called a marked quasi-polarized K3 surface of degree $2n$. Let

$$\Lambda_n := \langle 2n \rangle^{\perp} = \langle -2n \rangle \oplus U^{\oplus 2} \oplus E_8(-1)^{\oplus 2} \ \subset \Lambda_{K3},$$

$$\mathcal{D}_n := \{[\omega] \in \mathbf{P}(\Lambda_n \otimes \mathbf{C}) | (\omega, \omega) = 0, (\omega, \overline{\omega}) > 0\}.$$

The space $\mathcal{D}_n$ is a bounded symmetric domain with two connected components. By the surjectivity of the period mapping and the uniqueness of primitive embedding, $\mathcal{D}_n$ consists of the period points $[\mu(\omega_X)]$ of marked quasi-polarized K3 surfaces (X, H, μ).

Let $\iota : O(\Lambda_n) \to Aut(\mathcal{D}_n)$ be the natural homomorphism induced by $\Lambda_n \to \Lambda_n \otimes \mathbf{C}$. Consider the groups

$$\tilde{\Gamma}_n := Im\ \iota, \quad \Gamma_n := \{g \in \tilde{\Gamma}_n | \exists f \in O(\Lambda_{K3}) \text{ s.t. } f(h) = h,\ g = \iota(f|_{\Lambda_n})\},$$

and the quotient spaces

$$\tilde{\mathcal{P}}_n := \mathcal{D}_n/\tilde{\Gamma}_n\,,\ \mathcal{P}_n := \mathcal{D}_n/\Gamma_n\,.$$

Since $\tilde{\Gamma}_n$ and Γ_n are arithmetic groups acting on the bounded symmetric domain $\mathcal{D}_n$, the quotient spaces $\tilde{\mathcal{P}}_n$ and $\mathcal{P}_n$ are both quasi-projective by [BB]. The second space $\mathcal{P}_n$ is the period domain which parameterizes the isomorphism classes of quasi-polarized K3 surfaces (X, H) of degree $2n$. Here $(X, H) \simeq (X', H')$ if and only if there is $\sigma : X \simeq X'$ such that $\sigma^* H' = H$.

Lemma 3.1. *The group Γ_n is a normal subgroup of $\tilde{\Gamma}_n$ and*

$$\tilde{\Gamma}_n/\Gamma_n \simeq (\mathbf{Z}/2)^{\oplus(\tau(n)-1)}\,.$$

Proof. By [Ni, Theorem 1.14.2] we have an exact sequence

$$1 \to O(\Lambda_n)^* \to O(\Lambda_n) \to O(A_{\Lambda_n}) \to 1$$

where $O(\Lambda_n)^*$ denotes the kernel of the third arrow. We have

$$O(A_{\Lambda_n}) = O(\Lambda_n^*/\Lambda_n) \simeq O(\mathbf{Z}h^*/\mathbf{Z}h) \simeq (\mathbf{Z}/2)^{\oplus\tau(n)}.$$

Moding out the exact sequence above by $\pm id$ (given that $n > 1$), we get

$$1 \to O(\Lambda_n)^* \to O(\Lambda_n)/\pm id \to (\mathbf{Z}/2)^{\oplus(\tau(n)-1)} \to 1.$$

Now our assertion follows from the natural isomorphisms

$$\tilde{\Gamma}_n \simeq O(\Lambda_n)/\pm id, \quad \Gamma_n \simeq O(\Lambda_n)^*. \quad \square$$

Therefore the natural morphism

$$\varphi : \mathcal{P}_n = \mathcal{D}_n/\Gamma_n \to \tilde{\mathcal{P}}_n = \mathcal{D}_n/\tilde{\Gamma}_n$$

is Galois with Galois group $(\mathbf{Z}/2)^{\oplus(\tau(n)-1)}$.

Let $\mathcal{D}_n^1 \subset \mathcal{D}_n$ denote the subset consisting of the period points $[\mu(\omega_X)]$ of marked polarized K3 surfaces (X, μ) of degree $2n$ and of $\rho(X) = 1$. Note that H is uniquely determined by X and is ample when $\rho(X) = 1$. This subset $\mathcal{D}_n^1$ is the complement of a countable union of of proper closed subsets in $\mathcal{D}_n$ [Og2].

Consider the restriction of φ to the Picard number 1 part

Proposition 3.2. *There is a natural 1-1 correspondence between the sets* $\tilde{\mathcal{P}}_n^1$ *and* $\{FM(X)|\rho(X)=1,\ deg\ X=2n\}$.

Proof. By Lemma 1.5, if X is of $\rho(X)=1$ and of degree $2n$, then so are the FM partners Y of X. Given a marked K3 surface (X,μ) with a period point $[\mu(\omega_X)] \in \mathcal{D}_n^1$, we associate to it the set $FM(X)$. Let $g \in \tilde{\Gamma}_n$. Then the period point $g[\mu(\omega_X)]$ represents a second marked K3 surface (X',μ') in $\mathcal{D}_n^1$ with $g[\mu(\omega_X)] = [\mu'(\omega_{X'})]$. But $\mu(T(X)) = \Lambda_n = \mu'(T(X'))$. It follows that $(T(X), \mathbf{C}\omega_X) \simeq (T(X'), \mathbf{C}\omega_{X'})$, hence $FM(X) = FM(X')$. Thus we have a well-defined map

$$\tilde{\mathcal{P}}_n^1 \to \{FM(X)|\rho(X)=1,\ deg\ X=2n\},\ [\mu(\omega_X)] \mapsto FM(X).$$

This map is surjective.

We now show that the map is injective. Let (X,μ), (X',μ') be two marked K3 surfaces of degree $2n$ and of Picard number 1. Assume that $FM(X) = FM(X')$. Then, there is a Hodge isometry $f : (T(X), \mathbf{C}\omega_X) \to (T(X'), \mathbf{C}\omega_{X'})$. This means that $g' := \mu' \circ f \circ \mu^{-1} : \Lambda_n \to \Lambda_n$ is an element of $O(\Lambda_n)$ with $\iota(g')[\mu(\omega_X)] = [\mu'(\omega_{X'})]$. Hence $\iota(g') \in \tilde{\Gamma}_n$. This shows that the period points of (X,μ) and (X',μ') are in the same $\tilde{\Gamma}_n$-orbit in $\mathcal{D}_n$. □

Proposition 3.3. *The map* $\varphi : \mathcal{P}_n \to \tilde{\mathcal{P}}_n$ *is unramified over* $\tilde{\mathcal{P}}_n^1$ *and the fiber of* φ *at the point* $\tilde{\Gamma}_n[\mu(\omega_X)] \in \tilde{\mathcal{P}}_n^1$, *corresponding to* $FM(X)$, *is naturally isomorphic to the set* $FM(X)$.

Proof. The Galois group $\tilde{\Gamma}_n/\Gamma_n$ acts transitively on each fiber naturally. Thus each fiber, as a set, has cardinality at most $|\tilde{\Gamma}_n/\Gamma_n| = 2^{\tau(n)-1}$. Now suppose that $\rho(X)=1$. Define the map $FM(X) \to \varphi^{-1}\tilde{\Gamma}_n[\mu(\omega_X)]$ as follows.

Let $X' \in FM(X)$. One can choose an admissible marking μ' of X'. Then $\mu'(T(X)) = \Lambda_n$. We set $X' \mapsto \Gamma_n[\mu'(\omega_{X'})] \in \mathcal{P}_n$. If $\tilde{\mu}$ is another such marking of X', then we have $f := \mu' \circ \tilde{\mu}^{-1} \in O(\Lambda_{K3})$ with $f(h) = h$. Since $\mu'(T(X')) = \tilde{\mu}(T(X')) = \Lambda_n$, and $[\mu'(\omega_{X'})], [\tilde{\mu}(\omega_{X'})] \in \mathcal{D}_n$, it follows that $g = f|_{\Lambda_n} \in \Gamma_n$. This shows that $\Gamma_n[\tilde{\mu}(\omega_{X'})] = \Gamma_n[\mu'(\omega_{X'})]$. Thus $\Gamma_n[\mu'(\omega_{X'})]$ is independent of the choice of marking μ'.

Since $(T(X), \mathbf{C}\omega_X) \simeq (T(X'), \mathbf{C}\omega_{X'})$, the previous injectivity argument shows that $g[\mu(\omega_X)] = [\mu'(\omega_{X'})]$ for some $g \in \tilde{\Gamma}_n$. Thus $\tilde{\Gamma}_n[\mu(\omega_X)] = \tilde{\Gamma}_n[\mu'(\omega_{X'})]$. Hence for all $X' \in FM(X)$, the orbits $\Gamma_n[\mu'(\omega_{X'})] \in \mathcal{D}_n/\Gamma_n$ are in the same fiber $\varphi^{-1}[\mu(\omega_X)]$. Thus the map $FM(X) \to \varphi^{-1}(\tilde{\Gamma}_n[\mu(\omega_X)])$ is well-defined.

Next we shall show that this map is injective. Suppose that $\Gamma_n[\mu'(\omega_{X'})] = \Gamma_n[\mu(\omega_X)]$, i.e. $[\mu'(\omega_{X'})] = g[\mu(\omega_X)]$ for some $g \in \Gamma_n$. Then $g = \iota(f|_{\Lambda_n})$ for some $f \in O(\Lambda_{K3})$ with $f(h) = h$. This shows that $\mu^{-1} \circ f \circ \mu' : H^2(X', \mathbf{Z}) \to H^2(X, \mathbf{Z})$ is an isomorphism sending $(T(X'), \mathbf{C}\omega_{X'})$, $NS(X') = \mathbf{Z}\mu'^{-1}(h)$ to $(T(X), \mathbf{C}\omega_X)$, $NS(X) = \mathbf{Z}\mu^{-1}(h)$ respectively. Hence $X' \simeq X$ by the Torelli Theorem ([PSS], see also [BPV]) . Recall that $|\varphi^{-1}(\tilde{\Gamma}_n[\mu(\omega_X)])| \le 2^{\tau(n)-1}$ by Lemma 3.1 and $|FM(X)| = 2^{\tau(n)-1}$ by Corollary 1.6 (2). So, the map above is surjective as well.

Now we have:

$$|\varphi^{-1}[\mu(\omega_X)]| = 2^{\tau(n)-1} = deg\,\varphi\,.$$

Thus, φ is unramified over $\tilde{\mathcal{P}}_n^1$. □

Combining all together, we obtain the following:

Corollary 3.4. *The morphism $\varphi : \mathcal{P}_n^1 \to \tilde{\mathcal{P}}_n^1$ defines a principal homogeneous space structure of the 2-elementary abelian group $\tilde{\Gamma}_n/\Gamma_n \simeq (\mathbf{Z}/2)^{\oplus(\tau(n)-1)}$ on $FM(X)$ of a K3 surface X with $\rho(X) = 1$ and $\deg X = 2n$.*

As it is remarked before, the quotient spaces $\mathcal{P}_n$ and $\tilde{\mathcal{P}}_n$ are quasi-projective. Thus there exist projective compactifications $\overline{\mathcal{P}_n}$, $\overline{\tilde{\mathcal{P}}_n}$, and a morphism

$$\varphi : \overline{\mathcal{P}_n} \to \overline{\tilde{\mathcal{P}}_n}.$$

Question 3.5. *What is the ramification locus like? Note that this locus is Zariski closed subset which lies in the complement of the dense subset $\tilde{\mathcal{P}}_n^1$.*

§4. Second Application

Let X be a K3 surface. We shall call a Mukai vector $(r, H, s) \in \tilde{NS}(X)$ *special* if H is ample and *primitive*, and that $r > 0$, $(r, s) = 1$, and $(H^2) = 2rs$, hold.

As our second application of Theorem 2.1, we shall show the following refinement of Theorem 1.3:

Proposition 4.1. *Let X be K3 surface and $Y \in FM(X)$. Then there is a special Mukai vector (r, H, s) such that $Y \simeq M_H((r, H, s))$.*

Proof. By Theorem 1.3, we have $Y \simeq M_{H'}((a, H', b))$ where H' is an ample line bundle and a and b are positive integers such that $(a, b) = 1$ and $({H'}^2) = 2ab$. Write $H' = mH$ where H is primitive and m a positive integer. Consider the Kuranishi family of the polarized K3 surface (X, H), namely a smooth projective family

$$\pi : (\mathcal{X}, \mathcal{H}) \to \mathcal{K}$$

over some 19-dimensional polydisk $\mathcal{K}$ with central fiber $(\mathcal{X}_0, \mathcal{H}_0) = (X, H)$. As it is well known, the set

$$\mathcal{K}^1 := \{t \in \mathcal{K} | NS(\mathcal{X}_t) = \mathbf{Z}\mathcal{H}_t\}$$

is everywhere dense in $\mathcal{K}$. (See eg. [Og2]). We fix a marking $R^2\pi_*\mathbf{Z}_{\mathcal{X}} \simeq \Lambda_{K3} \times \mathcal{K}$ so that $\mathcal{H}_t \mapsto (h, t)$, for some constant vector $h \in \Lambda_{K3}$ with $(h^2) = 2n$.

Let

$$f : \mathcal{Y} \to \mathcal{K},$$

be the relative moduli space of semi-stable sheaves with respect to $\mathcal{H}$ with Mukai vectors $(a, m\mathcal{H}_t, b) \in \widetilde{NS}(\mathcal{X}_t)$. This f is projective and $\mathcal{Y}_t = M_{m\mathcal{H}_t}((a, m\mathcal{H}_t, b)) \in$ $FM(\mathcal{X}_t)$ by Lemma 1.4.

For $t \in \mathcal{K}^1$, the fiber $\mathcal{X}_t$ has Picard number 1. In this case, we know that

$$\mathcal{Y}_t \simeq \mathcal{M}_{\mathcal{H}_t}((r_t, \mathcal{H}_t, s_t))$$

for some Mukai vector $(r_t, \mathcal{H}_t, s_t)$ which is special, by Theorem 2.1.

Since $2r_t s_t = (\mathcal{H}_t^2) = (H^2) = 2n$ is constant and $\mathcal{K}^1$ is dense in $\mathcal{K}$, there are positive integers r,s and a sequence $\{t_k\}$ in $\mathcal{K}^1$ such that $r_{t_k} = r$ and $s_{t_k} = s$ are constant and $\lim_{k\to\infty} t_k = 0$.

Now consider the moduli space $Y' := M_H((r, H, s)) = \overline{M}_H((r, H, s))$ over $X = \mathcal{X}_0$. Y' is a FM partner of X by Theorem 1.3. Once again, by Lemma 1.4, we can extend this to a relative moduli space of stable sheaves over $\mathcal{K}$:

$$f' : \mathcal{Y}' \to \mathcal{K}$$

which is projective over $\mathcal{K}$, and such that $\mathcal{Y}'_0 = Y'$, that $\mathcal{Y}'_t = M_{\mathcal{H}_t}((r, \mathcal{H}_t, s))$ is a FM partner of $\mathcal{X}_t$, and that there exists an f'-ample primitive invertible sheaf $\mathcal{L}'$ of degree $2n$.

Since $\mathcal{K}$ is a polydisk, we can choose markings

$$R^2 f_* \mathbf{Z}_{\mathcal{Y}} \simeq \Lambda_{K3} \times \mathcal{K}, \qquad R^2 f'_* \mathbf{Z}_{\mathcal{Y}'} \simeq \Lambda_{K3} \times \mathcal{K}$$

so that $\mathcal{L}_t \mapsto (l, t)$ and $\mathcal{L}'_t \mapsto (l, t)$ respectively, for some constant primitive vector $l \in \Lambda_{K3}$ with $(l^2) = 2n$. Here we used the fact that $O(\Lambda_{K3})$ acts on the set of primitive vectors l with $(l^2) = 2n$ transitively. Corresponding to f and f', we have then two period maps:

$$\eta : \mathcal{K} \to \mathcal{P}_n\,, \qquad \eta' : \mathcal{K} \to \mathcal{P}_n\,.$$

Here $\mathcal{P}_n$ is the period domain defined in Section 3. Recall that this space is quasi-projective, hence Hausdorff.

At t_k, we have $\mathcal{Y}_{t_k} \simeq \mathcal{Y}'_{t_k}$. Since both sides have Picard number 1, we also have $(\mathcal{Y}_{t_k}, \mathcal{L}_{t_k}) \simeq (\mathcal{Y}'_{t_k}, \mathcal{L}'_{t_k})$. It follows that $\eta(t_k) = \eta'(t_k)$ for all k.

Since $\mathcal{P}_n$ is Hausdorff, it follows that $\eta(0) = \eta'(0)$. This means $(\mathcal{Y}_0, \mathcal{L}_0) \simeq (\mathcal{Y}'_0, \mathcal{L}'_0)$. Therefore $\mathcal{Y}_0 \simeq \mathcal{Y}'_0$, i.e. $Y \simeq Y' = M_H((r, H, s))$. $\square$

References

[BB] W. Baily and A. Borel, *Compactification of arithmetic quotients of bounded symmetric domains*, Annals of Math. **84** (1966), 442–528.

[BPV] W. Barth, C. Peters, A. Van de Ven, *Compact complex surfaces*, Springer-Verlag (1984).

[Cs] J. W. S. Cassels, *Rational quadratic forms*, Academic Press (1978).

[GM] S.I. Gelfand, Y.I. Manin, *Methods of homological algebra*, Springer-Verlag (1991).

[HLOY] S. Hosono, B. Lian, K. Oguiso, S.T. Yau, *Counting Fourier-Mukai partners and applications, math.AG/0202014, submitted.*

[HL] D. Huybrechts and M. Lehn, *The Geometry of Moduli Spaces of Sheaves*, Vieweg (1997).

[Ma1] M. Maruyama, *Stable vector bundles on an algebraic surface*, Nagoya Math. J. **58** (1975), 25–68.

[Ma2] M. Maruyama, *Moduli of stable sheaves, II*, J. Math. Kyoto Univ. **18** (1978), 557–614.

[Mu1] S. Mukai, *Symplectic structure of the moduli space of sheaves on an abelian or K3 surface*, Invent. Math. **77** (1984), 101–116.

[Mu2] S. Mukai, *On the moduli space of bundles on K3 surfaces I, in Vector bundles on algebraic varieties*, Oxford Univ. Press (1987), 341–413.

[Ni] V. V. Nikulin, *Integral symmetric bilinear forms and some of their geometric applications*, Math. USSR Izv. **14** (1980), 103–167.

[Og1] K. Oguiso, *K3 surfaces via almost-primes, math.AG/0110282*, Math. Res. Lett. **9** (2002), 47–63.

[Og2] K. Oguiso, *Local families of K3 surfaces and applications, (math.AG/0011258 and math.AG/0104049)*, J. Alg. Geom. to appear.

[Or] D. Orlov, *Equivalences of derived categories and K3 surfaces, math.AG/9606006*, J. Math.Sci. **84** (1997), 1361–1381.

[PPS] I. Pjatetskii-Shapiro, I. Shafarevich, *A Torelli theorem for algebraic surfaces of type K3*, Math. USSR-Izv **5** (1971), 547–587.

Shinobu Hosono
Graduate School of
Mathematical Sciences
University of Tokyo
Komaba 3-8-1, Meguro-ku
Tokyo 153-8914, JAPAN
Email: hosono@ms.u-tokyo.ac.jp

Bong H. Lian
Department of mathematics
Brandeis University
Waltham, MA 02154
U.S.A.
Email: lian@brandeis.edu

Keiji Oguiso
Graduate School of
Mathematical Sciences
University of Tokyo
Komaba 3-8-1, Meguro-ku
Tokyo 153-8914, JAPAN
Email: oguiso@ms.u-tokyo.ac.jp

Shing-Tung Yau
Department of mathematics
Harvard University
Cambridge, MA 02138
U.S.A.
Email: yau@math.harvard.edu

Contemporary Mathematics
Volume **322**, 2003

Moduli spaces associated to a singular variety and the moduli of bundles over universal curves

Jun Li

1. Introduction

Since the discovery of Donaldson invariants of four manifolds, the research on moduli spaces has become an integral part of the study of geometry, topology and algebra. This emphasis on (associated) moduli problems is reinforced recently by the research in Mathematical Physics, especially in Super-String theories.

Of the several technical tools used in studying moduli problems, the method of degeneration probably is one of the oldest and an effective one. During the early days of algebraic geometry research, the Italian school, aided by their strong geometric intuition, solved many problems of plane curves via degenerations. In 70-80's of the last century, using degeneration method Gieseker studied moduli of curves and of vector bundles over curves [**5, 6**]. There are other cases where degenerations were used successfully in moduli problem study [**9**].

The method of degeneration rose to its prominence during the study of Donaldson invariants. The gluing formula, or the Donaldson-Floer theory, is the differential geometric analogue of the degeneration method in algebraic geometry. Many people worked on this theory, including Kronheimer-Morwka's work on the general formula of Donaldson invariants of four manifolds of simple type [**16**]. Recently, parallel theory was developed for moduli of stable maps to symplectic manifolds, by A.Li-Ruan [**19**] and Ionel-Parker [**10, 11, 12**]. Part of the field theory of Eliashberg-Givental-Hofer [**4**] is also a research in this direction.

Algebraic geometers also studied the moduli of stable bundles using degeneration method. Among them were the work of O'Grady-Morgan in their work on the Donaldson's invariants of elliptic surfaces [**24**]; Gieseker and the author's work on the geometry of the moduli of stable bundles over surfaces [**7**]; And the work of Muñov on the gluing formula of the Donaldson invariants of algebraic surfaces [**22**].

One feature that distinguish these moduli spaces to the moduli of curves and the moduli of bundles over curves is that these moduli spaces usually are singular, due to the fact that the deformations of the objects under study may be obstructed.

2000 *Mathematics Subject Classification.* Primary 14D20; Secondary 14D06, 14D21.

This is the case for sheaves on surfaces, and is the case for stable morphisms to smooth varieties. It is worth noticing that in differential geometry one usually can perturb the background structures[1] to make the moduli spaces smooth, at least away from high codimension subsets. Unfortunately, such method is not available to algebraic geometers. It is to compensate the deficiency caused by the rigidity of algebraic geometry that Tian and the author, and later Behrend-Fantechi and others [**20, 21, 2, 1, 14**], developed the notion of virtual moduli cycles for moduli spaces with perfect obstruction theories. One special feature of this machinery is that the existence of virtual moduli cycles requires knowing the obstruction theories of the whole (compact) moduli spaces. Hence to combine the technique of degeneration with that of the virtual moduli cycles one needs to construct degenerations so that the obstruction theory are perfect everywhere. Such construction was found recently by the author [**17, 18**].

In this note, we will explain briefly the main idea behind this new construction. This will occupy the section 3. In retrospect, what we did is a high dimensional generalization of Gieseker's degeneration of moduli of stable bundles on curves and of Harris-Mumford's construction of admissible covers. In Section 2, we will review Gieseker's degeneration of moduli of bundles, which serves as a warm up. In Section 4 we will apply this idea to give a simple construction of the compactification of the moduli of stable bundles over the universal curves, recently constructed by [**28**].

2. Gieseker's degeneration

In this section we will review Gieseker's degeneration of moduli of stable vector bundles over nodal curves.

Let $\mathcal{X}/C$ be a family of projective curves over a parameter curve $0 \in C$ of which all fibers X_t of $\mathcal{X}/C$ over $t \neq 0 \in C$ are smooth and X_0 is an irreducible curve with one node. We fix an odd integer d and form the moduli of rank two degree d stable vector bundles over X_t, denote by $\mathfrak{M}_{X_t}(2,d)$. Since X_t is a smooth family of projective curves (for $t \neq 0$), the moduli spaces $\mathfrak{M}_{X_t}(2,d)$ form a smooth family of projective varieties over $C^\circ = C - 0$. The question is how to fill in the central fiber of the family $\mathfrak{M}_{\mathcal{X}^\circ/C^\circ}(2,d)$, where $\mathcal{X}^\circ = \mathcal{X} \times_C C^\circ$.

The simple minded approach is to take the moduli of all stable sheaves on X_0. Since X_0 is a curve with one node, this moduli space contains stable sheaves that are not locally free. Hence the total space (after filling in the central fiber) is not smooth near the points associated to these sheaves.

Gieseker took a different approach. Instead of considering all stable sheaves on X_0, he only consider all stable locally free sheaves on X_0. To make the moduli space proper, he also included all Hilbert stable locally free stable sheaves on X_n, where the later is the semistable curve that contains a chain of n-rational curves and its stabilization is X_0. (We will comment on the Hilbert stability momentarily.) The moduli space of all such sheaves form a proper moduli space. Here is the heuristic argument. Let $\mathcal{E}_s$ be a family of locally free sheaves on $X_{\rho(s)}$, where $\rho\colon (0,S) \to (0,C)$ is a morphism, so that it specialize to a non-locally free stable sheaf on X_0. To obtain a locally free sheaf as its specialization, we proceed as follows. First we form the surface $\mathcal{X}_S = \mathcal{X} \times_C S$, and let $\tilde{\mathcal{X}}_S$ be the desingularization of $\mathcal{X}_S$. Note that the special fiver of $\tilde{\mathcal{X}}_S$ over $0 \in S$ is an X_n for some n. We then let

[1]In case of ASD connections the background structure is the Riemannian metric and in case of stable maps it is the almost complex structure.

$\tilde{\mathcal{E}}_0$ be the specialization of $\tilde{\mathcal{E}}_s \equiv \mathcal{E}_{\rho(t)}$. As can be argued, we can find a specialization so that $\tilde{\mathcal{E}}_0$ is stable[2] and hence a locally free sheaf on X_n[3]. Then there is a smallest partial contraction $\pi: X_n \to X_m$ so that the push forward $\pi_* \tilde{\mathcal{E}}_0$ is still locally free. We define this to be the specialization of $\tilde{\mathcal{E}}_s$, and it is in the moduli space introduced by Gieseker.

We denote the new moduli space (of Hilbert stable locally free sheaves on X_n for some n, modulo an equivalence to be specified in Section 4) by $\mathfrak{M}^{\mathrm{Gi}}_{X_0}(2,d)$. We let

$$\mathfrak{M}^{\mathrm{Gi}}_{\mathcal{X}/C}(2,d) = \mathfrak{M}^{\mathrm{Gi}}_{X_0}(2,d) \cup \mathfrak{M}_{\mathcal{X}^\circ/C^\circ}(2,d).$$

This is the Gieseker's degeneration of the moduli of stable bundles over nodal curves.

A wonderful property of this degeneration is that *the total space of the new degeneration is smooth, its central fiber* $\mathfrak{M}^{Gi}_{X_0}(2,d)$ *has normal crossing singularities* and *the central fiber is itself a moduli space.* The proof of the first two follow from the deformation theory of sheaves on X_n, and its proof will be revisited in Section 4.

Let us indicate the importance of the *three characteristics* of this degeneration by reviewing Gieseker's proof of Newstead's conjecture on the vanishing of the top Chern classes of $\mathfrak{M}_{X_t}(2,d)$. These three characteristics are the guidance for our search of the generalization to other moduli problems and possibly over high dimensional bases. The details of the following sketch can be found in Gieseker's original paper [**6**].

First, for a family $\mathcal{X}/C$ of smooth genus g curves specializing to an irreducible nodal curve with one node, we take the Gieseker's degeneration $\mathfrak{M}^{\mathrm{Gi}}_{\mathcal{X}/C}(2,d)$. Newstead conjectured that for $k \le 2(g-1)$ the Chern classes

$$c_{m-k}(\mathfrak{M}_{X_t}(2,d)) = 0, \quad \text{where } m = \dim \mathfrak{M}_{X_t}(2,d).$$

Since the central fiber $\mathfrak{M}^{\mathrm{Gi}}_{X_0}(2,d)$ of $\mathfrak{M}^{\mathrm{Gi}}_{\mathcal{X}/C}(2,d)$ has normal crossing singularity, an easy argument shows that the above vanishing is equivalent to

$$c_{m-k}(\Omega_{\mathbf{M}}(\log \mathbf{D})) = 0, \quad k \le 2(g-1) \tag{2.1}$$

where for simplicity we denote by $\mathbf{M}$ the central fiber $\mathfrak{M}^{\mathrm{Gi}}_{\mathcal{X}/C}(2,d)_0$ and denote by $\mathbf{D}$ the singular locus of $\mathbf{M}$. Let $\tilde{\mathbf{M}}$ be the normalization of $\mathbf{M}$ and $\tilde{\mathbf{D}}$ be the preimage of $\mathbf{D}$. Then an application of the theory of mixed Hodge structure shows that (2.1) is equivalent to

$$c_{m-k}(\Omega_{\tilde{\mathbf{M}}}(\log \tilde{\mathbf{D}})) = 0, \quad k \le 2(g-1). \tag{2.2}$$

The next step is the genus reduction. Let X be the normalization of X_0 with $p_1 + p_2 \subset X$ the preimage of the node of X_0. Let (E,ρ) be a pair of a rank 2 degree d stable vector bundle on X and an isomorphism $\rho: E|_{p_1} \to E|_{p_2}$. Then by identifying $E|_{p_1}$ with $E|_{p_2}$ via ρ we obtain a stable vector bundle E^ρ on X_0. Let $\mathbf{N}^0$ be the set of all such pairs, which is naturally a smooth quasi-projective variety. We let $\pi: \mathbf{N}^0 \to \mathfrak{M}_X(2,d)$ be the obvious projection. Then π is a Zariski fiber bundle with fibers isomorphic to the general Lie group $GL(2)$. Further, there is a tautological family of stable locally sheaves $\mathcal{E}^0$ on $\mathbf{N}^0 \times X_0$ by gluing the universal bundle on $\mathfrak{M}_X(2,d)$ via the tautological family of isomorphisms on $\mathbf{N}^0$. It is apparent that $\mathbf{N}^0$

with the help of this tautological family over $\mathbf{N}^0$, embeds into $\mathfrak{M}^{\rm Gi}_{X_0}(2,d)$. Gieseker then constructed directly a compactification $\mathbf{N}$ of $\mathbf{N}^0$ whose fiber over $\mathfrak{M}_X(2,d)$ are a compactification of $GL(2)$; He constructed a family $\mathfrak{X}/\mathbf{N}$ (étale locally) of nodal curves, consisting of semistable models of X_0, that extends the constant family $\mathbf{N}^0\times X_0$ and finally he constructed a locally free sheaf $\mathcal{E}$ on $\mathfrak{X}$ extending the family $\mathcal{E}^0$ on $\mathbf{N}^0\times X_0$.

As can be seen from the construction, for a dense open subset of $s\in\mathbf{N}$ the restriction of $\mathcal{E}$ to fibers of $\mathfrak{X}$ over s are (Hilbert) stable, thus associates to members in $\mathbf{M}$, indeed in $\tilde{\mathbf{M}}$. it turns out that not all members of $\mathcal{E}$ are stable. Thus the above correspondence only defines a birational map $\tilde{\mathbf{M}}\sim\mathbf{N}$. Gieseker showed that there are smooth closed subsets $\Sigma^-\subset\tilde{\mathbf{M}}$ and $\Sigma^+\subset\mathbf{N}$ so that

$$\tilde{\mathbf{M}}-\Sigma^-\cong\mathbf{N}-\Sigma^+$$

and that the birational map can be resolved by a single flip along Σ^- and Σ^+:

$$\mathrm{bl}_{\Sigma^-}\tilde{\mathbf{M}}\cong\mathrm{bl}_{\Sigma^+}\mathbf{N}.$$

Here we will not go to the details of this flip, suffices to say that the analysis of this flip was aided by the fact that the central fiber is a moduli space itself.

Finally, based on the fiber bundle structure of $\mathbf{N}\to\mathfrak{M}_X(2,d)$ Gieseker showed that

$$c_{m-k}(\Omega_{\mathbf{N}}(\log\mathbf{D}'))=0,\quad k\le 2(g-1)$$

for some divisor $\mathbf{D}'\subset\mathbf{N}$. He then used the explicit information of the flips demonstrated that this vanishing result implies the vanishing (2.2).

The key point of this proof is that the degeneration $\mathfrak{M}^{\rm Gi}_{\mathcal{X}/C}(2,d)$ relates the moduli of stable bundles over a genus g curve to the moduli $\mathfrak{M}^{\rm Gi}_{X_0}(2,d)$, and its normalization; The space $\mathbf{N}$ is a fiber bundle over the moduli of stable bundles over a genus $g-1$ curve and finally the two spaces, the normalization of $\mathfrak{M}^{\rm Gi}_{X_0}(2,d)$ and the $\mathbf{N}$, are related by a simple flip along smooth loci. From this description, it is clear that the fact that *the new degeneration $\mathfrak{M}_{\mathcal{X}/C}(2,d)$ has smooth total ambient space*, that *the central fiber has normal crossing singularity* and that one can describe explicitly the flip, which is possible in part because *the central fiber is a moduli space itself*, are essential to the proof of the Newstead conjecture.

We now comment on the technical part of the original construction of $\mathfrak{M}^{\rm Gi}_{X_0}(2,d)$. As we know, the Geometric Invariant theory (in short GIT) was developed to attack moduli problems in general. This theory is very successful in constructing moduli spaces of sheaves and pairs of sheaves. Gieseker used GIT to construct his degeneration. There is a new feature about his construction. The usual moduli of sheaves concerns sheaves over a fixed base variety. In case of $\mathfrak{M}^{\rm Gi}_{X_0}(2,d)$, its elements are stable vector bundles $E\to X_n$ where their bases X_n are part of the data of the objects being parameterized. This made the application of GIT tricky. The idea of Gieseker was to use bundle $E\to X_n$ to embed X_n to a projective space $\mathbf{P}^N$, and define Hilbert stability of E to be the stability of the embedded curve in $\mathbf{P}^N$. This is a marvellous idea, which resolved the issue of not having a fixed bases. However, the way the embedding was constructed made the analysis of stability quite involved, especially when the base of E is a singular curve. Such analysis was done by Gieseker-Morrison for the case of rank two [**8**], and in high rank case done (with slightly modified Hilbert stability) recently by [**28**]. I should point out that Nagaraj-Seshadri has constructed the Gieseker's degeneration $\mathfrak{M}^{\rm Gi}_{\mathcal{X}/C}(r,d)$ for all rank, with a mixture of sheaves on X_0 and on X_n, bypassing the analysis of Hilbert stability.

It is worth noticing that in gauge theory the Floer-Donaldson theory was developed in parallel line. In differential geometry, one obtains a family of smooth Riemannian manifolds degenerating to a singular one by stretching a neck in a four manifold M. Here a neck in a four manifold M is an ϵ-tubular neighborhood $N_\epsilon\Sigma$ of a three manifold $\Sigma \subset M$; stretching the neck means to change the Riemannian metric g of M within $N_\epsilon\Sigma$ so that metrically $g_t|_{N_\epsilon\Sigma}$ converges to a disjoint union of

$$[0,\infty)\times\Sigma \cup (-\infty,0]\times\Sigma. \tag{2.3}$$

Globally, (M, g_t) will then converge to two manifolds (M_-, g_-) and (M_+, g_+), each has an end that is isomorphic to one of the component in (2.3). Now consider a sequence of ASD connections A_i on a fixed G-bundle over (M, g_{t_i}), with $t_i \to \infty$. A subsequence of $\{A_i\}$ will then converge to an ASD connection A_- on M_- and an ASD connection A_+ on M_+. In the normal circumstance, we will have

$$\int_{M_-} \|A_-\|^2 + \int_{M_+} \|A_+\|^2 = \int_M \|A_i\|^2. \qquad \text{(The right is independent of } i\text{)}$$

However, sometimes a strict inequality "<" holds. In this case we understand that some of the energy escaped at infinity. To recapture this, one needs to use dilation scheme to obtain an ASD connection on the normal bundle $N_{\Sigma/M}$, and sometimes a sequence of ASD connections on $N_{\Sigma/M}$. Comparing to Gieseker's construction, we see that (M, g_t) plays the role of X_t, that the pair X_- and X_+ plays the role of the two irreducible components of an analytic neighborhood of the node of X_0, the vanishing cycle in X_t of the family $\mathcal{X}$ plays the role of Σ while the inserted rational curves in X_n plays the role of the normal bundle $N_{\Sigma/M}$. Thus the Floer-Donaldson theory is a differential geometric analogue of Gieseker's degeneration. This indicates strongly that there should be an algebraic geometric approach to the Floer-Donaldson theroy.

This is what has been done recently in [**17, 18**]. We will review the key to this new theory in the next section.

3. The new approach to degeneration of moduli spaces

In this section, we will review the new theory of degenerations of moduli spaces [**17, 18**].

We begin with the basic set up. We let C be a connected smooth curve, $0 \in C$ be a point in a smooth curve and let $W \to C$ be a family of projective schemes so that the fibers W_t of W over $t \neq 0 \in C$ are smooth and the central fiber W_0 is the union of two smooth varieties Y_1 and Y_2 intersecting transversally along a smooth irreducible divisor. We denote by $D_i \subset Y_i$ the divisor $Y_1 \cap Y_2 \subset Y_i$ and denote by $D \subset W_0$ the divisor $Y_1 \cap Y_2$. Of course $D = D_1 = D_2$. We say W_0 is the gluing of (D_1, Y_1) and (Y_2, D_2) along $D_1 = D_2$. We call $Y_i^{\mathrm{rel}} = (Y_i, D_i)$ the relative pair.

Before going into details, let us make some general comment on the degeneration we will study momentarily. Suppose we are interested in certain properties of a moduli space of some objects defined over a smooth variety (called the base moduli spaces $\mathcal{M}_t$ associated to each base variety W_t. We expect that the relative moduli space $\mathcal{M}_{W^\circ/C^\circ} \triangleq \coprod_{t\in C^\circ} \mathcal{M}_t$ form a flat (or "virtually") family[4]. Usually, there is a way to define the corresponding moduli space $\mathcal{M}_0$ of the singular base W_0, and the relative moduli space $\mathcal{M}_{W/C}$ will be an obvious choice to study in algebraic geometry. However, as in the case studied by Gieseker for vector bundles over curves, the moduli space $\mathcal{M}_0$ usually contains objects that are degenerate along the nodal divisor $D \subset W_0$. In such cases the deformations of such objects have local obstruction, and consequently the central fiber of the relative moduli space $\mathcal{M}_{W/C}$ becomes worse than normal crossing (or the obstruction theory is not perfect).

Recall that the three criteria in search of a "good" degeneration of moduli spaces: *The total space is smooth (or "virtually"), the central fiber is normal crossing (or "virtually")* and *the central fiber is a moduli space.* With the family $\mathcal{M}_{W/C}$, we can certainly blow it up to obtain a new family that satisfies the first two criteria, but not the third one since after the blowing up it is usually difficult to trace its points. The alternative, as emphasized by Gieseker, is to find a "geometric" blow up, where you in effect construct a blow up that itself is a relative moduli space. The Gieseker's degeneration of vector bundles over curves is such an example.

We now demonstrate how this "geometric resolution" is constructed in case of the moduli of stable maps to varieties.

We let W/C be as before with a relative ample line bundle $H \to W$ fixed. We fix integers g, k and a family of algebraic classes $d_t \in H_2(W_t)$. We form the moduli space

$$\mathcal{M}_{g,n}(W_t, d) = \{f : X \to W_t \mid X = (X, p_1, \cdots, p_k),\ g(X) = g,\ \deg f = d_t\}/\sim .$$

Here we require X to be a connected nodal curve and f be stable. We know $\mathcal{M}_{g,n}(W_t, d)$ is a proper, separated DM-stack, of expected dimension

$$\text{exp.}\dim \mathcal{M}_{g,n}(W_t, d) = (3 - \dim W_t)(g-1) + k + d_t \cdot c_1(W_t).$$

We let

$$\text{ev} : \mathcal{M}_{g,n}(W_t, d) \longrightarrow W_t^k$$

be the evaluation map. Then the GW-invariants of W_t are

$$\Phi^{W_t}_{g,k,d}(\beta) = \int_{[\mathcal{M}_{g,n}(W_t,d)]^{\text{vir}}} \text{ev}^* \beta, \quad \beta \in H^*(X^k).$$

When $\mathcal{M}_{g,n}(W_t, d)$ has pure dimension and is equal to the expected dimension, $[\mathcal{M}_{g,n}(W_t, d)]^{\text{vir}}$ is the fundamental class of $\mathcal{M}_{g,n}(W_t, d)$. In general, we need to use the notion of virtual cycles $[\mathcal{M}_{g,n}(W_t, d)]^{\text{vir}}$, constructed first by Li-Tian [**20, 21**] and also by Behrend-Fantechi [**1, 2**].

The question we like to address is the following: *How to relate the Gromov-Witten invariants of W_t with that of W_0, and then with that of the pairs (Y_1, D_1) and (Y_2, D_2).*

This is a typical problem in studying modul spaces via degenerations. As before, we form the relative moduli space $\mathcal{M}_{g,n}(W/C, d)$[5] and study the associated

[4]When local deformations of objects in the moduli space are non vanishing, the family may not be flat. By "virtual" flat we mean that the total moduli space admits a perfect obstruction theory and the induced obstruction theory of any fiber is isomorphic to its natural obstruction theory.

[5]Her d represents the family $d_t \in H_2(W_t)$.

GW-invariants $\Phi^{W_t}_{g,k,d}$. However, because of the singularity of W_0, the deformation theory of $\mathcal{M}_{g,n}(W_0, d)$ usually is not perfect, and hence we do not have virtual cycle $[\mathcal{M}_{g,n}(W_0, d)]^{\text{vir}}$. Worse still, when there is a stable morphism $F: X \to W_0$ so that an irreducible component of X is mapped entirely into the nodal divisor $D \subset W_0$ then the task of relating to "relative" Gromov-Witten invariants of (Y_i, D_i) becomes almost untractable.

So as explained we need to perform a "geometric resolution" of $\mathcal{M}_{g,n}(W/C, d)$ so that the new space on one hand has the desired "regularity" property and on the other hand is itself a moduli space. Similar to Gieseker's construction and the Floer-Donaldson theory, the "resolution" will be the moduli space of *stable* morphisms to all possible semistable models of W/C.

We now construct the class of expanded degenerations (semistable models) of W. We let Δ be the projective bundle over D: $\Delta = \mathbb{P}(\mathbf{1}_D \oplus N_{D_2/Y_2})$, where $\mathbf{1}_D$ is the trivial holomorphic line bundle on D and N_{D_2/Y_2} is the normal bundle of D_2 in Y_2. Δ has two distinguished divisors: $D_- = \mathbb{P}(\mathbf{1}_D \oplus 0)$ called the left distinguished divisor and $D_+ = \mathbb{P}(0 \oplus N_{D_2/Y_2})$ called the right distinguished divisor. Using this identification, we can *glue* an ordered chain of n Δ's by identifying the right distinguished divisor (i.e. D_+) in the k-th Δ with the left distinguished divisor (i.e. D_-) of the $(k+1)$-th Δ, for $k = 1, \cdots, n-1$. We then *glue* Y_1 to this chain by identifying the distinguished divisor D_1 in Y_1 with the D_- in the first Δ of the chain, and then *glue* Y_2 to this scheme by identifying the right distinguished divisor D_+ of the last Δ with D_2 in Y_2. We denote the resulting scheme by $W[n]_0$. Note that $W[n]_0$ has $(n+2)$-irreducible components, the Y_1 and Y_2 at its two ends and a subchain of n-ruled varieties Δ in the middle.

We let $G[n]$ be the group $(\mathbb{C}^*)^{\times n}$. The $\mathbb{C}^*$ acts on $\mathbf{1}_D \oplus N_{D_2/Y_2}$ via $(a, b)^\sigma = (a, \sigma b)$, which induces an action on Δ. Clearly, this action fixes the two divisors $D_\pm \subset \Delta$. Hence by making the l-th copy of $(\mathbb{C}^*)^{\times n}$ acts on the l-th copy of the Δ in $W[n]_0$, we obtain a $G[n]$-action on $W[n]_0$. We call the collection $\{W[n]_0\}_n$ the semistable models of W_0.

Following the example of Gieseker, the new degeneration should be the moduli space of all *stable* morphisms to $W[n]_0$ for some n. Let $f: X \to W[n]_0$ be a morphism, with $p_1, \cdots, p_k$ its marked points of X. We say f is *non-degenerate* if no irreducible components of X are mapped entirely into the singular locus of $W[n]_0$ and no marked points are mapped into the singular locus of $W[n]_0$. For some technical reason, we need to impose *pre-deformable* condition on f as well. In short, a *non-degenerate* map f is *pre-deformable* if every $x \in X$ that is mapped to the singular locus of $W[n]_0$ is a node of X and the contact (tangential) orders of the restrictions $f|_{U_1}$ and $f|_{U_2}$ along the nodal divisor are identical. Here U_1 and U_2 are the two irreducible components of the germ of X at z. We impose this condition because if f is a member of a flat family of stable morphisms whose general members are maps into W_t than f is automatically *pre-deformable.*

The key to the construction is the new equivalence relation and the associated automorphism group of an $f: X \to W[n]_0$.

We say two $f_i: X_i \to W[n]_0$, where $i = 1$ and 2, are equivalent if there is an isomorphism $\varphi: X_1 \to X_2$ (as marked curves) and an isomorphism $\sigma: W[n]_0 \to W[n]_0$ induced by $\sigma \in G[n]$ so that $\sigma \circ f_1 = f_2 \circ \varphi$.

DEFINITION 3.1. *Let $f : X \to W[n]_0$ be an ordinary stable morphism. We say f is a stable morphism to $W[n]_0/G[n]$ if f is non-degenerate, is pre-deformable and if the automorphism group $\mathfrak{Aut}(f)$ is finite.*

What is surprising is that the set of equivalence classes of all degree d stable morphisms from k pointed genus g curves to $W[n]_0/G[n]$, for all possible n, form a proper, separated Deligne-Mumford stack with perfect obstruction theory.

In the remainder of this section, we will sketch a proof of the existence theorem, emphasizing the new idea that makes the generalization of Gieseker's degeneration to high dimension possible.

As in the construction of moduli spaces, the first task is to make sense of the families of stable morphisms to $W[n]_0/G[n]$ for all possible n. This is done by the introduction of an Artin stack $\mathfrak{W}/C$, of all semistable models of W/C.

For simplicity, we assume $C = \mathbf{A}^1$. We let $G[n]$ acts on $\mathbf{A}^{n+1}$ via

$$\mathbf{t}^\sigma = (\sigma_1 t_1, \sigma_1^{-1}\sigma_2 t_2, \cdots, \sigma_{n-1}^{-1}\sigma_n t_n, \sigma_n^{-1} t_{n+1}). \tag{3.1}$$

Note that if we let

$$\mathbf{p} : \mathbf{A}^{n+1} \to \mathbf{A}^1, \quad \mathbf{p}(\mathbf{t}) = t_1 \times \cdots \times t_{n+1},$$

then $\mathbf{p}$ is $G[n]$-equivariant with the trivial $G[n]$-action on $\mathbf{A}^1$. The standard model $W[n]$ will be constructed as a desingularization of

$$W \times_{\mathbf{A}^1} \mathbf{A}^{n+1}.$$

We construct $W[n]$ by induction on n. For $n = 0$, $W[0] = W$. For $n = 1$, $W_1 = W \times_{\mathbf{A}^1} \mathbf{A}^2$ is smooth except along the locus $D \times_{\mathbf{A}^1} 0$, which is a smooth codimension 3 subscheme in W_1. It is easy to see that there are two ways to small resolve the singularity of W_1. We let $W[1]$ be one of the small resolution that has the following property: Since W_1 is over $\mathbf{A}^2$ the resolution $W[1]$ is also over $\mathbf{A}^2$; The fiber of $W[1]$ over $0 \in \mathbf{A}^2$ is the $W[1]_0$ introduced before; We choose the resolution so that the fibers of $W[1]$ over the first and the second coordinate line of $\mathbf{A}^2$ is a smoothing of the first and the second singular divisor D_i of $W[1]_0$. (Here we order the singular loci of $W[1]$ consecutively so that D_1 is attached to Y_1.)

The $W[2]$ is constructed similarly. We let W_2 be the fiber product $W[1] \times_{\mathbf{A}^2} \mathbf{A}^3$ with $\mathbf{A}^3 \to \mathbf{A}^2$ the morphism $(t_1, t_2, t_3) \mapsto (t_1, t_2 t_3)$. We let $W[2] \to \mathbf{A}^3$ be a small resolution of W_2. We choose the small resolution so that the fiber of $W[2]$ over $0 \in \mathbf{A}^3$ is $W[2]_0$, that the fibers over the i-th coordinate line is a smoothing of the i-th singular divisor $D_i \subset W[2]_0$.

We construct $W[n] \to \mathbf{A}^{n+1}$ inductively following the same line.

The $G[n]$ action on $\mathbf{A}^{n+1}$ lifts uniquely to a $G[n]$ action on $W[n] \to \mathbf{A}^{n+1}$. Its induced action on $W[n]_0 \equiv W[n] \times_{\mathbf{A}^{n+1}} 0$ is the $G[n]$-action on Δ introduced before.

We now define the stack of semistable models of W/C. We first define the groupoid $\mathfrak{W}$ to be the category of pairs $(W_S/S, \pi)$, where S are C-schemes, W_S are families of projective schemes over S and $\pi : W_S \to W$ are C-projections, such that there are open coverings U_α of S and $\rho_\alpha : U_\alpha \to \mathbf{A}^{n_\alpha+1}$ so that

$$W_S \times_S U_\alpha \cong W[n_\alpha] \times_{\mathbf{A}^{n_\alpha+1}} U_\alpha,$$

which is compatible to the projection $W_S \to W$. Two families W_S and W'_S are isomorphic if there is an S-isomorphism $f : W_S \to W'_S$ compatible to the tautological projections $W_S \to W$ and $W'_S \to W$. By our construction, $\mathfrak{W}$ forms a stack.

With the stack $\mathfrak{W}$, a family of semistable model W_S in $\mathfrak{W}$ over an C-scheme S is merely a family $W_S \in \mathfrak{W}(S)$. An S-family of stable morphisms to $\mathfrak{W}$ then consists of a family $W_S \in \mathfrak{W}(S)$, an S-family of ordinary stable morphisms $f : \mathcal{X} \to W_S$ so that for each closed $s \in S$ the fiber of f over s is stable in the sense of the Definition 3.1[6]. In short we will call such family a family of stable morphisms to $\mathfrak{W}$. Given two S-families of stable morphisms to $\mathfrak{W}$, say given by $f_i : \mathcal{X}_i \to W_{S,i}$, an arrow (equivalence) from f_1 to f_2 over 1_S consists of an S-isomorphism $\varphi : \mathcal{X}_1 \to \mathcal{X}_2$ and an arrow (equivalence) $\sigma : W_{S,1} \to W_{S,2}$ so that $\sigma \circ f_1 = f_2 \circ \varphi$.

It is now straightforward to define the groupoid of all families of stable morphisms to $\mathfrak{W}$, of the prescribed topology. We denote this groupoid by $\mathcal{M}_{g,n}(\mathfrak{W}, d)$.

THEOREM 3.2. *The groupoid $\mathcal{M}_{g,n}(\mathfrak{W}, d)$ forms a separated, C-proper Deligne-Mumford stack. The stack $\mathcal{M}_{g,n}(\mathfrak{W}, d)$ admits a natural perfect obstruction theory.*

We now comment on the key ideas on the proof of the Theorem.

First, we look at the separatedness. Let $0 \in S$ be a curve with $S^\circ = S - 0$, let $f' : \mathcal{X}' \to W'_S$ and $f'' : \mathcal{X}'' \to W''_S$ be two S-families in $\mathfrak{W}(S^\circ)$ so that $f'|_{S^\circ}$ is equivalent to $f''|_{S^\circ}$. By valuation criteria for separatedness, we need to show that f' is equivalent to f''. Since $f'|_{S^\circ} \simeq f''|_{S^\circ}$, $W'_S \times_S S^\circ \cong W''_S \times_S S^\circ$. For simplicity, we consider the case where W'_S has general smooth fibers. We let $\tilde{W}_S$ be a desingularization of $W'_S \times_S W''_S$. Possibly after a base change, we can assume $\tilde{W}_S$ is smooth, its fibers over 0 is one of $W[n]_0$. By using the blocks $W[n]$ we can show that the family $\tilde{W}_S$ is a family in $\mathfrak{W}(S)$.

We now let $\tilde{f} : \tilde{\mathcal{X}} \to \tilde{W}_S$ be a family of ordinary stable morphisms that is the extension of $f^\circ|_{S^\circ} \cong f''|_{S^\circ}$. We let $\tilde{f}_0$ be the restriction of $\tilde{f}$ to the central fiber. Now let $\Delta \subset \tilde{W}_{S,0}$ be any irreducible component of the ruled variety. We say that this Δ is redundant if to each $c \in \mathbb{C}^*$ there is an isomorphism $\varphi : \tilde{f}_0 \to \tilde{f}_0$ so that $\sigma_c \circ \tilde{f}_0 = \tilde{f}_0 \circ \varphi$, where $\sigma_c : \tilde{W}_{S,0} \to \tilde{W}_{S,0}$ is the isomorphism that is the identity away from Δ and is a dilation along fibers of $\Delta \to D$ by c. Since f' is already a family of stable morphisms in $\mathcal{M}_{g,n}(\mathfrak{W}, d)$, under the contraction $\tilde{W}_S \to W'_S$, f' must be the contraction of $\tilde{f}$. Hence Δ is redundant if and only if Δ is contracted under $\tilde{W}_S$. For the same reason, Δ is redundant if and only if it is contracted under the map $\tilde{W}_S \to W''_S$. Hence $W'_S \cong W''_S$ and f' is equivalent to f''. This proves that $\mathcal{M}_{g,n}(\mathfrak{W}, d)$ is separated.

The properness is similar, and will not be repeated here [**17**].

The boundedness of the stack $\mathcal{M}_{g,n}(\mathfrak{W}, d)$ follows from a universal estimate that whenever $f : X \to W[n]_0$ is a stable morphism to $\mathfrak{W}$ then the number n is bounded from above by the sum of the genus g, the number of the marked points k and the degree d_0 of f. Here the degree d_0 is equal to $\deg(f \circ \pi)^* c_1(H)$, where H is the fixed relative ample line bundle on W/C and $\pi : W[n]_0 \to W$ is the projection.

We now sketch the proof that $\mathcal{M}_{g,n}(\mathfrak{W}, d)$ is étale covered by a DM-stack. Let $\xi \in \mathcal{M}_{g,n}(\mathfrak{W}, d)$ be a point associated to a stable morphism $f_0 : X_0 \to W[n]_0$. We need to construct a DM-stack U and an étale morphism $U \to \mathcal{M}_{g,n}(\mathfrak{W}, d)$ so that its image contains ξ. We take the family $W[n]$ and consider the moduli space $\mathcal{M}_{g,n}(W[n], d)$ of (ordinary) stable morphisms to $W[n]$ of the prescribed topological type. The space $\mathcal{M}_{g,n}(W[n], d)$ is a DM-stack. In [**17**], the author showed that there is a locally closed substack $\mathcal{V} \subset \mathcal{M}_{g,n}(W[n],d)$ parameterizing all maps $f: X \to W[n]$ in $\mathcal{M}_{g,n}(W[n],d)$ that are also stable morphisms to $\mathfrak{W}$. Namely, f are non-degenerate, pre-deformable and $\mathfrak{Aut}(f)$ are finite. Since $G[n]$ acts on $W[n]$ and since $\mathcal{V}$ is canonically a DM-substack of $\mathcal{M}_{g,n}(W[n],d)$, $G[n]$ acts on $\mathcal{V}$ as well. Now it follows from the definition that to each $[f] \in \mathcal{V}$ its stabilizer $\mathrm{Stab}_{[f]}$ is a subgroup of $\mathfrak{Aut}(f)$, which we know is finite. Thus $G[n]$ acts on $\mathcal{V}$ with finite stabilizer everywhere, and hence $\mathcal{V}/G[n]$ is a DM-stack. By the definition of $\mathcal{M}_{g,n}(\mathfrak{W},d)$, the universal family on $\mathcal{V}$ defines a tautological morphism $\mathcal{V}/G[n] \to \mathcal{M}_{g,n}(\mathfrak{W},d)$ whose image contains ξ. It is easy to see that the morphism $\mathcal{V}/G[n] \to \mathcal{M}_{g,n}(\mathfrak{W},d)$ is étale.

We now summarize the reason why the set of morphisms to $W[n]_0$ with the stability condition and the equivalence relation defined forms a Deligne-Mumford stack. First, we know that to construct a desirable degeneration we need to enlarge the category of stable morphisms to W_0, to include also morphisms to $W[n]_0$, for $n > 0$. This force us to look at the set

$$\mathfrak{W}^{\mathrm{set}} = \{W_t \mid t \in C^\circ\} \cup \{W[n]_0\}_{n \in \mathbb{Z}^{\geq 0}}.$$

Following Grothendieck, the algebraic structure on this set is given by all algebraic families that contains elements in $\mathfrak{W}^{\mathrm{set}}$ as its members. This will include all (basic) families $W[n]$ mentioned before. As proved in [**17**], all other families can be derived by these basic families. The next thing we noticed is that the set of automorphisms of $W[n]_0$ (relative to W/C) is $G[n]$. These equivalence are part of the intrinsic structure of $\mathfrak{W}^{\mathrm{set}}$. With these structures added, $\mathfrak{W}$ becomes an Artin stack. The upside of $\mathfrak{W}$ is that it carries with it all the information about how $W[n]_0$ deforms as semistable models of W/C. The down-side, at least to those who feel uneasy with the Artin stacks, is that it is highly unseparated. Now if we look at the space of all (new) stable morphisms modulo the (new) equivalence relation, then we get to keep the upside of the $\mathfrak{W}$. Namely, we know how deformation of an $f: X \to W[n]_0$ looks like. But miraculously the downside of $\mathfrak{W}$ is gone. Since given a (new) stable morphism $f: X \to W[n]_0$ almost all but finitely many automorphisms $\sigma: W[n]_0 \to W[n]_0$ will move the map f and thus can not be part of automorphisms of $[f]$. Thus *the moduli space of stable morphisms to the Artin stack $\mathfrak{W}$ become a separated, C-proper Deligne-Mumford stack.*

The deformation theory of of $[f] \in \mathcal{M}_{g,n}(\mathfrak{W},d)$ can easily be worked out using the locally closed $\mathcal{V} \subset \mathcal{M}_{g,n}(W[n],d)$. Indeed, of the three requirements for stability: *non-degeneracy, pre-deformable and finite automorphism groups*, the first and the third are open conditions and the second is a closed condition. Thus the deformation of $\mathcal{V}$ is a combination of the deformation theory of $\mathcal{M}_{g,n}(W[n],d)$ and the restriction imposed on by the pre-deformable requirement. The deformation of $\mathcal{M}_{g,n}(W[n],d)$ is well-understood now. The requirement that a pre-deformable morphism stay pre-deformable is a local condition, and can be worked out by hand. Combined, we obtain the obstruction theory of $\mathcal{V}$. By construction, it is $G[n]$-equivariant and its $G[n]$-quotient is the obstruction theory of $\mathcal{V}/G[n]$. The obstruction theory of $\mathcal{M}_{g,n}(\mathfrak{W},d)$ is the descent of that of $\mathcal{V}/G[n]$.

In conclusion, we see that the key to the construction is the blocks $W[n]/G[n]$, and its associated Artin stack $\mathfrak{W}$. Using the Artin stack $\mathfrak{W}$, we can make sense of the families of moduli objects over semistable models of W/C, which by a tautological argument, shows that the moduli groupoid forms a stack. To show the moduli stack is a DM-stack, we need to construct its étale charts. To do this, we can look at the relative moduli stack of stable morphisms to $W[n]/\mathbf{A}^{n+1}$. Such stacks are usually DM-stack. Then the quotient of this stack by $G[n]$ forms a chart of the moduli stack. This quotient stack will be a DM-stack if the stabilizer in $G[n]$ of every point is finite, which is the case if we impose appropriate stability condition.

4. Compactification of a relative moduli space

In this section, we will show how to use the idea developed in [**17, 18**] to (re)construct a compactification of the relative moduli space of stable vector bundles over the universal curve over $\mathcal{M}_g$. Here we follow the convention that $\mathcal{M}_g$ is the stack of Deligne-Mumford stable curves and $\mathcal{M}_g^0$ is the substack of all smooth curves.

The relative moduli space of stable vector bundles over the universal curves $\mathcal{C}_g$ on $\mathcal{M}_g^0$ is a basic moduli space whose existence follows directly from the existence of the moduli of stable vector bundles over smooth curves. An important question for this moduli space is how to compactify this moduli space so that it becomes a relative moduli space over the whole stack $\mathcal{M}_g$. In the rank 1 case, this was done by Caparaso [**3**]. In high rank case, Pandharipande constructed a compactification using relative GIT theory [**25**]. The elements of his compactification consists of stable sheaves over stable curves. As explained in the first section, near the stable but non-locally free sheaves the compactification will be singular. Ideally, such compactification should be constructed along the line of Gieseker's based on Hilbert stability. Recently, Schmitt [**28**] has constructed a compactification, using a modified Hilbert stability. Since the compactification was constructed by GIT, it is projective over $\mathcal{M}_g$. We expect that this compactification will have smooth total space and have normal crossing fibers over $\mathcal{M}_g$.

In this section we will show how to construct a similar compactification directly, using the idea based on [**17, 18**]. Here is the strategy of our construction. We fix a genus $g \geq 2$, a rank $r \geq 1$ and a prime $\chi > 2rg$. We intend to construct a compactification of the relative moduli of stable vector bundles on $\mathcal{C}_g/\mathcal{M}_g^0$ of rank r and Euler characteristic χ. We begin with the (Artin) stack of all connected semistable[7] curves $\mathfrak{M}_g$ and its universal curve $\mathfrak{C}_g$. We then form the relative stack of locally free sheaves on $\mathfrak{C}_g/\mathfrak{M}_g$, of rank r and Euler characteristic χ. We denote this stack by $\mathfrak{V}_{r,\chi}$. We later introduce a stability of locally free sheaves on nodal curves, following the Simpson stability of sheaves. We will show that the substack $\mathfrak{V}^{\mathrm{st}}_{r,\chi} \subset \mathfrak{V}_{r,\chi}$ of all stable locally free sheaves form a proper, separated and smooth Deligne-Mumford stack so that the nature projection $\mathfrak{V}^{\mathrm{st}}_{r,\chi} \to \mathcal{M}_g$ has normal crossing fibers.

As one will see, although the stack $\mathfrak{W}$ used in the previous section never occur in the construction of $\mathfrak{V}^{\mathrm{st}}_{r,\chi}$, it is actually ubiquitous. In case W/C is a family of curves of arithmetic genus g then the stack $\mathfrak{W}$ is indeed a substack of $\mathfrak{M}_g$. In this sense, the stack $\mathfrak{M}_g$ is a more general version of $\mathfrak{W}$ when the singular locus of W_0 has more than one components.

We first define the stack $\mathfrak{V}_{r,\chi}$ more carefully. As we know, a stack is basically a groupoid that satisfies certain sheaf like axioms. For $\mathfrak{M}_g$, it is a groupoid whose objects over a scheme S, denoted by $\mathfrak{M}_g(S)$, is the collection of all flat family of connected semistable curves over S of arithmetic genus g. An arrow (isomorphism) between two members $\mathcal{X}_S$ and $\mathcal{X}'_S \in \mathfrak{M}_g(S)$ that lies over 1_S consists of an S-isomorphism

$$\sigma : \mathcal{X}_S \xrightarrow{\cong} \mathcal{X}'_S.$$

We define $\mathfrak{V}_{r,\chi}$ be a groupoid over $\mathfrak{M}_g$ as follows: For each $\mathcal{X} \in \mathfrak{M}_g(S)$, we define $\mathfrak{V}_{r,\chi}(\mathcal{X})$ be the set of all locally free sheaves $\mathcal{E}$ on $\mathcal{X}$, of rank r everywhere and so that the Euler characteristic of $\mathcal{E}|_{\mathcal{X}_s}$ is χ for all closed $s \in S$. Here as usual for $s \in S$ we denote $\mathcal{X}_s$ the fiber of $\mathcal{X}$ over s. Given two members $\mathcal{E}$ and $\mathcal{E}'$ over $\mathcal{X}$ and $\mathcal{X}' \in \mathfrak{M}_g(S)$ respectively, an equivalence between $\mathcal{E}$ and $\mathcal{E}'$ is a pair (φ, σ) where $\sigma : \mathcal{X} \to \mathcal{X}'$ is an arrow in $\mathfrak{M}_g$ and φ is an isomorphism

$$\sigma^*\mathcal{E}' \cong \mathcal{E} \otimes_{\mathcal{O}_{\mathcal{X}}} \pi_S^*\mathcal{L}, \tag{4.1}$$

where $\mathcal{L}$ is an invertible sheaf on S and $\pi_S : \mathcal{X} \to S$ is the projection. We say two equivalences (φ_1, σ_1) and (φ_2, σ_2) of $\mathcal{E}$ and $\mathcal{E}'$ are equivalent if $\sigma_1 \equiv \sigma_2$ and there is an isomorphism $\psi : \mathcal{L}_1 \cong \mathcal{L}_2$ so that the diagram

$$\begin{array}{ccc}
\sigma_1\mathcal{E}' & \xrightarrow{\varphi_1} & \mathcal{E} \otimes_{\mathcal{O}_{\mathcal{X}}} \pi_S^*\mathcal{L}_1 \\
\Vert & & \downarrow{\scriptstyle 1_{\mathcal{E}} \otimes \pi_S^*\psi} \\
\sigma_2\mathcal{E}' & \xrightarrow{\varphi_2} & \mathcal{E} \otimes_{\mathcal{O}_{\mathcal{X}}} \pi_S^*\mathcal{L}_2
\end{array}$$

is commutative. We define an arrow between $\mathcal{E}$ and $\mathcal{E}' \in \mathfrak{V}_{r,\chi}(S)$ be an equivalence class of the equivalences between $\mathcal{E}$ and $\mathcal{E}'$. Here we use the equivalence classes of equivalences instead of equivalences directly because then the automorphism group of a stable E over X, defined by viewing $[E]$ as a member of $\mathfrak{V}_{r,\chi}$, will become finite.

The following existence theorem holds by tautology.

PROPOSITION 4.1. *The groupoid $\mathfrak{V}_{r,\chi}$ forms a stack. Further there is a tautological morphism $\mathfrak{V}_{r,\chi} \to \mathfrak{M}_g$ sending $\mathcal{E} \in \mathfrak{V}_{r,\chi}$ to the base curve $\mathcal{X}$ of $\mathcal{E}$.*

Our next step is to define the stability condition. Let X be a semistable curve in $\mathfrak{M}_g$ and let $\rho : X \to X^0$ be its stabilization. Since $g \geq 2$, the dualizing sheaf ω_{X^0} is ample on X^0. We let $\mathcal{R}_X$ be the set of all irreducible components of X contracted by $\rho : X \to X^0$. We fix a map $\eta : \mathcal{R}(X) \to \mathbb{Z}^+$. We then let H_η be a $\mathbb{Q}$-line bundle on X so that its restriction to a non-contracted component $D \subset X$ is $\rho^*\omega_{X^0}|_D$ and its restriction to a contracted component $D \subset X$ has degree $\rho(D)\epsilon$, where $\epsilon \sim 0^+$ is a sufficiently small positive rational number. Clearly, H_η is $\mathbb{Q}$-ample on X.

DEFINITION 4.2. *Let E be a locally free sheaf on X. We say E is weakly stable (with respect to H_η) if for any proper subsheaf $0 \neq F \subset E$ we have*

$$\frac{\chi(F)}{l.c.\chi(F(n))} < \frac{\chi(E)}{l.c.\chi(E(n))}, \quad \text{for } n \gg 0. \tag{4.2}$$

Here $E(n) = E \otimes H_\eta^{\otimes n}$ and $l.c.\chi(E(n))$ is the leading coefficient of $\chi(F(n))$ in n. We say E is stable if for each rational component $D \subset X$ contracted by stabilization the restriction $E|_D$ is not $\mathcal{O}_D^{\oplus r}$.

We remark that the weakly stable is the Simpson stability [**27**] associated to ample divisor H_η.

It is known that being stable is an open condition. Hence our next step is to define $\mathfrak{V}_{r,\chi}^{\mathrm{st}}$ be the open substack of $\mathfrak{V}_{r,\chi}$ of all stable locally free sheaves in $\mathfrak{V}_{r,\chi}$. There is one technical problem. In defining the stability we need to use the polarization H_η, which depends on the choice of $\eta : \mathcal{R}_X \to \mathbb{Z}^+$. Further there is unlikely a choice of $\mathbb{Q}$-line bundle over the universal family $\mathfrak{C}_g$ whose restriction to each fiber is H_η for some η. Fortunately, the stability so defined does not depend on the choice of η, at least when χ is prime as chosen.

LEMMA 4.3. *Let $\chi > 2rg$ be a prime as chosen. Then for every semistable curve X the stability of vector bundles E over X of rank r and Euler characteristic χ do not depend on the choice of $\eta\colon \mathcal{R}_X \to \mathbb{Z}^+$.*

PROOF. Let X be fixed and let η_1 and η_2 be two such maps. Suppose E is a locally free sheaf on X as stated so that E is H_{η_1}-stable but not H_{η_2}-stable. Then there must be a proper subsheaf $0 \neq F \subset E$ so that (4.2) holds with η replaced by η_1 and does not hold with η replaced by η_2. Since E is H_{η_1}-stable, since $\chi > 2rg$ and ϵ is sufficiently small, it is proved in [**15**] that the restriction of E to each contracted component $D \subset X$ has

$$E|_D \cong \mathcal{O}_D^{\oplus a} \oplus \mathcal{O}_D(1)^{\oplus b}$$

and there is a non-contracted component $D' \subset X$ so that $F|_{D'}$ is not a torsion sheaf. From the second property, we know that $\text{l.c.}\chi(F \otimes H_{\eta_i}^{\otimes n})$ is very close to the positive integer

$$l(F) \triangleq \text{l.c.}\chi(F \otimes \rho^* \omega_{X^0}^{\otimes n}) \in \mathbb{Z}^+.$$

Since the numerators $\chi(F)$ and $\chi(E)$ are independent of η_i, we must have

$$\frac{\chi(F)}{l(F)} = \frac{\chi}{2r(g-1)}.$$

Further, since $F \subset E$ is a subsheaf, $l(F) \leq 2r(g-1)$. Hence because χ is prime, the above identity is possible only if $l(F) = 2r(g-1)$, namely the quotient E/F is only supported on the contracted components of X. Then apply the fact that the restriction of E to contracted components are non-negative, $\chi(E/F) > 0$. This contradicts to the identity above. This proves that the stability is independent of the choice of η. □

In the proof of the above Lemma, we used the fact that ϵ is sufficiently small, depending on X. Indeed, a single ϵ can be found for our purpose, thanks to the following Lemma.

LEMMA 4.4. *Let E be a vector bundle over a semistable curve $X \in \mathfrak{M}_g$. Suppose ϵ is sufficiently small and E is H_η-stable as defined in Definition 4.2, the number of the contracted components of X is no more than $r(g+1)$.*

PROOF. We let $\mathcal{D} \subset X$ be the union of all irreducible components contracted by stabilization and let $\mathcal{D}_1 \cdots, \mathcal{D}_k$ be the connected components of $\mathcal{D}$. First, each $\mathcal{D}_i$ is mapped to a node of X^0, and no two $\mathcal{D}_i$ are mapped to the same node. Thus k is bounded from above by the number of the nodes of X^0, which is at most $g+1$. Secondly, because of the H_η-stability of E, the number of irreducible components in each $\mathcal{D}_i$ can not exceed the rank r [**15**]. The reason is the following. As we mentioned in the proof of the previous Lemma, H_η-weakly stable implies that for

$H^0(E|_{\mathcal{D}_i})$ that vanishes at both p_1 and p_2 are the zero section[8]. Thus by Riemann-Roch theorem the number of rational curves in $\mathcal{D}_i$ is at most the rank of E, which is r. This proves the bound in the Lemma. □

We define the sub-groupoid $\mathfrak{V}^{\mathrm{st}}_{r,\chi} \subset \mathfrak{V}_{r,\chi}$ be the one consisting of all families $\mathcal{E} \to \mathcal{X}$, $\mathcal{X} \in \mathfrak{M}_g(S)$, so that for each closed $s \in S$ the restriction $\mathcal{E}|_{\mathcal{X}_s}$ is stable with respect to an $\eta: \mathcal{R}_{\mathcal{X}_s} \to \mathbb{Z}^+$. By the Lemmas before, $\mathfrak{V}^{\mathrm{st}}_{r,\chi}$ is a well-defined sub-groupoid of $\mathfrak{V}_{r,\chi}$, and thus is a substack of $\mathfrak{V}_{r,\chi}$.

The previous Lemma immediately shows that $\mathfrak{V}^{\mathrm{st}}_{r,\chi}$ is bounded.

COROLLARY 4.5. *The stack* $\mathfrak{V}^{st}_{r,\chi}$ *is bounded.*

PROOF. We let $\mathfrak{B}$ be the set of all semistable curves in $\mathfrak{M}_g$ that are the images of $\mathfrak{V}^{\mathrm{st}}_{r,\chi} \to \mathfrak{M}_g$. By the previous Lemma, $\mathfrak{B}$ is bounded. Then because we can choose one sufficiently small ϵ to work for all stability condition, and also because Simpson stable bundles on individual X are bounded, we conclude that $\mathfrak{V}^{\mathrm{st}}_{r,\chi}$ is bounded. This proves the Corollary. □

The next step is to show that the automorphism group of every $E \in \mathfrak{V}^{\mathrm{st}}_{r,\chi}$ is finite. Here as usual we define the automorphism group of $[E]$, denoted by $\mathfrak{Aut}([E])$, be the set of all arrows $[E] \to [E]$ in the stack $\mathfrak{V}_{r,\chi}$.

LEMMA 4.6. *Let* $E \to X$ *be any member in* $\mathfrak{V}^{st}_{r,\chi}$. *Then the automorphism group* $\mathfrak{Aut}([E])$ *is a subgroup of* $\mathrm{Aut}(X^0)$, *thus is finite.*

PROOF. Let E over X be as in the Lemma. Let $\xi \in \mathfrak{Aut}(E)$ be any element. By definition, ξ is a pair of an automorphism $\sigma : X \to X$ and an isomorphism $\psi: E \cong \sigma^* E$. Clearly, the automorphism σ induces an automorphism $\sigma^0: X^0 \to X^0$. This defines a homomorphism

$$\mathfrak{Aut}([E]) \longrightarrow \mathrm{Aut}(X^0).$$

We claim that it is injective. Namely, if $\sigma^0 = \mathrm{id}_{X^0}$ then σ must be the identity id_X and $\psi = c1_E$ for some $c \in \mathbb{C}^*$. By our definition of arrows in $\mathfrak{V}^{\mathrm{st}}_{r,\chi}$, such equivalence is the identity arrow in $\mathfrak{V}^{\mathrm{st}}_{r,\chi}$. This will show that $\mathfrak{Aut}([E]) \to \mathrm{Aut}(X^0)$ is injective.

We now prove the claim. Since E is H_η-stable for ϵ sufficiently small, it is proved in [**15**] that the push forward $\rho_* E$ is a sheaf on X^0 and that it has no element supported on points of X^0. Further it is direct to show that the push forward is Simpson stable with respect to the polarization ω_{X^0}. Here we used the fact that χ is a large prime. However, since $\sigma^0 = \mathrm{id}_{X^0}$, ψ induces an isomorphism of $\rho_* E$. Because $\rho_* E$ is stable, this induced isomorphism must be $c1_{\rho_* E}$, for some $c \in \mathbb{C}^*$. Finally, using the fact that the restriction of E to each chain of contracted rational curves (i.e. the $\mathcal{D}_i$ as before) is regular (see [**15**]), one proves that σ must be an identity and the isomorphism ψ must also be $c1_E$ (see [**15**] for details). This proves the claim. □

We next show that $\mathfrak{V}^{\mathrm{st}}_{r,\chi}$ is separated and proper. For this, an easy fact helps us technically.

LEMMA 4.7. *The tautological morphism* $\mathfrak{V}^{st}_{r,\chi} \to \mathfrak{M}_g$ *is a smooth morphism.*

[8] A vector bundle on $\mathcal{D}_i$ satisfying this and the condition $E|_D \cong \mathcal{O}_D^{\oplus a} \oplus \mathcal{O}_D(1)^{\oplus b}$ are called regular vector bundle in [**15**].

PROOF. Let X be any semistable curve and $E \to X$ a stable locally free sheaf in $\mathfrak{V}^{\mathrm{st}}_{r,\chi}$. Let $\mathcal{X}$ be any flat deformation of X. Then since E is locally free and $\dim X = 1$, we can extend E to a family of locally free sheaves $\mathcal{E}$ over $\mathcal{X}$. Since being stable is an open condition, there is a Zariski open subset $S' \subset S$ containing $[X]$ so that the fibers of $\mathcal{E}$ over $s \in S'$ are stable. Thus there is a family in $\mathfrak{V}^{\mathrm{st}}_{r,\chi}$ extending the element $[E]$ and is a lift of the family $\mathcal{X} \in \mathfrak{M}_g$. This proves that $\mathfrak{V}^{\mathrm{st}}_{r,\chi}$ is smooth over $\mathfrak{M}_g$. □

LEMMA 4.8. *The stack $\mathfrak{V}^{st}_{r,\chi}$ is separated.*

PROOF. We apply the valuation criteria of separatedness. Let $\xi \in S$ be a closed point in an affine smooth curve S, let $S^\circ = S - \xi$, let $\mathcal{X}_{S^\circ}$ be a family of semistable curves in $\mathfrak{M}_g(S^\circ)$ and let $\mathcal{E}^\circ$ be a family of stable locally free sheaves on $\mathcal{X}_{S^\circ}$ in $\mathfrak{V}^{\mathrm{st}}_{r,\chi}(S^\circ)$. To verify the separatedness criterion, we need to show that there is at most one extension of $\mathcal{E}^\circ$ to families of stable sheaves in $\mathfrak{V}^{\mathrm{st}}_{r,\chi}(S)$. Suppose there are two families $\mathcal{E}' \to \mathcal{X}'_S$ and $\mathcal{E}'' \to \mathcal{X}''_S$ in $\mathfrak{V}^{\mathrm{st}}_{r,\chi}(S)$, both are extensions of $\mathcal{E}^\circ \to \mathcal{X}_{S^\circ}$. We need to show that there is an isomorphism $\sigma: \mathcal{X}'_S \to \mathcal{X}''_S$ extending the identity on $\mathcal{X}_{S^\circ}$ and an isomorphism $\mathcal{E}' \cong \sigma^*\mathcal{E}''$.

We now make a reduction. Since $\mathfrak{V}^{\mathrm{st}}_{r,\chi} \to \mathfrak{M}_g$ is smooth, and since the set of smooth curves in $\mathfrak{M}_g$ is dense in $\mathfrak{M}_g$, to prove the separatedness we only need to consider the case where all fibers of $\mathcal{X}_{S^\circ}$ over S° are smooth. Possibly after a base change of S, we can assume that there is a family $\mathcal{X}_S$ so that

- $\mathcal{X}_S \times_S S^\circ \equiv \mathcal{X}_{S^\circ}$;
- $\mathcal{X}_S$ *has smooth total space and its fiber over $\xi \in S$ has normal crossing singularities, thus a semistable curve;*
- $\mathcal{X}_S$ *are desingularizations of $\mathcal{X}'_S$ and $\mathcal{X}''_S$, and the tautological projections π' : $\mathcal{X}_S \to \mathcal{X}'_S$ and $\pi'': \mathcal{X}_S \to \mathcal{X}''_S$ commute with the birational $\mathcal{X}'_S \sim \mathcal{X}''_S$.*

We fix a $\mathbb{Q}$-ample line bundle H_η on $\mathcal{X}_{S,\xi}$ and extend it to $\mathcal{H}$ over the family $\mathcal{X}_S$. Then it can be shown easily [**15**] that both $\pi'^*\mathcal{E}'$ and $\pi''^*\mathcal{E}''$ are families of locally free Simpson stable vector bundles over $\mathcal{X}_S$ with respect to $\mathcal{H}$. Since they are isomorphic over the general fibers of $\mathcal{X}_S$, they are isomorphic by the separatedness of the moduli of Simpson stable sheaves. Further, we observe that a rational curve $D \subset \mathcal{X}_{S,\xi}$ is contracted by π' if and only if $\pi'^*\mathcal{E}'|_D$ is trivial. Because $\pi'^*\mathcal{E}' \cong \pi''^*\mathcal{E}''$, both π' and π'' contracts *identical* set of rational curves in $\tilde{\mathcal{X}}_S$. Hence $\mathcal{X}'_S \cong \mathcal{X}''_S$ and under this isomorphism

$$\mathcal{E}' \equiv \pi'_*\pi'^*\mathcal{E}' \cong \pi''_*\pi''^*\mathcal{E}'' \equiv \mathcal{E}''.$$

This verifies the valuation criterion for separatedness. □

LEMMA 4.9. *The stack $\mathfrak{V}^{st}_{r,\chi}$ is proper.*

PROOF. We will prove the properness using valuation criterion. Let $\xi \in S$ be a closed point in a smooth curve S. We let $S^\circ = S - \xi$ and let $\mathcal{E}^\circ \in \mathfrak{V}^{\mathrm{st}}_{r,\chi}(S^\circ)$ be a family over $\mathcal{X}_{S^\circ}$. We need to show that we can extend $\mathcal{E}^\circ \to \mathcal{X}_{S^\circ}$ to a family $\mathcal{E} \to \mathcal{X}_S$ in $\mathfrak{V}^{\mathrm{st}}_{r,\chi}(S)$. Again, since $\mathfrak{V}^{\mathrm{st}}_{r,\chi}$ is flat over $\mathfrak{M}_g$ and since smooth curves in $\mathfrak{M}_g$ are dense in it, we can assume without lose of generality that all fibers of $\mathcal{X}_{S^\circ}$

it to a $\mathbb{Q}$-line bundle $\mathcal{H}$ on $\mathcal{X}_S$. Clearly, the family $\mathcal{E}^\circ$ is a family of Simpson $\mathcal{H}$-stable locally free sheaves on $\mathcal{X}_{S^\circ}/S^\circ$. Then by [**27**], there is an extension of $\mathcal{E}^\circ$ to a sheaves of $\mathcal{O}_{\mathcal{X}_S}$-modules $\mathcal{E}$, flat over S, so that $\mathcal{E}\otimes_{\mathcal{O}_{\mathcal{X}_S}}\mathcal{O}_{\mathcal{X}_{S,\xi}}$ is semistable with respect to the polarization H_η. In particular there are no elements in $\mathcal{E}\otimes_{\mathcal{O}_{\mathcal{X}_S}}\mathcal{O}_{\mathcal{X}_{S,\xi}}$ supported at point. Since $\mathcal{X}_S$ is a smooth surface, $\mathcal{E}$ is a subsheaf of its double dual $\mathcal{E}^{\vee\vee}$ with cokernel $\mathcal{E}^{\vee\vee}/\mathcal{E}$ supported at points along the central fiber $\mathcal{X}_{S,\xi}$. In case the cokernel is non-trivial, then the exact sequence

$$0=\mathrm{Tor}(\mathcal{E}^{\vee\vee},\mathcal{O}_{\mathcal{X}_{S,\xi}})\longrightarrow \mathrm{Tor}(\mathcal{E}^{\vee\vee}/\mathcal{E},\mathcal{O}_{\mathcal{X}_{S,\xi}})\longrightarrow \mathcal{E}\otimes_{\mathcal{O}_{\mathcal{X}_S}}\mathcal{O}_{\mathcal{X}_{S,\xi}}\longrightarrow$$

shows that $\mathcal{E}\otimes_{\mathcal{O}_{\mathcal{X}_S}}\mathcal{O}_{\mathcal{X}_{S,\xi}}$ has torsion supported at points, so can not be stable. This proves that $\mathcal{E}$ is locally free.

Because $\mathcal{E}$ is locally free, by a Lemma proved in [**15**], for any rational curve $D_i\subset\tilde{\mathcal{X}}_{S,\xi}$ in the contracted locus (of the stabilization of $\mathcal{X}_{S,\xi}$) the restriction $\mathcal{E}|_{D_i}$ must be of the form $\mathcal{O}_{D_i}^{\oplus a}\oplus\mathcal{O}_{D_i}(1)^{\oplus b}$. We let $\tilde{\mathcal{X}}_S$ be the contraction of $\mathcal{X}_S$ along those D_i so that $\mathcal{E}|_{D_i}$ is trivial. Let $\pi:\mathcal{X}_S\to\tilde{\mathcal{X}}_S$ be the contraction morphism and let $\tilde{\mathcal{E}}=\pi_*\mathcal{E}$. It is direct to check that $\tilde{\mathcal{E}}$ is locally free and its restriction to the central fiber $\tilde{\mathcal{X}}_{S,\xi}$ is a member in $\mathfrak{V}^{\mathrm{st}}_{r,\chi}$. Thus $\tilde{\mathcal{E}}\to\tilde{\mathcal{X}}_S$ is a desired extension. This verifies the valuation criterion for properness. □

Since $\mathfrak{V}^{\mathrm{st}}_{r,\chi}$ is bounded, it is easy to see that there is a scheme of finite type that surjects onto $\mathfrak{V}^{\mathrm{st}}_{r,\chi}$. Then because $\mathfrak{V}^{\mathrm{st}}_{r,\chi}$ is separated, proper and because the automorphism group of all of its members are finite, we can find a scheme of finite type that is étale surjects onto $\mathfrak{V}^{\mathrm{st}}_{r,\chi}$. This proves that

THEOREM 4.10. *The stack $\mathfrak{V}^{st}_{r,\chi}$ is a proper, separated Deligne-Mumford stack. Further, it is smooth and the fibers of the tautological morphism $\mathfrak{V}^{st}_{r,\chi}\to\mathcal{M}_g$ have only normal crossing singularities.*

PROOF. We only need to prove the second part. The smoothness of $\mathfrak{V}^{\mathrm{st}}_{r,\chi}$ follows directly from the proof that $\mathfrak{V}^{\mathrm{st}}_{r,\chi}\to\mathfrak{M}_g$ is smooth. Now we show that fibers of $\mathfrak{V}^{\mathrm{st}}_{r,\chi}\to\mathcal{M}_g$ have only normal crossing singularities. Let $E\to X$ be a member in $\mathfrak{V}^{\mathrm{st}}_{r,\chi}$ and let $\hat{V}$ be the germ of infinitesimal deformations of X with $\hat{\mathcal{X}}\to\hat{V}$ its the tautological family. Let $\hat{V}\to\mathfrak{M}_g$ be the tautological map and

$$\hat{\mathfrak{V}}\triangleq\mathfrak{V}^{\mathrm{st}}_{r,\chi}\times_{\mathfrak{M}_g}\hat{V}$$

be the fiber product. Since $\mathfrak{V}^{\mathrm{st}}_{r,\chi}$ is smooth over $\mathfrak{M}_g$, $\hat{\mathfrak{V}}$ is smooth. Then because $\hat{V}$ to $\mathcal{M}_g$ has normal crossing fibers, $\hat{\mathfrak{V}}\to\mathcal{M}_g$ has normal crossing fibers.

On the other hand, we let $\mathfrak{Aut}_0(X)$ be the connected component of $\mathfrak{Aut}(X)$ containing 1_X. Then $\mathfrak{Aut}_0(X)\cong(\mathbb{C}^*)^{\times n}$, where n is the number of contracted components of X under stabilization, and $\mathfrak{Aut}_0(X)$ acts on $\hat{V}$ and its universal family $\hat{\mathcal{X}}$, and thus induces an action on the Cartesian product $\hat{\mathfrak{V}}$. Because the automorphism group of each $[E\to X]\in\mathfrak{V}^{\mathrm{st}}_{r,\chi}$ is a subgroup of $\mathfrak{Aut}(X)$, it is direct to check that the group $\mathfrak{Aut}_0(X)$ acts on $\hat{\mathfrak{V}}$ freely. Thus the quotient $\hat{\mathfrak{V}}/\mathfrak{Aut}_0(X)$ is smooth, and hence the tautological map $\hat{\mathfrak{V}}/\mathfrak{Aut}_0(X)\to\mathcal{M}_g$ has normal crossing fibers. Finally, it is direct to check that the tautological map $\hat{\mathfrak{V}}/\mathfrak{Aut}_0(X)\to\mathfrak{V}^{\mathrm{st}}_{r,\chi}$ is étale. This proves that $\mathfrak{V}^{\mathrm{st}}_{r,\chi}\to\mathcal{M}_g$ has normal crossing fibers everywhere. □

Since $\mathfrak{V}^{\mathrm{st}}_{r,\chi}$ is a DM-stack, we will denote it by

$$\mathcal{M}_{r,\chi}(\mathcal{C}_g/\mathfrak{M}_g).$$

Since it is proper, it is a compactification of the relative moduli of stable vector bundles on smooth curves over $\mathcal{M}_g^0$.

We now comment on this compactification. It is known that the the moduli $\mathcal{M}_X(r,d)$ of stable vector bundles over a smooth curve is isomorphic to $\mathcal{M}_X(r,d+kr)$ for any integer $k \in \mathbb{Z}$. However, the same might not be true for non-irreducible curves since the choices of ample line bundles will affect the stability of bundles over them. In our construction, we choose the ample line bundle on X to be sufficiently close to the pull back of ω_{X^0}. In this sense, the compactification $\mathcal{M}_{r,\chi}(\mathcal{C}_g/\mathfrak{M}_g)$ constructed here is "fairly" canonical. The other choice we made is the degree, a prime $\chi > 2rg$. It is not clear that if both χ and $\chi' > 2rg$ are primes and satisfying $\chi - \chi' \equiv 0 \mod r$, then are $\mathcal{M}_{r,\chi}(\mathcal{C}_g/\mathfrak{M}_g)$ and $\mathcal{M}_{r,\chi'}(\mathcal{C}_g/\mathfrak{M}_g)$ isomorphic? Of course they are birational. In this note, the compactification is constructed as a Deligne-Mumford stack, over the Deligne-Mumford stack $\mathcal{M}_g$. If we let $\mathbf{M}_{r,\chi}(\mathcal{C}_g/\mathcal{M}_g)$ be its coarse moduli space, then it is a connected, algebraic space with only finite quotient singularities. If we let $\mathbf{M}_g$ be the coarse moduli space of $\mathcal{M}_g$, then the fiber of the tautological projection

$$\mathbf{M}_{r,\chi}(\mathcal{C}_g/\mathcal{M}_g) \longrightarrow \mathbf{M}_g$$

over $[X] \in \mathbf{M}_g$ is the quotient of $\mathbf{M}_X^{\mathrm{Gi}}(r,\chi)$ by $\mathrm{Aut}(X)$. Here $\mathbf{M}_X^{\mathrm{Gi}}(r,\chi)$ is the coarse moduli space of the fiber product

$$\mathcal{M}_{r,\chi}(\mathcal{C}_g/\mathfrak{M}_g) \times_{\mathcal{M}_g} [X].$$

Here by abuse of notation we view $[X]$ as an ordinary point and $[X] \to \mathcal{M}_g$ be the tautological map. Note that $\mathbf{M}_X^{\mathrm{Gi}}(r,\chi)$ is smooth when X is smooth and has normal crossing singularity when X is nodal.

One important question remain open is whether there is a universal family over the moduli stack $\mathcal{M}_{r,\chi}(\mathcal{C}_g/\mathfrak{M}_g)$. Since in the definition of the moduli groupoid an arrow is an equivalence class of isomorphisms, it is not clear that whether a universal family exists. The existence of universal family will be useful to study the degeneration of moduli of stable vector bundles on smooth curves of fixed determinants (c.f. [**29**]).

References

[1] K. Behrend, *Gromov-Witten invariants in alebraic geometry*, Invent. Math. **127**(1997), no. 3, 601-617.

[2] K. Behrend and B. Fantechi, *The intrinsic normal cone*, Invent. Math. **128**(1997), no. 1, 45-88.

[3] L. Caporaso, *A compactification of the universal Picard variety over the moduli space of stable curves*, J. Amer. Math. Soc. **7**(1994), no. 3, 589-660.

[4] Y. Eliashberg, A. Givental and H. Hofer, *Introduction to symplectic field theory*, GAFA 2000 (Tel Aviv, 1999). Geom. Funct. Anal. 2000, Special Volume, Part II, 560-673.

[5] D. Gieseker, *Stable curves and special divisors: Petri's conjecture*, Invent. Math. **66**(1982), no. 2, 251-275.

[6] D. Gieseker, *A degeneration of the moduli space of stable bundles*, J. Differential Geom. **19**(1984), no. 1, 173-206.

[7] D. Gieseker, and J. Li, *Irreducibility of moduli of rank two vector bundles*. J. Diff. Geom. **40**(1994), 23-104; *High rank*. J. Dif.

[8] D. Gieseker and I. Morrison, *Hilbert stability of rank two bundles on curves*, J. Differential

[10] E-N. Ionel and T. Parker, *Gromov-Witten Invariants of Symplectic Sums*, Math. Res. Lett. **5**(1998), no. 5, 563-576.

[11] E-N. Ionel and T. Parker, *Relative Gromov-Witten Invariants*, preprint, math.SG/9907155.

[12] E-N. Ionel and T. Parker, *The Symplectic Sum Formula for Gromov-Witten Invariants*, preprint, math.SG/0010217.

[13] I. Kausz, *A Gieseker type degeneration of moduli stacks of vector bundles on curves*, preprint, math.AG/0201197.

[14] B. Kim, A. Kresch, and T. Pantev,*Functoriality in intersection theory and a conjecture of Cox, Katz and Lee*, preprint.

[15] Y-H. Kiem and J. Li, *Vanishing of the top Chern classes of the moduli of vector bundles*, in preparation.

[16] P.B. Kronheimer and T.S. Morwka, *Embedded surfaces and the structure of Donaldson's polynomial invariants*, J. Diff. Geom. **41**(1995), no. 3, 573-734.

[17] J. Li, *Stable morphisms to singular schemes and relative stable morphisms*, J. Diff. Geom, **57** (2001) 509-578.

[18] J. Li, *A Degeneration formula of GW-invariants*, to appear in J. Diff. Geom.

[19] A-M. Li and Y. Ruan, *Symplectic surgery and Gromov-Witten invariants of Calabi-Yau 3-folds I*, Invent. Math. **145**(2001), no. 1, 151-218.

[20] J. Li and G. Tian, *Algebraic and symplectic geometry of virtual moduli cycles*, Algebraic geometry–Santa Cruz 1995, 143-170, Proc. Sympos. Pure Math., 62, Part 2, Amer. Math. Soc., Providence, RI, 1997;

[21] J. Li and G. Tian, *Virtual moduli cycles and Gromov-Witten invariants of algebraic varieties*, J. Amer. Math. **11**(1998), no. 1, 119-174.

[22] V. Muñoz, *Donaldson invariants for connected sums along surfaces of genus 2*, Topology Appl. **107**(2000), no. 3, 215-232.

[23] D.S. Nagaraj and C.S Seshadri, *Degenerations of the moduli spaces of vector bundles on curves. I*, Proc. Indian Acad. Sci. Math. Sci. **107**(1997), no. 2, 101-137; *Degenerations of the moduli spaces of vector bundles on curves. II. Generalized Gieseker moduli spaces*, Proc. Indian Acad. Sci. Math. Sci. **109**(1999), no. 2, 165-201.

[24] K.G. O'Grady and J.W Morgan, *Differential topology of complex surfaces. Elliptic surfaces with $p_g = 1$: smooth classification*, Lecture Notes in Mathematics, 1545. Springer-Verlag, Berlin, 1993.

[25] R. Pandharipande, *A compactification over $\overline{M}_g$ of the universal moduli space of slope-semistable vector bundles*, J. Amer. Math. Soc. **9**(1996), no. 2, 425-471.

[26] C.S. Seshadri, *Degenerations of the moduli spaces of vector bundles on curves*, School on Algebraic Geometry (Trieste, 1999), 205-265, ICTP Lect. Notes, 1, Abdus Salam Int. Cent. Theoret. Phys., Trieste, 2000.

[27] C.T. Simpson, *Moduli of representations of the fundamental group of a smooth projective variety. II*, Inst. Hautes tudes Sci. Publ. Math. No. **80**(1994), 5-79.

[28] A. Schmitt, *The Hilbert Compactification of the Universal Moduli Space of Semistable Vector Bundles over Smooth Curves*, preprint.

[29] X.T. Sun, *Degeneration of SL(n)-bundles on a reducible curve*, preprint, math.AG/0112072.

DEPARTMENT OF MATHEMATICS, STANFORD UNIVERSITY, STANFORD, CA 94305

Contemporary Mathematics
Volume **322**, 2003

Convolution on homology groups of moduli spaces of sheaves on K3 surfaces

Hiraku Nakajima

ABSTRACT. Quiver varieties are noncompact hyper-Kähler manifolds, and have many analogy with moduli spaces of sheaves on a $K3$ surface. The author constructed representations of Lie algebras on the homology groups of the former spaces by convolution operators. In this paper, we give a similar construction on the homology groups of the latter spaces.

1. Moduli spaces of sheaves on $K3$ surfaces

Let X be a $K3$ surface over $\mathbf{C}$. Let $(H^*(X,\mathbf{Z}), \langle\ ,\ \rangle)$ be its Mukai lattice, i.e., the integral cohomology group together with the pairing

$$\langle \mathbf{v}, \mathbf{v}' \rangle = \int_X v_1 \cup v_1' - v_0 \cup v_2' - v_2 \cup v_0',$$

where v_i, v_i' are degree $2i$ part of $\mathbf{v}, \mathbf{v}' \in H^*(X,\mathbf{Z})$. It is a lattice of signature $(4,20)$.

For a coherent sheaf E on X, we define its Mukai vector $\mathbf{v}(E)$ by $\mathbf{v}(E) = \operatorname{ch}(E)\sqrt{\operatorname{td} X} \in H^*(X,\mathbf{Z})$, where $\operatorname{ch}(E)$ is the Chern character of E, $\operatorname{td} X$ is the Todd genus of X. The Riemann-Roch theorem says

$$\langle \mathbf{v}(E), \mathbf{v}(F) \rangle = -\chi(E,F) = -\sum_i (-1)^i \dim \operatorname{Ext}^i(E,F)$$

for coherent sheaves E, F on X.

A vector $\mathbf{v} \in H^*(X,\mathbf{Z})$ is called *primitive* if it is not written as $\mathbf{v} = m\mathbf{v}'$ for some $\mathbf{v}' \in H^*(X,\mathbf{Z})$, $m > 1$.

A vector $\mathbf{v} \in H^*(X,\mathbf{Z})$ is called a (-2)-*vector* if $\langle \mathbf{v}, \mathbf{v} \rangle = -2$. A coherent sheaf E on X is called *exceptional* if

$$\dim \operatorname{Ext}^i(E,E) = \begin{cases} 1 & \text{if } i = 0, 2, \\ 0 & \text{if } i = 1. \end{cases}$$

2000 *Mathematics Subject Classification.* Primary 14D20, 17B10; Secondary 14J28, 44A35.

In particular, its Mukai vector is a (-2)-vector. For example, the structure sheaf $\mathcal{O}_X$ and the structure sheaf $\mathcal{O}_C$ of a (-2)-curve C are exceptional. A (-2)-vector $\mathbf{v}$ defines a *reflection* on the Mukai lattice by

$$R_{\mathbf{v}}\colon x \longmapsto x + \langle x, \mathbf{v}\rangle \mathbf{v}, \qquad \text{for } x \in H^*(X, \mathbf{Z}).$$

A vector $\mathbf{v} \in H^*(X, \mathbf{Z})$ is called *isotropic* if $\langle \mathbf{v}, \mathbf{v}\rangle = 0$.

Let H be an ample line bundle over X. For a vector $\mathbf{v} \in H^*(X, \mathbf{Z})$, let $M_H(\mathbf{v})$ denote the moduli space of H-stable sheaves E on X with $\mathbf{v}(E) = \mathbf{v}$. A fundamental result of Mukai [**17**] says that it is a nonsingular quasi-projective variety with

$$\dim M_H(\mathbf{v}) = 2 + \langle \mathbf{v}, \mathbf{v}\rangle, \tag{1.1}$$

and it has a holomorphic symplectic structure. Let us give examples of low dimensional moduli spaces due to Mukai [**18**].

(1) If v is a (-2)-vector, $M_H(\mathbf{v})$ is either empty or a single point [loc. cit., Corollary 3.6].
(2) If v is a primitive isotropic vector and H is generic, $M_H(\mathbf{v})$ is a $K3$ surface, provided it is nonempty [loc. cit., Theorem 1.4].

2. Quiver varieties

In this section we give a definition of quiver varieties and their properties [**20, 23**]. Our definition of quiver varieties is different from the original one in [loc. cit.], but is equivalent. We want to emphasize analogy between quiver varieties and moduli spaces of sheaves on $K3$ surfaces. In fact, quiver varieties are moduli spaces of sheaves on ALE spaces (minimal resolutions of simple singularities) [**13**], and the author's starting point of the study of quiver varieties was to prove analog of the above Mukai's result [**19**].

2.1. Definition. Let Γ be a nontrivial finite subgroup of $\mathrm{SL}_2(\mathbf{C})$. Let $R(\Gamma)$ be its representation ring. As a free abelian group, it has a base indexed by isomorphism classes of irreducible representations $\rho_0, \rho_1, \dots, \rho_r$. As a convention, we assume that ρ_0 is the trivial representation.

Let Q be the 2-dimensional representation of Γ given by the inclusion $\Gamma \subset \mathrm{SL}_2(\mathbf{C})$. The representation ring $R(\Gamma)$ has a pairing $\langle\ ,\ \rangle$ given by

$$\langle \mathbf{v}, \mathbf{v}'\rangle = -(\mathbf{v}, \mathbf{C} \otimes \mathbf{v}'),$$

where $\mathbf{C}$ is the virtual representation $\bigwedge^0 Q - \bigwedge^1 Q + \bigwedge^2 Q$, and $(\ ,\ )$ is the natural inner product on $R(\Gamma)$ such that $\{\rho_i\}$ is an orthonormal base. The so-called McKay correspondence (see e.g., [**24**, Chapter 4]) says that $(R(\Gamma), \langle\ ,\ \rangle)$ is isomorphic to the root lattice of an affine Lie algebra $\widehat{\mathfrak{g}}$ of type $A_n^{(1)}$, $D_n^{(1)}$, $E_n^{(1)}$ after the sign change. We call it the *McKay lattice*. The McKay lattice has 1-dimensional kernel with the generator δ whose ρ_0-component is positive. It is known that δ is identified with the isomorphism class of the regular representation of Γ. The set $\{\rho_i\}$ is identified with the set of simple roots of $\widehat{\mathfrak{g}}$. As in the case of Mukai lattice, we can define (-2)-vector (resp. isotropic vector) $\mathbf{v}$ in $R(\Gamma)$. In fact, it is known that (see [**8**, Chap. 5])

(1) a (-2)-vector is a real root (and defines a reflection on $R(\Gamma)$ by the same formula in the previous section).
(2) an isotropic vector is an imaginary root, i.e., $n\delta$ for some $n \in \mathbf{Z} \setminus \{0\}$.

Let $\mathrm{Hilb}^n(\mathbf{C}^2)$ be the Hilbert scheme of n points on $\mathbf{C}^2$. As a set, it consists of ideals I of the polynomial ring $\mathbf{C}[x, y]$ of two variables satisfying $\dim \mathbf{C}[x, y]/I = n$. It is known that

(1) $\mathrm{Hilb}^n(\mathbf{C}^2)$ is a resolution of singularities of the nth symmetric product $S^n(\mathbf{C}^2)$ (Fogarty).

(2) $\mathrm{Hilb}^n(\mathbf{C}^2)$ has a holomorphic symplectic structure (Beauville, Mukai).

See [**24**] for more details on Hilbert schemes of points on surfaces.

The natural action of Γ on $\mathbf{C}^2$ induces an action on $\mathrm{Hilb}^n(\mathbf{C}^2)$. We consider the fixed point set $\left(\mathrm{Hilb}^n(\mathbf{C}^2)\right)^\Gamma$ consisting of Γ-invariant ideals. Let $\mathbf{v}$ be an isomorphism class of a representation of Γ. We consider it as an element of $R(\Gamma)$. For such $\mathbf{v}$, let

$$M(\mathbf{v}) \overset{\text{def.}}{=} \left\{ I \in \left(\mathrm{Hilb}^n(\mathbf{C}^2)\right)^\Gamma \,\middle|\, [\mathbf{C}[x, y]/I] = \mathbf{v} \right\},$$

where $n = \dim \mathbf{v}$ and $[\mathbf{C}[x, y]/I]$ is the class of the Γ-module $\mathbf{C}[x, y]/I$. For $\mathbf{v} \in R(\Gamma)$, which is not a class of an *actual* representation, we set $M(\mathbf{v})$ to be empty. $M(\mathbf{v})$ is an example of quiver varities of affine type.

To readers familiar with quiver varieties : The other vector $\mathbf{w}$ is the 0th fundamental weight Λ_0. Quiver varieties for arbitrary $\mathbf{w}$ can be realized as Γ-fixed point sets of framed moduli spaces of higher rank torsion free sheaves on $\mathbf{P}^2$. See [**27**].

It can be shown that $M(\mathbf{v})$, if nonempty, has a holomorphic symplectic structure (in fact, a hyper-Kähler structure). We have the following dimension formula:

$$\dim M(\mathbf{v}) = 2v_0 + \langle \mathbf{v}, \mathbf{v} \rangle, \tag{2.1}$$

where v_0 is the ρ_0-component of $\mathbf{v}$, i.e., $v_0 = (\rho_0, \mathbf{v})$. This is very similar to (1.1).

When $v_0 = 1$ and other components of $\mathbf{v}$ are 0, i.e., $\mathbf{v}$ is the class of the trivial 1-dimensional representation, $M(\mathbf{v})$ is a single point consisting of the maximal ideal at the origin. (More generally, if $v_0 = 1$ and $\langle \mathbf{v}, \mathbf{v} \rangle = -2$, it is known that $M(\mathbf{v})$ is a single point. This can be proved as a simple application of the main result in [**20**]. Note that such $\mathbf{v}$ is $\delta + \alpha$ where α is a root of the finite dimensional Lie algebra $\mathfrak{g}$.)

As a next example, let $\mathbf{v} = \delta$. A typical point in $M(\mathbf{v})$ is an ideal consisting of functions vanishing at a free Γ-orbit. Therefore $M(\mathbf{v})$ contains $\left(\mathbf{C}^2 \setminus \{0\}\right)/\Gamma$. In fact, it is not difficult to show that $M(\mathbf{v})$ is the minimal resolution of $\mathbf{C}^2/\Gamma$. This result was obtained by Ginzburg-Kapranov (unpublished) and Ito-Nakamura [**7**] independently, but is also a re-interpretation of Kronheimer's construction [**12**]. (See [**24**, Chapter 4] for detail.)

These two examples are analogs of Mukai's results mentioned in the previous section.

Furthermore, a general $M(\mathbf{v})$, provided that it is nonempty, is a resolution of singularities of $S^m(\mathbf{C}^2/\Gamma)$ for some m. If $m = 0$, it means a single point. (See [**27**].)

Now the following analogy are obvious:

Mukai lattice $(H^*(X, \mathbf{Z}), \langle\ ,\ \rangle)$	McKay lattice $(R(\Gamma), \langle\ ,\ \rangle)$
a moduli space $M_H(\mathbf{v})$	a quiver variety $M(\mathbf{v})$

2.2. Lie algebra action on homology groups of quiver varieties. In this subsection, we recall the author's construction of a structure of a representation of the affine Lie algebra $\widehat{\mathfrak{g}}$ on the direct sum of homology groups $\bigoplus_{\mathbf{v}} H_*(M(\mathbf{v}), \mathbf{C})$.

For each simple root α_i, there exist e_i, f_i, h_i in the affine Lie algebra $\widehat{\mathfrak{g}}$ satisfying a relation similar to that for $\mathfrak{sl}_2$:

$$[h_i, e_i] = 2e_i, \quad [h_i, f_i] = -2f_i, \quad [e_i, f_i] = h_i. \tag{2.2}$$

The affine Lie algebra $\widehat{\mathfrak{g}}$ is generated by e_i, f_i, h_i for each simple roots α_i, together with the so-called degree operator d. We do not recall other defining relations among those generators, but can be found in [**8**].

We now define operators e_i, f_i, h_i on the direct sum of homology groups of quiver varieties by *convolutions*. Let

$$\mathfrak{P}_i(\mathbf{v}) \overset{\text{def.}}{=} \{(I_1, I_2) \in M(\mathbf{v} - \rho_i) \times M(\mathbf{v}) \mid I_1 \supset I_2\},$$

or more generally

$$\mathfrak{P}_i^{(n)}(\mathbf{v}) \overset{\text{def.}}{=} \{(I_1, I_2) \in M(\mathbf{v} - n\rho_i) \times M(\mathbf{v}) \mid I_1 \supset I_2\}.$$

It is known that $\mathfrak{P}_i^{(n)}(\mathbf{v})$ is a nonsingular lagrangian subvariety in the product $M(\mathbf{v} - n\rho_i) \times M(\mathbf{v})$, provided it is nonempty. The fundamental class of $\mathfrak{P}_i^{(n)}(\mathbf{v})$ defines an operator

$$H_*(M(\mathbf{v} - n\rho_i), \mathbf{C}) \ni c \longmapsto p_{2*}(p_1^* c \cap \left[\mathfrak{P}_i^{(n)}(\mathbf{v})\right]) \in H_*(M(\mathbf{v}), \mathbf{C}),$$

where p_1, p_2 are the first and second projections in the product $M(\mathbf{v}-n\rho_i) \times M(\mathbf{v})$. Here the smoothness of the product is used to consider the intersection $\cap$. (Since $M(\mathbf{v})$ is noncompact, we must study the support of $p_1^* c \cap \left[\mathfrak{P}_i^{(n)}(\mathbf{v})\right]$ in order to check that its pushforward with respect to p_2 is well-defined. But the study is not difficult.) Moreover, since $\mathfrak{P}_i^{(n)}(\mathbf{v})$ is a half-dimensional subvariety, the degrees are changed as

$$H_{d(\mathbf{v}-n\rho_i)+s}(M(\mathbf{v} - n\rho_i), \mathbf{C}) \longrightarrow H_{d(\mathbf{v})+s}(M(\mathbf{v}), \mathbf{C}),$$

where $d(\mathbf{v}) = \dim M(\mathbf{v})$. Moving $\mathbf{v}$, we consider the above as an operator on $\bigoplus_{\mathbf{v},s} H_{d(\mathbf{v})+s}(M(\mathbf{v}), \mathbf{C})$ and denote it by $f_i^{(n)}$. We set $f_i = f_i^{(1)}$. In fact, one can show

$$f_i^{(n)} = \frac{1}{n!} f_i^n.$$

So $f_i^{(n)}$ is the so-called the *divided power*, and the notation is compatible with the standard one.

The operators e_i, $e_i^{(n)}$ are defined by reversing the role of two factors up to sign:

$$H_*(M(\mathbf{v}), \mathbf{C}) \ni c \longmapsto (-1)^{nr(\mathbf{v})} p_{1*}(p_2^* c \cap \left[\mathfrak{P}_i^{(n)}(\mathbf{v})\right]) \in H_*(M(\mathbf{v} - n\rho_i), \mathbf{C}),$$

where $r(\mathbf{v}) = \frac{1}{2}(\dim M(\mathbf{v} - \rho_i) - \dim M(\mathbf{v}))$. Finally we define the operators h_i, d by

$$h_i|_{H_*(M(\mathbf{v}),\mathbf{C})} = (\delta_{0i} + \langle \rho_i, \mathbf{v}\rangle) \operatorname{id},$$
$$d|_{H_*(M(\mathbf{v}),\mathbf{C})} = -(\rho_0, \mathbf{v}) \operatorname{id}.$$

THEOREM 2.3. *For each s, the direct sum* $\bigoplus_{\mathbf{v}} H_{d(\mathbf{v})+s}(M(\mathbf{v}), \mathbf{C})$ *has a structure of an integrable representation of* $\widehat{\mathfrak{g}}$ *given by above operators. Each summand* $H_{d(\mathbf{v})+s}(M(\mathbf{v}), \mathbf{C})$ *is a weight space of the representation. If* $s = 0$*, it is irreducible and highest weight (isomorphic to the so-called basic representation.)*

Let

$$L(\mathbf{v}) \overset{\text{def.}}{=} \{I \in M(\mathbf{v}) \mid \operatorname{Supp} \mathbf{C}[x,y]/I \subset \{0\}\}.$$

It is known that $L(\mathbf{v})$ is a lagrangian subvariety in $M(\mathbf{v})$ and homotopically equivalent to $M(\mathbf{v})$ [**20**]. It follows that $H_{d(\mathbf{v})}(M(\mathbf{v}), \mathbf{C}) \cong H_{d(\mathbf{v})}(L(\mathbf{v}), \mathbf{C})$ and irreducible components of $L(\mathbf{v})$ is a base of this space. This base is related to Kashiwara's crystal base. We will explain the relation briefly in §4.1.

3. $\mathfrak{sl}_2$-action on homology groups of moduli spaces of sheaves on a $K3$ surface

Thanks to the analogy between the moduli spaces $M_H(\mathbf{v})$ of sheaves on $K3$ surfaces and quiver varieties, it is natural to expect the construction in the previous section has a counter part for $M_H(\mathbf{v})$. It will be given in this section.

Let E_0 be a H-stable exceptional sheaf on a polarized $K3$ surface (X, H). Let $\mathbf{v}_0 = \mathbf{v}(E_0)$. For a vector $\mathbf{v} \in H^*(X, \mathbf{Z})$, we consider a moduli space of coherent systems:

$$\mathfrak{P}_{E_0}^{(n)}(\mathbf{v}) \overset{\text{def.}}{=} \{(E, U) \mid E \in M_H(\mathbf{v}), U \in \operatorname{Gr}_n(\operatorname{Hom}(E_0, E))\},$$

where $\operatorname{Gr}_n(\operatorname{Hom}(E_0, E))$ is the Grassmann manifold consisting of n-dimensional subspaces of $\operatorname{Hom}(E_0, E)$. For $(E, U) \in \mathfrak{P}_{E_0}^{(n)}(\mathbf{v})$, we have an evaluation homomorphism

$$U \otimes E_0 \to E. \tag{3.1}$$

For $\mathbf{v} \in H^*(X, \mathbf{Z})$, we define its twisted degree by

$$\deg_{E_0}(\mathbf{v}) \overset{\text{def.}}{=} (v_1 \operatorname{rank} E_0 - v_0 c_1(E_0), H).$$

We put the following technical conditions on $\mathbf{v}$, E_0:

(1) E_0 is a vector bundle.
(2) $\deg_{E_0}(\mathbf{v}) = \min\{L \cdot H \mid L \in \operatorname{Pic}(X), L \cdot H > 0\}$.

Under these assumptions, Yoshioka [**33**], Markman [**16**], Kawai-Yoshioka [**10**] showed that

PROPOSITION 3.2. (1) *Suppose that the degree* 0*-part of* $\mathbf{v} - n\mathbf{v}_0$ *is nonnegative. Then* (3.1) *is injective and its cokernel is a* H*-stable sheaf. Therefore, we have two projections*

$$M_H(\mathbf{v} - n\mathbf{v}_0) \leftarrow \mathfrak{P}_{E_0}^{(n)}(\mathbf{v}) \to M_H(\mathbf{v}).$$

(2) *Suppose that the degree* 0*-part of* $\mathbf{v} - n\mathbf{v}_0$ *is negative. Then we have two projections*

$$M_H(n\mathbf{v}_0^* - \mathbf{v}^*) \leftarrow \mathfrak{P}_{E_0}^{(n)}(\mathbf{v}) \to M_H(\mathbf{v}),$$

Roughly speaking, the second projection in (3) is given by the 'dual' of the kernel of (3.1). See [**16, 10**] for more detail. As in [**16**], it is natural to set $M_H(\mathbf{v}) \overset{\text{def.}}{=} M_H(-\mathbf{v}^*)$ if the degree 0-part of $\mathbf{v}$ is negative. And we use this convention hereafter.

We consider the image of the fundamental class of $\mathfrak{P}_{E_0}^{(n)}(\mathbf{v})$ in the homology group $M_H(\mathbf{v}) \times M_H(\mathbf{v} - n\mathbf{v}_0)$, and define operators exactly as in the case of quiver varieties:

$$H_{d(\mathbf{v})+s}(M_H(\mathbf{v}), \mathbf{C}) \leftrightarrows H_{d(\mathbf{v}-n\mathbf{v}_0)+s}(M_H(\mathbf{v} - n\mathbf{v}_0), \mathbf{C}).$$

Fixing $\mathbf{v}$, we consider these as operators $e^{(n)}$, $f^{(n)}$ on

$$\bigoplus_{\substack{\mathbf{v}'=\mathbf{v}-n\mathbf{v}_0\\ n\in\mathbf{Z}}} H_{d(\mathbf{v}')+s}(M_H(\mathbf{v}'), \mathbf{C})$$

(up to sign as in the case of quiver varieties). Note that the above condition 2 holds for $\mathbf{v}' = \mathbf{v} - n\mathbf{v}_0$ if it holds for $\mathbf{v}$. We set $e^{(1)} = e$, $f^{(1)} = f$. We define h as

$$h|_{H_*(M_H(\mathbf{v}'),\mathbf{C})} = \langle \mathbf{v}_0, \mathbf{v}' \rangle \operatorname{id}.$$

THEOREM 3.3. (1) $e^{(n)} = \frac{1}{n!}e^n$, $f^{(n)} = \frac{1}{n!}f^n$.
(2) *The operators e, f, h satisfy the defining relations of $\mathfrak{sl}_2$.*

The proof will be given in the next section.

When we have a collection of exceptional sheaves $\{E_1, E_2, \dots, E_n\}$ on a $K3$ surface X, we expect that a Kac-Moody Lie algebra with Cartan matrix

$$-\left(\langle \mathbf{v}(E_i), \mathbf{v}(E_j)\rangle\right)$$

has a representation on the homology groups $\bigoplus_{\mathbf{v}} H_{d(\mathbf{v})+s}(M_H(\mathbf{v}), \mathbf{C})$, where $\mathbf{v}$ runs over an appropriate set of Mukai vectors. However, the technical condition 2 does not fit with this setting. So we need a further study of moduli spaces.

As an application of Theorem 3.3, we have an isomorphism of homology groups

$$H_*(M_H(\mathbf{v}), \mathbf{C}) \cong H_*(M_H(R_{\mathbf{v}_0}(\mathbf{v})), \mathbf{C})$$

given by $r = (\exp f)(\exp(-e))(\exp f)$, a well-known result in the representation theory of $\mathfrak{sl}_2$. An isomorphism between the same homology groups was given by Markman [**16**]. He also showed that it is a ring isomorphism. It should not be difficult to check that his isomorphism is the same as above isomorphism (or its variant).

4. Brill-Noether theory

In this section, we recall Brill-Noether theory for quiver varieties and moduli spaces of sheaves on $K3$. The first case was found originally by Lusztig [**14**] in a slightly different setting, and by the author [**20, 23**] in this setting. The second case had been studied by Yoshioka [**33**], Markman [**16**] and Kawai-Yoshioka [**10**]. We present the first one in terms of ideals, not in terms of representations of a quiver, so that the analogy to the second one becomes clear.

It is worthwhile to mention that our Brill-Noether theory gives a natural generalization of a *Mukai flop*, which is a certain birational map between holomorphic symplectic manifold [**18**]. It is a birational map between $M_H(\mathbf{v})$ and $M_H(R_{\mathbf{v}_0}(\mathbf{v}))$. This generalization was called a *stratified Mukai flop* by Markman.

4.1. The case of quiver varieties. Let $A\#\Gamma$ be the smash product of Γ with the coordinate ring $A = \mathbf{C}[x,y]$ of $\mathbf{C}^2$. Let $A_0 = A/\mathfrak{m}$, where $\mathfrak{m}$ is the maximal ideal at the origin. This is a module of $A\#\Gamma$, where Γ acts trivially. More generally, if I is a Γ-invariant ideal of A, in particular if $I \in M(\mathbf{v})$, A/I is a module of $A\#\Gamma$. Let

$$M(\mathbf{v})_{i;n} \overset{\text{def.}}{=} \{I \in M(\mathbf{v}) \mid \dim \operatorname{Hom}_{A\#\Gamma}(A_0 \otimes \rho_i, A/I) = n\}.$$

It is a locally closed subvariety. One can show that

$$M(\mathbf{v})_{i;n} = \{I \in M(\mathbf{v}) \mid \dim \operatorname{Hom}_{A\#\Gamma}(I, A_0 \otimes \rho_i) = n + \delta_{0i} + \langle \rho_i, \mathbf{v}\rangle\}. \tag{4.1}$$

See [**26**, §4.7] for detail.

Take $I \in M(\mathbf{v})_{i;n}$. Let

$$\operatorname{Hom}_{A\#\Gamma}(A_0 \otimes \rho_i, A/I) \otimes (A_0 \otimes \rho_i) \to A/I$$

be the evaluation homomorphism, which is an analog of (3.1). This is obviously injective. Let us define I' so that A/I' is the cokernel. It is clear that $I' \in M(\mathbf{v} - n\rho_i)_{i;0}$ and give us a map

$$p\colon M(\mathbf{v})_{i;n} \to M(\mathbf{v} - n\rho_i)_{i;0}.$$

It is also clear that $(p(I), I)$ is contained in $\mathfrak{P}_i^{(n)}(\mathbf{v})$ and we have an isomorphism

$$\mathfrak{P}_i^{(n)}(\mathbf{v}) \cap p_1^{-1}(M(\mathbf{v} - n\rho_i)_{i;0}) = \mathfrak{P}_i^{(n)}(\mathbf{v}) \cap p_2^{-1}(M(\mathbf{v})_{i;n}) \cong M(\mathbf{v})_{i;n}.$$

Here p_1, p_2 are the projections to factors of $M(\mathbf{v} - n\rho_i) \times M(\mathbf{v})$.

More generally we have a diagram

$$M(\mathbf{v} - n\rho_i)_{i;r-n} \xleftarrow{p_1} \mathfrak{P}_i^{(n)}(\mathbf{v})_r \xrightarrow{p_2} M(\mathbf{v})_{i;r},$$

where

$$\mathfrak{P}_i^{(n)}(\mathbf{v})_r \overset{\text{def.}}{=} p_2^{-1}(M(\mathbf{v})_{i;r}).$$

($r = n$ is the above special case.)

PROPOSITION 4.2. (1) p_1 *is the Grassmann bundle of* n*-planes associated with a vector bundle* $\operatorname{Hom}_{A\#\Gamma}(I', A_0 \otimes \rho_i)$ *of rank* $r + m$ *over* $\{I' \in M(\mathbf{v} - n\rho_i)_{i;r-n}\}$, *where* $m = n + \delta_{0i} + \langle \rho_i, \mathbf{v}\rangle$.

(2) p_2 *is the Grassmann bundle of* n*-planes associated with a vector bundle* $\operatorname{Hom}_{A\#\Gamma}(A_0 \otimes \rho_i, A/I)$ *of rank* r *over* $\{I \in M(\mathbf{v})_{i;r}\}$.

(3) *We have*

$$\frac{1}{2}\dim M(\mathbf{v} - n\rho_i) + \dim(\text{fiber of } p_1) = \frac{1}{2}\dim M(\mathbf{v}) + \dim(\text{fiber of } p_2).$$

These follows from a translation of results for quiver varieties [**20**, 10.10]. But it is also easy to show directly as follows (cf. [**26**, §7.2]). For $I' \in M(\mathbf{v} - n\rho_i)_{i;r-n}$, the fiber $p_1^{-1}(I')$ consists of Γ-invariant ideals I arising in an exact sequence

$$0 \to I \to I' \to A_0 \otimes \rho_i^{\oplus n} \to 0.$$

Note that $\operatorname{Ext}^1_A(A_0 \otimes \rho_i, A_0 \otimes \rho_i)$ has no Γ-invariant part and the quotien I'/I must be of above form. These I are parametrized by inclusions

$$\mathbf{C}^n \cong \operatorname{Hom}_{A\#\Gamma}(A_0 \otimes \rho_i^{\oplus n}, A_0 \otimes \rho_i) \subset \operatorname{Hom}_{A\#\Gamma}(I', A_0 \otimes \rho_i).$$

These I' are parametrized by inclusions

$$\mathbf{C}^n \cong \operatorname{Hom}_{A\#\Gamma}(A_0 \otimes \rho_i, A_0 \otimes \rho_i^{\oplus n}) \subset \operatorname{Hom}_{A\#\Gamma}(A_0 \otimes \rho_i, A/I).$$

Finally the formula (3) follows from the dimension formula (2.1).

For example, consider the case $\mathbf{v} = \delta$. As we mentioned in §2, $M(\mathbf{v})$ is the minimal resolution of $\mathbf{C}^2/\Gamma$. The dimension formula says $M(\mathbf{v} - \rho_i)$ $(i \neq 0)$ is 0-dimensional. As an application of the main result in [**20**], it is easy to see that $M(\mathbf{v} - \rho_i)$ $(i \neq 0)$ is nonempty, hence is a single point. Therefore $\mathfrak{P}_i(\mathbf{v})$ can be considered as a subvariety of $M(\mathbf{v})$. Furthermore, Proposition 4.2, in this situation, says that $\mathfrak{P}_i(\mathbf{v})$ is a projective line, and $\mathfrak{P}_i(\mathbf{v})$ is $M(\mathbf{v})_{i;1}$. It is also easy to see that $\mathfrak{P}_i(\mathbf{v})$'s $(i \neq 0)$ are the irreducible components of the exceptional set of the minimal resolution $M(\mathbf{v}) \to \mathbf{C}^2/\Gamma$ [**22**]. It is easy to check that this description of the irreducible components is the same as Ito-Nakamura's description [**7**] rediscovered later (see [**26**, 6.3]).

Let us briefly explain a 'combinatorial structure' on the disjoint union $\bigsqcup_{\mathbf{v}} \operatorname{Irr} L(\mathbf{v})$ of the set of irreducible components of $L(\mathbf{v})$. We use the above diagram with $r = n$. (This construction is due to Lusztig.)

Let $Y \in \operatorname{Irr} L(\mathbf{v})$. We define $\varepsilon_i(Y)$ by

$$\varepsilon_i(Y) = \min\{n \mid M(\mathbf{v})_{i;n} \cap Y \neq \emptyset\}.$$

Then $M(\mathbf{v})_{i;\varepsilon_i(Y)} \cap Y$ is a nonempty open subset of Y. Let us consider its image under p. Since the supports of $\mathbf{C}[x,y]/I$ and $\mathbf{C}[x,y]/p(I)$ differ only 0, we have $p(I) \in L(\mathbf{v} - n\rho_i)$ for $I \in Y$. Moreover, the formula (3) above gives the dimension of the image is equal to $\frac{1}{2}\dim M(\mathbf{v} - n\rho_i) = \dim L(\mathbf{v} - n\rho_i)$. Therefore the closure of the image is again an irreducible component of $L(\mathbf{v} - n\rho_i)$. It satisfies $\varepsilon_i = 0$. Conversely if Y' is an irreducible component of $L(\mathbf{v} - n\rho_i)$ with $\varepsilon_i(Y') = 0$, then the closure of $p^{-1}(Y' \cap M(\mathbf{v} - n\rho_i)_{i;0})$ is an irreducible component of $L(\mathbf{v})$ with $\varepsilon_i = n$. It is easy to show that for any $Y \in \operatorname{Irr} L(\mathbf{v})$, there exists i such that $\varepsilon_i(Y) > 0$ unless $\mathbf{v} = 0$. Therefore, any irreducible component of $L(\mathbf{v})$ can be constructed from $L(0) = M(0) = pt$ by the above procedure.

This is not the end of the story. This procedure gives us a certain structure on $\bigsqcup_{\mathbf{v}} \operatorname{Irr} L(\mathbf{v})$, which is known as Kashiwara's *crystal*. (We do not recall the definition of the crystal here. See [**26**, §7].) Kashiwara introduced the notion of the crystal by abstracting the properties of his crystal bases of representations of quantized enveloping algebras. Kashiwara-Saito showed that the above crystal is isomorphic to the crystal of basic representation of the corresponding quantum affine algebra [**9**].

4.2. Sheaves on $K3$. Let us consider the second case, i.e., sheaves on $K3$ surfaces. In stead of $A_0 \otimes \rho_i$, we consider a rigid stable sheaf E_0 on $K3$.

Let E_0 be a H-stable exceptional sheaf on a polarized $K3$ surface (X, H) with Mukai vector $\mathbf{v}_0$. We assume the conditions in §3. Let

$$M_H(\mathbf{v})_{E_0;n} \overset{\text{def.}}{=} \{E \in M_H(\mathbf{v}) \mid \dim \operatorname{Hom}(E_0, E) = n\},$$

where $\mathbf{v}$ is the Mukai vector of E. Serre duality, the stability of E and the condition imply $\operatorname{Ext}^2(E_0, E) = 0$. Therefore, we have

$$M_H(\mathbf{v})_{E_0;n} = \{E \in M_H(\mathbf{v}) \mid \dim \operatorname{Ext}^1(E_0, E) = n + \langle \mathbf{v}_0, \mathbf{v}\rangle\}.$$

This is an analog of (4.1). We also a natural morphism

$$p\colon M_H(\mathbf{v})_{E_0;n} \to M_H(\mathbf{v}-n\mathbf{v}_0)_{E_0;0}$$

and an isomorphism

$$\begin{aligned}&\mathfrak{P}^{(n)}_{E_0}(\mathbf{v}) \cap p_1^{-1}\left(M_H(\mathbf{v}-n\mathbf{v}_0)_{E_0;0}\right) = \mathfrak{P}^{(n)}_i(\mathbf{v}) \cap p_2^{-1}\left(M_H(\mathbf{v})_{E_0;n}\right)\\ &\cong M_H(\mathbf{v})_{E_0;n}.\end{aligned}$$

More generally we have a diagram

$$M_H(\mathbf{v}-n\mathbf{v}_0)_{E_0;r-n} \xleftarrow{p_1} \mathfrak{P}^{(n)}_{E_0}(\mathbf{v})_r \xrightarrow{p_2} M_H(\mathbf{v})_{E_0;r},$$

where

$$\mathfrak{P}^{(n)}_{E_0}(\mathbf{v})_r \overset{\text{def.}}{=} p_2^{-1}(M_H(\mathbf{v})_{E_0;r}).$$

We have

PROPOSITION 4.3. (1) *p_1 is the Grassmann bundle of n-planes associated with a 'vector bundle'* $\operatorname{Ext}^1(E', E_0)$ *of rank $r+m$ over $\{E' \in M_H(\mathbf{v}-n\mathbf{v}_0)_{E_0;r-n}\}$, where $m = n + \langle \mathbf{v}_0, \mathbf{v}\rangle$.*

(2) *p_2 is the Grassmann bundle of n-planes associated with a 'vector bundle'* $\operatorname{Hom}(E_0, E)$ *of rank r over $\{E \in M_H(\mathbf{v})_{E_0;r}\}$.*

(3) *We have*

$$\frac{1}{2}\dim M_H(\mathbf{v}-n\mathbf{v}_0) + \dim(\text{fiber of } p_1) = \frac{1}{2}\dim M_H(\mathbf{v}) + \dim(\text{fiber of } p_2).$$

See [**33**, Lemma 2.4], [**16**, Theorem 33], [**10**, 5.113] for proof of (1),(2). (3) is a direct consequence of the dimension formula (1.1) together with $\langle \mathbf{v}_0, \mathbf{v}_0\rangle = -2$. Here we put vector bundles in the quotation mark ' ' since universal bundles may not exist.

A recent result by Onishi-Yoshioka [**29**] can be considered as analog of the example after Proposition 4.2. They found an *ADE* configuration of (−2)-curves in a 2-dimensional moduli space of sheaves on $K3$ and irreducible components are given by coherent systems [loc. cit., Propositions 2.7,2.8].

We can mimick the inductive construction of irreducible components of the lagrangian subvariety for the quiver variety. For an irreducible lagrangian subvariety Y of $M_H(\mathbf{v})$, let

$$\varepsilon_{E_0}(Y) \overset{\text{def.}}{=} \min\{n \mid M_H(\mathbf{v})_{E_0;n} \cap Y \neq \emptyset\}.$$

Mimicking the construction for quiver varieities, we consider

$$Y' \overset{\text{def.}}{=} \overline{p(M_H(\mathbf{v})_{E_0;n} \cap Y)}.$$

This is an irreducible lagrangian subvariety of $M_H(\mathbf{v}-n\mathbf{v}_0)$ with $\varepsilon_{E_0}(Y') = 0$. (We must check that it is isotropic. But it is easy.) We can recover Y from Y' by

$$Y = \overline{p^{-1}(M_H(\mathbf{v}-n\mathbf{v}_0)_{E_0;0} \cap Y')}.$$

This can be used for a construction of lagrangian subvarieties, starting from a known lagrangian subvariety, say a fiber of a lagrangian fibration.

It is obvious that we have $[h,e]=2e$, $[h,f]=-2f$. We study the convolution products

$$(4.4)\qquad \begin{aligned} &p_{13*}\left(p_{12}^*\left[\mathfrak{P}_{E_0}(\mathbf{v}+\mathbf{v}_0)\right]\cap p_{23}^*\left[\omega\mathfrak{P}_{E_0}(\mathbf{v}+\mathbf{v}_0)\right]\right),\\ &q_{13*}\left(q_{12}^*\left[\omega\mathfrak{P}_{E_0}(\mathbf{v})\right]\cap q_{23}^*\left[\omega\mathfrak{P}_{E_0}(\mathbf{v})\right]\right),\end{aligned}$$

where p_{ij} and q_{ij} are projections to the product of ith and jth factors in

$$M_H(\mathbf{v})\times M_H(\mathbf{v}+\mathbf{v}_0)\times M_H(\mathbf{v}),\qquad M_H(\mathbf{v})\times M_H(\mathbf{v}-\mathbf{v}_0)\times M_H(\mathbf{v})$$

respectively, and ω is the exchange of the factor. The both products have degree $\dim M_H(\mathbf{v})$.

Consider

$$Z_n=\{(E_1,E_3)\in M_H(\mathbf{v})_{E_0;n}\times M_H(\mathbf{v})_{E_0;n}\mid p(E_1)=p(E_3)\},$$

where $p\colon M_H(\mathbf{v})_{E_0;n}\to M_H(\mathbf{v}-n\mathbf{v}_0)_{E_0;0}$ is as before. It is a fiber product of two copies of the Grassmann bundle of n-planes in a vector bundle of rank $2n+\langle\mathbf{v}_0,\mathbf{v}\rangle$ over $M_H(\mathbf{v}-n\mathbf{v}_0)_{E_0;0}$ by Proposition 4.3. Therefore

$$\dim Z_n=\dim M_H(\mathbf{v}).$$

Moreover, we have

$$(4.5)\qquad \left.\begin{aligned} &p_{13}\left(p_{12}^{-1}\left(\mathfrak{P}_{E_0}(\mathbf{v}+\mathbf{v}_0)\right)\cap p_{23}^{-1}\left(\omega\mathfrak{P}_{E_0}(\mathbf{v}+\mathbf{v}_0)\right)\right)\\ &q_{13}\left(q_{12}^{-1}\left(\omega\mathfrak{P}_{E_0}(\mathbf{v})\right)\cap q_{23}^{-1}\left(\omega\mathfrak{P}_{E_0}(\mathbf{v})\right)\right)\end{aligned}\right\}\subset\bigsqcup_n Z_n.$$

Therefore the cycles (4.4) are, in fact, the classes in $H_{\dim M_H(\mathbf{v})}(Z_n,\mathbf{C})$. Thus these are linear combinations of fundamental classes of irreducible components of Z_n. (In fact, Z_n is irreducible since $M_H(\mathbf{v}-n\mathbf{v}_0)$ is irreducible. But we do not need this fact.) We claim

$$\begin{aligned} p_{13*}\left(p_{12}^*\left[\mathfrak{P}_{E_0}(\mathbf{v}+\mathbf{v}_0)\right]\cap p_{23}^*\left[\omega\mathfrak{P}_{E_0}(\mathbf{v}+\mathbf{v}_0)\right]\right)&=\pm\langle\mathbf{v}_0,\mathbf{v}\rangle[Z_0]+[Z_1]\\ q_{13*}\left(q_{12}^*\left[\omega\mathfrak{P}_{E_0}(\mathbf{v})\right]\cap q_{23}^*\left[\omega\mathfrak{P}_{E_0}(\mathbf{v})\right]\right)&=[Z_1],\end{aligned}$$

if $\langle\mathbf{v}_0,\mathbf{v}\rangle\ge 0$, and

$$\begin{aligned} p_{13*}\left(p_{12}^*\left[\mathfrak{P}_{E_0}(\mathbf{v}+\mathbf{v}_0)\right]\cap p_{23}^*\left[\omega\mathfrak{P}_{E_0}(\mathbf{v}+\mathbf{v}_0)\right]\right)&=\left[Z_{1-\langle\mathbf{v}_0,\mathbf{v}\rangle}\right]\\ q_{13*}\left(q_{12}^*\left[\omega\mathfrak{P}_{E_0}(\mathbf{v})\right]\cap q_{23}^*\left[\omega\mathfrak{P}_{E_0}(\mathbf{v})\right]\right)&=\mp\langle\mathbf{v}_0,\mathbf{v}\rangle\left[Z_{-\langle\mathbf{v}_0,\mathbf{v}\rangle}\right]+\left[Z_{1-\langle\mathbf{v}_0,\mathbf{v}\rangle}\right]\end{aligned}$$

otherwise. If $n=0$ or $-\langle\mathbf{v}_0,\mathbf{v}\rangle$, the Grassmann bundle is trivial, i.e., it is the bundle of the 0-subspace or the whole space. And $[Z_0]$ or $\left[Z_{-\langle\mathbf{v}_0,\mathbf{v}\rangle}\right]$ is the class of the diagonal. Thus the claim implies the assertion.

We only consider the case $\langle\mathbf{v}_0,\mathbf{v}\rangle\ge 0$. The other case can be treated in a similar way. Let U be the complement of the diagonal in $M_H(\mathbf{v})\times M_H(\mathbf{v})$. The intersection $Z_n\cap U$ is a nonempty open subset of Z_n unless the Grassmann bundle p is an identity, i.e., n is either 0 or the rank of the vector bundle $=2n+\langle\mathbf{v}_0,\mathbf{v}\rangle$. By our assumption $\langle\mathbf{v}_0,\mathbf{v}\rangle\ge 0$, we only have the case $n=0$. In order to determine the coefficients of $[Z_n]$ with $n\neq 0$, it is enough to study on the open set U. Take $(E_1,E_3)\in U\cap Z_n$ and let S_1,S_3 be the n-planes in the $(2n+\langle\mathbf{v}_0,\mathbf{v}\rangle)$-dimensional vector space corresponding to E_1,E_3. Suppose that (E_1,E_3) is contained in either of two intersections of (4.5), and let (E_1,E_2,E_3) (resp. (E_1,E_2',E_3)) be the corresponding point in $p_{12}^{-1}\left(\mathfrak{P}_{E_0}(\mathbf{v}+\mathbf{v}_0)\right)\cap p_{23}^{-1}\left(\omega\mathfrak{P}_{E_0}(\mathbf{v}+\mathbf{v}_0)\right)$ (resp. $q_{12}^{-1}\left(\omega\mathfrak{P}_{E_0}(\mathbf{v})\right)\cap q_{23}^{-1}\left(\omega\mathfrak{P}_{E_0}(\mathbf{v})\right)$). Then E_2 (resp. E_2') corresponds to $S_1\cap S_3$ (resp. S_1+S_3). Thus the existences of E_2 and E_2' are equivalent, i.e., we must have $\dim(S_1\cap S_3)=n-1$, which is equivalent to $\dim(S_1+S_3)=n+1$. This happens for generic (S_1,S_3) if and only if $n=1$ (by our assumption $\langle \mathbf{v}_0, \mathbf{v}\rangle \geq 0$). Moreover, in this case, the intersection is transversal and the projections p_{13}, q_{13} are isomorphisms. Therefore the coefficients of $[Z_n]$ ($n \neq 0$) of products in (4.4) are as in our claim. In order to determine the coefficient of $[Z_0]$, we note that $M_H(\mathbf{v})_{E_0;0}$ is a nonempty open subset of $M_H(\mathbf{v})$ and we can restrict the cycle to this open subset. Then we can compute the products by the standard excess intersection formula (cf. [**23**, 8.5]) to get the claim. □

5. Conjectures

In §§3,4 we have studied the convolution operators induced by exceptional sheaves. It is natural to ask what happens if we consider more general sheaves. The author's Heisenberg algebra representation of Hilbert schemes of points is an example. Let us explain it briefly (see [**24**, §8] for more detail.) We consider the following subvariety in the triple product:

$$P[i] \overset{\text{def.}}{=} \{(E_1, E_2, x) \in \operatorname{Hilb}^{n-i} X \times \operatorname{Hilb}^n X \times X \mid E_1 \supset E_2, \operatorname{Supp} E_1/E_2 = \{x\}\}$$

for $i > 0$. We have a projection $\Pi\colon P[i] \to X$ to the third factor. For a class $\alpha \in H_*(X, \mathbf{C})$, $\Pi^*\alpha \cap [P[i]]$ defines operators

$$H_*(\operatorname{Hilb}^{n-i} X, \mathbf{C}) \to H_*(\operatorname{Hilb}^n X, \mathbf{C}), \qquad H_*(\operatorname{Hilb}^n X, \mathbf{C}) \to H_*(\operatorname{Hilb}^{n-i} X, \mathbf{C})$$

as before. We denote these by $P_\alpha[i]$ and $P_\alpha[-i]$. Then these operators satisfy the Heisenberg relation:

$$[P_\alpha[i], P_\beta[j]] = (-1)^{i-1} i \delta_{i+j,0} \int_X \alpha \cap \beta \cdot \operatorname{id}.$$

Let $\mathbf{v}_0$ be a primitive isotropic vector. As we mentioned, $M_H(\mathbf{v}_0)$ is a $K3$ surface. Suppose that there exists a universal family $\mathcal{E}$ on $X \times M_H(\mathbf{v}_0)$. Then we can define a Fourier-Mukai functor $\mathcal{F}_\mathcal{E}\colon D(X) \to D(M_H(\mathbf{v}_0))$:

$$\mathcal{F}_\mathcal{E}(\bullet) \overset{\text{def.}}{=} \mathbf{R}p_{2*}\left(\mathcal{E} \otimes p_1^*(\bullet)\right),$$

where $D(X)$, $D(M_H(\mathbf{v}_0))$ are bounded derived categories of coherent sheaves on X, $M_H(\mathbf{v}_0)$, and p_i is the projection of $X \times M_H(\mathbf{v}_0)$ into the ith factor. Let $\mathcal{F}_\mathcal{E}^H$ be a similar homomorphism $H^*(X, \mathbf{C}) \to H^*(M_H(\mathbf{v}_0), \mathbf{C})$ (see [**34**, 2.6]). Under some assumptions on $\mathbf{v}$, Yoshioka [loc. cit., §7] showed that $\mathcal{F}_\mathcal{E}$ induces an isomorphism

$$M_H(\mathbf{v}) \to M_{\widehat{H}}(-\mathcal{F}_\mathcal{E}^H(\mathbf{v})),$$

where $\widehat{H}$ is a certain naturally defined ample line bundle over $M_H(\mathbf{v}_0)$. Note that $\mathcal{F}_\mathcal{E}^H(\mathbf{v}_0^\vee)$ is the fundamental class $\omega_{M_H(\mathbf{v}_0)}$ of $M_H(\mathbf{v}_0)$. Therefore, if we choose $\mathbf{v}$ so that $-\mathcal{F}_\mathcal{E}(\mathbf{v})$ is the positive generator 1 of $H^0(M_H(\mathbf{v}_0), \mathbf{C})$, we have an isomorphism

$$M_H(\mathbf{v} + (1-n)\mathbf{v}_0^\vee) \to \operatorname{Hilb}^n(M_H(\mathbf{v}_0)).$$

This shows that we have a Heisenberg algebra representation on $\bigoplus_n H_*(M_H(\mathbf{v} + (1-n)\mathbf{v}_0^\vee), \mathbf{C})$. This observation was informed to us by Yoshioka, as an answer to author's question. It is probably interesting to explore a direct proof. Yoshioka also showed that variants of $\mathcal{F}_\mathcal{E}$ also induce isomorphisms under some (different) assumptions on $\mathbf{v}$ [loc. cit., §§7,8]. The same construction of the Heisenberg algebra representation works for these variants. Operators associated with more general

Let us comment on a generalization to other directions. In our construction of the representation of the affine Lie algebra $\widehat{\mathfrak{g}}$, we can replace homology groups of quiver varieties by equivariant K-homology groups, and then we get a representation of the quantum toroidal algebra [**25**]. The group acting on quiver varieties is $\mathbf{C}^*$, and the deformation parameter q of the quantum toroidal algebra is identified with the canonical generator of the representation ring of $\mathbf{C}^*$. The $\mathbf{C}^*$-action cannot be defined on moduli spaces of sheaves on a $K3$ surface, but it is still natural to expect that a toroidal Lie algebra, that is a loop Lie alebra of the affine Lie algebra $\widehat{\mathfrak{g}}\otimes\mathbf{C}[z,z^{-1}]$ has a representation on K-homology groups of moduli spaces. Loop generators $e_i\otimes z^r$, $f_i\otimes z^r$ are given by the r-th tensor power of the natural line bundle over $\mathfrak{P}_{E_0}(\mathbf{v})$. A quantum toroidal algebra contains a quantum affine algebra as a subalgebra. By a theory of Lusztig (see [**15**]), an affine braid group acts on the K-homology groups of quiver varieties. It is natural to expect similar result for moduli spaces of sheaves on $K3$.

Furthermore, it is natural to expect that the representation can be 'lifted' to derived categories of coherent sheaves on quiver varieties or moduli spaces of sheaves on $K3$. It is a philosophy of a 'categorification' which the author learned from I. Frenkel and Khovanov. ([**6**] is related to this paper.) For quiver varieties, the computation in [**25**] can be, more or less, generalized to derived categories. In the case of Hilbert schemes of points, the derived category of $\operatorname{Hilb}^n X$ is isomorphic to the derived category of S_n-equivariant coherent sheaves on X^n, thanks to works of Haiman [**4**] and Bridgeland-King-Reid [**2**]. By induction/restriction operators, the latter becomes a representation of the Heisenberg algebra by Segal [**30**] (see also [**32**] for detail). It is also interesting to study the categorification of the affine braid group representation. Special cases are studied by Seidel-Thomas [**31**] (when $M(\mathbf{v})$, $M_H(\mathbf{v})$ are 2-dimensional), and by Namikawa [**28**], Kawamata [**11**] (for Mukai flops). Under analogy between quiver varieties and cotangent bundles of flag varieties (see the introduction of [**25**]), it is also related to recent works of Bezrukavnikov and his collegues (see [**1**] and the reference therein).

References

[1] R. Bezrukavnikov, I. Mirković, and D. Rumynin, *Localization of modules for a semisimple Lie algebra in prime characteristic*, preprint, math.RT/0205144.

[2] T. Bridgeland, A. King and M. Reid, *The McKay correspondence as an equivalence of derived categories*, J. Amer. Math. Soc. **14** (2001), 535–554.

[3] N. Chriss and V. Ginzburg, Representation theory and complex geometry, Birkhähser, 1997.

[4] M. Haiman, *Hilbert schemes, polygraphs and the Macdonald positivity conjecture*, J. Amer. Math. Soc. **14** (2001), 941–1006.

[5] J.A. Harvey and G. Moore, *On the algebras of BPS states*, Comm. Math. Phys. **197** (1998), 489–519.

[6] R.S. Huerfano and M.Khovanov, *A category for the adjoint representation*, J. Algebra, **246** (2001), 514–542.

[7] Y. Ito and I. Nakamura, *Hilbert schemes and simple singularities*, in "Recent Trends in Algebraic Geometry – EuroConference on Algebraic Geometry (Warwick, July 1996)", ed. by K. Hulek and others, 151–233, Cambridge Univ. Press, 1999.

[8] V.G. Kac, Infinite dimensional Lie algebras (3rd Ed.), Cambridge Univ. Press 1990.

[9] M. Kashiwara and Y. Saito, *Geometric construction of crystal bases*, Duke Math. **89** (1997), 9–36.

[10] T. Kawai and K. Yoshioka, *String partition functions and infinite products*, Adv. Theor. Math. Phys. **4** (2000), 397–485.

[11] Y. Kawamata, *D-equivalence and K-equivalence*, preprint, math.AG/0205287.

[12] P.B. Kronheimer, *The construction of ALE spaces as a hyper-Kähler quotients*, J. Differential Geom. **29** (1989) 665–683.

[13] P.B. Kronheimer and H. Nakajima, *Yang-Mills instantons on ALE gravitational instantons*, Math. Ann. **288** (1990), 263–307.

[14] G. Lusztig, *Quivers, preverse sheaves, and quantized enveloping algebras*, J. Amer. Math. Soc. **4** (1991), 365–421.

[15] ———, Introduction to Quantum Groups, Birkhäuser, Boston, 1993.

[16] E. Markman, *Brill-Noether duality for moduli spaces of sheaves on K3 surfaces*, J. Algebraic Geom. **10** (2001), 623–694.

[17] S. Mukai, *Symplectic structure of the moduli space of sheaves on an abelian or K3 surface*, Invent. Math. **77** (1984), 101–116.

[18] ———, *On the moduli space of bundles on K3 surfaces. I*, in "Vector bundles on algebraic varieties (Bombay, 1984)", 341–413, Tata Inst. Fund. Res., Bombay, 1987.

[19] H. Nakajima, *Moduli spaces of anti-self-dual connections on ALE gravitational instantons*, Invent. Math. **102** (1990), 267–303.

[20] ———, *Instantons on ALE spaces, quiver varieties, and Kac-Moody algebras*, Duke Math. J. **76** (1994), 365–416.

[21] ———, *Homology of moduli spaces instantons on ALE spaces I*, J. Differential Geom. **40** (1994), 105–127.

[22] ———, *Varieties associated with quivers*, in "Representation theory of algebras and related topics", 139–157, CMS conference proceedings **19**, AMS, 1996.

[23] ———, *Quiver varieties and Kac-Moody algebras*, Duke Math. J. **91** (1998), 515–560.

[24] ———, *Lectures on Hilbert schemes of points on surfaces*, Univ. Lect. Ser. **18**, AMS, 1999.

[25] ———, *Quiver varieties and finite dimensional representations of quantum affine algebras*, J. Amer. Math. Soc., **14** (2001), 145–238.

[26] ———, *Quiver varieties and McKay correspondence*, in "Proceeding of 'Open Calabi-Yau manifolds – approaches from algebraic geometry and string theory' (2001 Dec., Hokkaido)", 1–49, 2002.

[27] ———, *Geometric construction of representations of affine algebras*, to appear in "Proceedings of the International Congress of Mathematicians, Vol. I (Beijing 2002)".

[28] Y. Namikawa, *Mukai flops and derived categories*, preprint, math.AG/0203287.

[29] N. Onishi and K. Yoshioka, *A note on the 2-dimensional moduli spaces of stable sheaves on K3 surfaces*, preprint, math.AG/0208241.

[30] G. Segal, *Equivariant K-theory and symmetric products*, Lecture at Cambridge, July 1996.

[31] P. Seidel and R. Thomas, *Braid group actions on derived categories of coherent sheaves*, Duke Math. J. **108** (2001), 37–108.

[32] W. Wang, *Equivariant K-theory, wreath products, and Heisenberg algebra*, Duke Math. **103** (2000), 1–23.

[33] K. Yoshioka, *Some examples of Mukai's reflections on K3 surfaces*, J. reine angew. Math. **515** (1999), 97–123.

[34] ———, *Moduli spaces of stable sheaves on abelian surfaces*, Math. Ann. **321** (2001), 817–884.

Department of Mathematics, Kyoto University, Kyoto 606-8502, Japan
E-mail address: nakajima@kusm.kyoto-u.ac.jp

Contemporary Mathematics
Volume **322**, 2003

Curves in the Hilbert schemes of points on surfaces

Wei-Ping Li, Zhenbo Qin, and Qi Zhang

ABSTRACT. We study certain algebraic curves in the Hilbert scheme $X^{[n]}$ of points on a simply-connected smooth projective surface X, and describe the moduli spaces of these curves. Moreover, when X is the complex projective plane $\mathbb{P}^2$, we determine the nef cone and effective cone of $X^{[n]}$, and study the contraction of the extremal ray on $X^{[n]}$.

1. Introduction

The Hilbert scheme $X^{[n]}$ of points in a smooth projective surface X is a desingularization of the n-th symmetric product of X (see [**Fo1**]). An element ξ in $X^{[n]}$ is a length-n 0-dimensional closed subscheme of X. Recently, there are two surprising discoveries, mainly due to the work of Göttsche [**Got**], Nakajima [**Na1**] and Grojnowski [**Gro**], that the Hilbert schemes $X^{[n]}$ $(n \geq 0)$ have relationships with modular forms on the one hand and with representations of infinite dimensional Lie algebras on the other hand. Their work basically deals with the cohomology groups (the Hodge numbers as well) of $X^{[n]}$. The geometry of the Hilbert scheme $X^{[n]}$ is another interesting area of research. An excellent example in this regard is Beauville's work [**Bea**] which says that the Hilbert scheme $X^{[n]}$ admits a hyperkähler structure when X is a $K3$ surface.

We are interested in curves contained in the Hilbert scheme $X^{[n]}$ when X is a simply-connected smooth projective surface. Throughout the paper, we assume $n \geq 2$. For a fixed point $x \in X$, let $M_2(x) = \{\xi \in X^{[2]} \,|\, \mathrm{Supp}(\xi) = \{x\}\}$. It is known that $M_2(x) \cong \mathbb{P}^1$. Fix distinct points $x, x_1, \ldots, x_{n-2} \in X$, and define

$$\beta_n = \big\{\, \xi + x_1 + \ldots + x_{n-2} \in X^{[n]} \,|\, \xi \in M_2(x) \big\}. \tag{1.1}$$

Using well-known results about punctual Hilbert schemes, we characterize all the curves γ in the Hilbert scheme $X^{[n]}$ homologous to β_n (Theorem 3.2). It turns out that the moduli space $\mathfrak{M}(\beta_n)$ of all these curves has dimension $(2n-2)$ and its top stratum consists of all the curves γ of the form (1.1). Moreover, the normal bundles of the curves γ of the form (1.1) are determined (Proposition 3.4).

2000 *Mathematics Subject Classification.* Primary 14C05; Secondary 14E05, 14J26.
Key words and phrases. Hilbert schemes, algebraic surfaces, moduli space of curves.
Partially supported by the grant HKUST6170/99P.

Next, we let $X = \mathbb{P}^2$. Fix a line $\ell \subset X$, and distinct points $x_1, \ldots, x_{n-1} \in X$ such that $x_i \notin \ell$ for $i = 1, \ldots, (n-1)$. Define the curve β_ℓ in $X^{[n]}$ by

$$\beta_\ell = \big\{ x + x_1 + \ldots + x_{n-1} \in X^{[n]} \,|\, x \in \ell \big\}. \tag{1.2}$$

We prove that the effective cone of the Hilbert scheme $X^{[n]}$ is spanned by β_n and $\beta_\ell - (n-1)\beta_n$, and determine the nef cone of $X^{[n]}$ (Theorem 4.1). In addition, we show that a curve γ in $X^{[n]}$ is homologous to $\beta_\ell - (n-1)\beta_n$ if and only if there exists a line C in $X = \mathbb{P}^2$ such that γ is a line in $C^{[n]} \subset X^{[n]}$ (Theorem 5.1). So all these curves γ in $X^{[n]}$ is parameterized by the Grassmannian bundle $\mathrm{Grass}(\mathrm{Sym}^n((T_{(\mathbb{P}^2)^*})^*), 2)$ over the dual space $(\mathbb{P}^2)^*$.

Using a result of Beltrametti and Sommese [**BSG**] on k-very ampleness, we construct a morphism $\varphi_1 \colon X^{[n]} \to \mathrm{Grass}\big(H^0(X, \mathcal{O}_X(n-1)), n\big)$ mapping $\xi \in X^{[n]}$ to the n-dimensional quotient space $H^0\big(X, \mathcal{O}_\xi \otimes \mathcal{O}_X(n-1)\big)$ of $H^0\big(X, \mathcal{O}_X(n-1)\big)$. When $n \geq 3$, φ_1 is the contraction of an extremal ray. Moreover, $\varphi_1 : X^{[n]} \to \varphi_1(X^{[n]})$ is a small contraction when $n \geq 4$. The contraction locus of φ_1 consists of all the points on the curves γ in $X^{[n]}$ homologous to $\beta_\ell - (n-1)\beta_n$. We prove that when $n \geq 4$, the flip for the small contraction φ_1 exists (Theorem 6.2).

Finally, we remark that our results here have been used to compute the 1-point Gromov-Witten invariants of the Hilbert scheme $X^{[n]}$, and to study the quantum cohomology of $X^{[n]}$. We refer to [**L-Q**] for details.

Conventions: Let $0 \leq k \leq n$ and V be an n-dimensional vector space. We use the Grassmannian $\mathrm{Grass}(V, k)$ to denote the set of all k-dimensional quotients of V, or equivalently, the set of all $(n-k)$-dimensional subspaces of V. Also, we take $\mathbb{P}(V) = \mathrm{Grass}(V, 1)$. So the set of lines in $\mathbb{P}(V)$ is the Grassmannian $\mathrm{Grass}(V, 2)$.

Acknowledgments: The authors thank D. Edidin, L. Ein, R. Friedman, S. Katz, Y. Kawamata and Y. Ruan for stimulating discussions. In addition, the first author thanks S. Katz and the Mathematics Department of Oklahoma State University for their hospitality during his visit there in June of 1999. This work is partially supported by a grant from the American Institute of Mathematics.

2. Basic facts about the Hilbert scheme of points on a surface

Let X be a simply-connected smooth projective surface. For $\xi \in X^{[n]}$, let I_ξ and $\mathcal{O}_\xi$ be the corresponding sheaf of ideals and structure sheaf respectively. For a point $x \in X$, let ξ_x be the component of ξ supported at x and $I_{\xi,x} \subset \mathcal{O}_{X,x}$ be the stalk of I_ξ at x. It is known from [**Fo1**] that $X^{[n]}$ is smooth. In $X^{[n]} \times X$, we have the universal codimension-2 subscheme:

$$\mathcal{Z}_n = \{(\xi, x) \subset X^{[n]} \times X \,|\, x \in \mathrm{Supp}(\xi)\} \subset X^{[n]} \times X. \tag{2.1}$$

Define the incidence variety $X^{[n-1,n]} = \{(\xi, \eta) \in X^{[n-1]} \times X^{[n]} \,|\, \xi \subset \eta\} \subset X^{[n-1]} \times X^{[n]}$. It is known that $X^{[n-1,n]}$ is smooth and of dimension $2n$. Let X^n and $X^{(n)}$ be the n-th Cartesian product and symmetric product respectively. Then, we have the Hilbert-Chow morphism ρ_n and the quotient map υ_n:

$$\rho_n \colon X^{[n]} \to X^{(n)}, \qquad \upsilon_n \colon X^n \to X^{(n)}. \tag{2.2}$$

For a subset $Y \subset X$, we define the subset $M_n(Y)$ in the Hilbert scheme $X^{[n]}$:

$$M_n(Y) = \{\xi \in X^{[n]} | \mathrm{Supp}(\xi) \text{ is a point in } Y\} \subset X^{[n]}. \tag{2.3}$$

So for $x \in X$, $M_n(x)$ is just the punctual Hilbert scheme of points on X at x. It is known that $M_n(x)$ are isomorphic for all X and all $x \in X$. Let $\mathrm{Hilb}^n(\mathbb{C}^2, 0)$ be the punctual Hilbert scheme of points on $\mathbb{C}^2$ at the origin. We adopt the isomorphism

$$M_n(x) \cong \mathrm{Hilb}^n(\mathbb{C}^2, 0). \tag{2.4}$$

Given $Y \subset X^{[n-k]}$ and $\eta \in X^{[k]}$ with $\{\mathrm{Supp}(\xi) | \xi \in Y\} \cap \mathrm{Supp}(\eta) = \emptyset$, we use $Y + \eta$ to represent the subset in $X^{[n]}$ consisting of all the points $\xi + \eta$ with $\xi \in Y$.

Let f_n and g_n be the projections of $X^{[n-1,n]} \subset X^{[n-1]} \times X^{[n]}$ to $X^{[n-1]}$ and $X^{[n]}$ respectively. Let p_n and q_n be the projections of $X^{[n]} \times X$ to $X^{[n]}$ and X respectively. Define the morphisms $\pi_n \colon \mathcal{Z}_n \to X^{[n]}$ by $\pi_n(\eta, x) = \eta$ and $q \colon X^{[n-1,n]} \to X$ by $q(\xi, \eta) = \mathrm{Supp}(I_\xi / I_\eta)$. In addition, we put $\psi_n = (g_n, q) \colon X^{[n-1,n]} \to \mathcal{Z}_n$ and $\phi_n = (f_n, q) \colon X^{[n-1,n]} \to X^{[n-1]} \times X$. It is known (see [**E-S**]) that $\psi_n : X^{[n-1,n]} \to \mathcal{Z}_n$ is canonically isomorphic to $\mathbb{P}(\omega_{\mathcal{Z}_n})$ where $\omega_{\mathcal{Z}_n}$ is the dualizing sheaf of $\mathcal{Z}_n$. So the fiber $(\psi_n)^{-1}(\eta, x)$ is isomorphic to the projective space $\mathbb{P}(\omega_{\eta_x} \otimes \mathbb{C}_x)$.

Next, we review some results on the homology groups of $X^{[n]}$. For $n \geq 2$, let

$$B_n = \{\xi \in X^{[n]} \,|\, |\mathrm{Supp}(\xi)| < n\}$$

which is known as the boundary divisor. For a real surface C in X, define

$$D_C = \{\xi \in X^{[n]} \,|\, C \cap \mathrm{Supp}(\xi) \neq \emptyset\}.$$

Note that the homology class of D_C depends only on the homology class of C. When C is a complex curve in X, D_C is a divisor in $X^{[n]}$. It follows from [**Got, Gro, Na1**] that if the real surfaces $C_1, \ldots, C_s$ form a basis of $H_2(X, \mathbb{C})$, then $B_n, D_{C_1}, \ldots, D_{C_s}$ form a basis of $H_{4n-2}(X^{[n]}, \mathbb{C})$.

3. Curves homologous to β_n

Let X be a simply-connected smooth projective surface. First of all, we study curves γ in the punctual Hilbert scheme $M_n(x) \subset X^{[n]}$ such that $\gamma \sim \beta_n$, where "$\sim$" denotes "homologous". Identify $M_n(x)$ with $\mathrm{Hilb}^n(\mathbb{C}^2, 0)$ as in (2.4). Let $R = \mathcal{O}_{\mathbb{C}^2, 0}$ be the local ring of $\mathbb{C}^2$ at the origin, and $\mathfrak{m} = (u, v)$ be the maximal ideal of R where u, v form a coordinate system of $\mathbb{C}^2$ at the origin. Let $\eta \in \mathrm{Hilb}^n(\mathbb{C}^2, 0)$. It is known (see [**Iar**]) that $I_{\eta,0} \supset \mathfrak{m}^n$ and there is an embedding $\tau \colon \mathrm{Hilb}^n(\mathbb{C}^2, 0) \to \mathrm{Grass}(R/\mathfrak{m}^n, n)$ where $R/\mathfrak{m}^n$ is considered as a $\mathbb{C}$-vector space of dimension $\binom{n+1}{2}$, and τ maps $\eta \in \mathrm{Hilb}^n(\mathbb{C}^2, 0)$ to the quotient $R/\mathfrak{m}^n \to R/I_{\eta,0} = \mathcal{O}_{\eta,0} \to 0$. For simplicity, put $\mathbb{G} = \mathrm{Grass}(R/\mathfrak{m}^n, n)$ and $\ell_1 = \binom{n+1}{2}$. Then over $\mathbb{G}$, there are a universal quotient bundle $\mathcal{Q}$, a universal subbundle $\mathcal{S}$ and an exact sequence:

$$0 \to \mathcal{S} \to (R/\mathfrak{m}^n) \otimes \mathcal{O}_{\mathbb{G}} \to \mathcal{Q} \to 0.$$

It is known that $\det(\mathcal{Q})$ induces the Plücker embedding $\mathfrak{p} : \mathbb{G} \to \mathbb{P}((\wedge^{\ell_1 - n}\mathbb{C}^{\ell_1})^*)$. So if $\mathcal{H}$ is the hyperplane line bundle over $\mathbb{P}((\wedge^{\ell_1 - n}\mathbb{C}^{\ell_1})^*)$, then $\det(\mathcal{Q}) = \mathfrak{p}^*\mathcal{H}$. Now regard the curve $\gamma \subset M_n(x)$ as a curve in $\mathrm{Hilb}^n(\mathbb{C}^2, 0)$ via the identification $M_n(x) \cong \mathrm{Hilb}^n(\mathbb{C}^2, 0)$. Then, we can prove that $\gamma \sim \beta_n$ if and only if $\gamma \cdot B_n = -2$ if and only if $(\mathfrak{p} \circ \tau)(\gamma)$ is a line. This enables us to obtain the following.

LEMMA 3.1. *Let γ be a curve in $M_n(x)$. Then, $\gamma \sim \beta_n$ if and only if $\gamma =$ $f_{n}(C)$ where C is a line in $(\psi_n)^{-1}(\eta, x) \cong \mathbb{P}(\omega_{\eta_x} \otimes \mathbb{C}_x)$ for some $\eta \in M_{n-1}(x)$*

$x \in X$, $\gamma_k \sim \beta_k$ in $X^{[k]}$, and $\xi \in X^{[n-k]}$ is a fixed point with $x \notin \mathrm{Supp}(\xi)$. In view of Lemma 3.1, we obtain a characterization of curves in $X^{[n]}$ homologous to β_n.

THEOREM 3.2. *Let $n \geq 2$, and X be a simply-connected smooth projective surface. Then, a curve γ in $X^{[n]}$ is homologous to β_n if and only if $\gamma = f_{n+1}(C)$ where C is a line in the projective space $(\psi_{n+1})^{-1}(\eta, x)$ for some $(\eta, x) \in \mathcal{Z}_{n+1}$. Moreover, in this case, (η, x) and C are uniquely determined by γ.*

Our next goal is to give a global description of the moduli space $\mathfrak{M}(\beta_n)$ of all the curves in $X^{[n]}$ homologous to β_n. Note that $\mathfrak{M}(\beta_n)$ is the union of certain irreducible components in the Hilbert scheme of curves in $X^{[n]}$. We shall use the concept of the relative Grassmannian scheme (see [**Mum, Sim**]). The following is proved by using the universal properties of Hilbert schemes and relative Grassmannians.

THEOREM 3.3. *Let $n \geq 2$, X be a simply-connected smooth projective surface, and $\omega_{\mathcal{Z}_{n+1}}$ be the dualizing sheaf of $\mathcal{Z}_{n+1} \subset X^{[n+1]} \times X$. Then, there exists a bijective morphism from the relative Grassmannian* $\mathrm{Grass}(\omega_{\mathcal{Z}_{n+1}}, 2)$ *over $\mathcal{Z}_{n+1}$ to the moduli space $\mathfrak{M}(\beta_n)$ of all the curves in $X^{[n]}$ homologous to β_n.*

Finally, we describe a top stratum in the moduli space $\mathfrak{M}(\beta_n)$. Let $x \in X$ and $\xi = x_1 + \ldots + x_{n-2} \in X^{[n-2]}$ be fixed such that $x, x_1, \ldots, x_{n-2}$ are distinct points. Then $M_2(x) + \xi \sim \beta_n$. The set of all such curves has dimension

$$\#(\text{moduli of } x) + \#(\text{moduli of } \xi) = 2 + 2(n-2) = 2n - 2.$$

Using the estimations in [**E-L, E-S**], we can prove that this set is the top stratum in the moduli space $\mathfrak{M}(\beta_n)$. In particular, the dimension of $\mathfrak{M}(\beta_n)$ is $(2n-2)$, one more than the expected dimension $c_1(T_{X^{[n]}}) \cdot \beta_n + \dim X^{[n]} - \dim \mathrm{Aut}(\mathbb{P}^1) = (2n-3)$. Therefore the moduli space of curves in $X^{[n]}$ homologous to β_n is obstructed. This can also be seen from the following proposition which deals with the normal bundles of the curves γ corresponding to the points in the top stratum.

PROPOSITION 3.4. *Let $\gamma = M_2(x) + x_1 + \ldots + x_{n-2}$ where $x, x_1, \ldots, x_{n-2}$ are distinct points in X. Let $N_{\gamma \subset X^{[n]}}$ be the normal bundle of γ in $X^{[n]}$. Then,*

$$N_{\gamma \subset X^{[n]}} \cong \mathcal{O}_\gamma^{\oplus(2n-2)} \oplus \mathcal{O}_\gamma(-2).$$

The proof of Proposition 3.4 uses the fact that in an analytic open neighborhood of γ, the Hilbert scheme $X^{[n]}$ splits into $X^{[2]} \times X^{[n-2]}$. So the result for a general n follows from the easy case of $n = 2$.

4. The nef cone and effective cone of $X^{[n]}$ when $X = \mathbb{P}^2$

First of all, recall from [**BSG**] that a line bundle L on a projective surface X is *k-very ample* if the restriction $H^0(X, L) \to H^0(X, \mathcal{O}_\xi \otimes L)$ is surjective for every $\xi \in X^{[k+1]}$. The concept of k-very ampleness relates $X^{[k+1]}$ to a Grassmannian as follows. The surjective restriction represents an element in $\mathrm{Grass}(H^0(X, L), k+1)$. So we get a morphism $\varphi_k(L)\colon\ X^{[k+1]} \to \mathrm{Grass}(H^0(X, L), k+1)$ when L is k-very ample. Letting $h = h^0(X, L)$ and $\mathfrak{P}\colon\ \mathrm{Grass}(\mathbb{C}^h, k+1) \to \mathbb{P}((\wedge^{h-(k+1)}\mathbb{C}^h)^*)$ be the Plücker embedding, then we see from the Appendix of [**BSG**] that

$$(\mathfrak{P} \circ \varphi_k(L))^* \mathcal{H} = \mathcal{O}_{X^{[k+1]}}\big(D_{c_1(L)} - B_{k+1}/2\big) \tag{4.1}$$

where $\mathcal{H}$ is the hyperplane line bundle over the projective space $\mathbb{P}((\wedge^{h-(k+1)}\mathbb{C}^h)^*)$.

In the rest of this paper, we let $X = \mathbb{P}^2$. By the Theorem 2.1 in [**BSG**], the line bundle $\mathcal{O}_X(n)$ is n-very ample. Note that for $k \geq 1$, k-very ampleness implies $(k-1)$-very ampleness. So we get the following two morphisms:

$$\varphi_1 = \varphi_{n-1}(\mathcal{O}_X(n-1))\colon X^{[n]} \to \mathrm{Grass}(H^0(X, \mathcal{O}_X(n-1)), n), \tag{4.2}$$

$$\varphi_2 = \varphi_{n-1}(\mathcal{O}_X(n))\colon X^{[n]} \to \mathrm{Grass}(H^0(X, \mathcal{O}_X(n)), n). \tag{4.3}$$

For $i = 1$ or 2, let $\ell_i = h^0(X, \mathcal{O}_X(n-2+i)) = \binom{n+i}{2}$, $N_i = \dim(\wedge^{\ell_i - n}\mathbb{C}^{\ell_i}) - 1$, $\mathfrak{P}_i\colon \mathrm{Grass}(\mathbb{C}^{\ell_i}, n) \to \mathbb{P}^{N_i}$ be the Plücker embedding, $\Phi_i = \mathfrak{P}_i \circ \varphi_i\colon X^{[n]} \to \mathbb{P}^{N_i}$, and $\mathcal{H}_i$ be the hyperplane line bundle over $\mathbb{P}^{N_i}$. By (4.1),

$$(\Phi_1)^*\mathcal{H}_1 = \mathcal{O}_{X^{[n]}}((n-1)D_\ell - B_n/2), \quad (\Phi_2)^*\mathcal{H}_2 = \mathcal{O}_{X^{[n]}}(nD_\ell - B_n/2).$$

In particular, $((n-1)D_\ell - B_n/2)$ is nef. Moreover, by the Main Theorem in [**C-G**], we conclude that φ_2 is an embedding. Thus $(nD_\ell - B_n/2)$ is very ample.

THEOREM 4.1. *Let $n \geq 2$ and ℓ be a line in $X = \mathbb{P}^2$.*

(i) *The nef cone of $X^{[n]}$ is the cone spanned by D_ℓ and $((n-1)D_\ell - B_n/2)$;*
(ii) *The effective cone of $X^{[n]}$ is the cone spanned by β_n and $(\beta_\ell - (n-1)\beta_n)$;*
(iii) *Let β be the homology class of an effective curve in the Hilbert scheme $X^{[n]}$. Then, $\beta \cdot (nD_\ell - B_n/2) = 1$ if and only if either $\beta = \beta_n$ or $\beta = (\beta_\ell - (n-1)\beta_n)$.*

The main idea to establish this theorem is as follows. By [**Fo2, BSG**], the Picard group of $X^{[n]}$ is generated by D_ℓ and $(-B_n/2)$. By the Proposition 2.2 in [**E-S**], the morphism ϕ_n introduced in Sect. 2 is the blowing-up of $X^{[n-1]} \times X$ along $\mathcal{Z}_{n-1}$. Together with the morphism g_n, we show that if C is a line in $X = \mathbb{P}^2$, then every line in the projective space $C^{[n]} \subset X^{[n]}$ is rationally equivalent to the 1-cycle $\beta_\ell - (n-1)\beta_n$ in the Chow group $A_1(X^{[n]})$. Finally, we use the intersection numbers $\beta_\ell \cdot D_\ell = 1$, $\beta_\ell \cdot (-B_n/2) = 0$, $\beta_n \cdot D_\ell = 0$, $\beta_n \cdot (-B_n/2) = 1$, and the fact that B_n and D_ℓ form a basis of $H_{4n-2}(X^{[n]}, \mathbb{C})$.

5. Curves homologous to $\beta_\ell - (n-1)\beta_n$

Now we study curves γ in $X^{[n]}$ homologous to $\beta_\ell - (n-1)\beta_n$ where ℓ is a line in $X = \mathbb{P}^2$. From the previous section, we know that if C is a line in $X = \mathbb{P}^2$, then every line in $C^{[n]} \subset X^{[n]}$ is homologous to $\beta_\ell - (n-1)\beta_n$. Conversely, let $\gamma \subset X^{[n]}$ be a curve homologous to $\beta_\ell - (n-1)\beta_n$. It can be showed that $p_1(\upsilon_n^{-1}(\rho_n(\gamma)))$ is a line C_γ together with some isolated points in X, where p_1 is the projection from X^n to the first factor, and ρ_n and υ_n are defined in (2.2). Furthermore, if p_γ is a linear homogeneous polynomial defining the line C_γ, then any degree-$(n-1)$ homogeneous polynomial divisible by p_γ must vanish at ξ for every $\xi \in \gamma$. From these observations, we deduce the following theorem.

THEOREM 5.1. *Let $n \geq 2$. Then, a curve $\gamma \subset X^{[n]}$ is homologous to $\beta_\ell - (n-1)\beta_n$ if and only if there is a line C in $X = \mathbb{P}^2$ such that γ is a line in $C^{[n]} \subset X^{[n]}$. Moreover, C is uniquely determined by the curve γ.*

Next, we discuss the normal bundle $N_{\gamma \subset X^{[n]}}$ of a curve γ in $X^{[n]}$ homologous to $\beta_\ell - (n-1)\beta_n$. By Theorem 5.1, there exists a unique line C in $X = \mathbb{P}^2$ such that γ is a line in $C^{[n]} \subset X^{[n]}$. In particular, N … $\cong \mathcal{O}$ … $(1)^{\oplus(n-1)}$ …

By the results in [**AIK**], $C^{[n]}$ is the locus of zeros of a transversally regular section of the rank-n locally free sheaf $\pi_{n*}(q_n^*\mathcal{O}_X(1)|_{\mathcal{Z}_n})$ over $X^{[n]}$. It follows that $N_{C^{[n]}\subset X^{[n]}} \cong \pi_{n*}(q_n^*\mathcal{O}_X(1)|_{\mathcal{Z}_n})|_{C^{[n]}} \cong \mathcal{O}_{C^{[n]}}^{\oplus 2} \oplus \mathcal{O}_{C^{[n]}}(-1)^{\oplus(n-2)}$. Combining with the above exact sequence, we are able to determine $N_{\gamma\subset X^{[n]}}$.

PROPOSITION 5.2. *Let $n \geq 2$, and ℓ be a line in $X = \mathbb{P}^2$. Let $\gamma \subset X^{[n]}$ be a curve homologous to $\beta_\ell - (n-1)\beta_n$. Then,*

(i) $N_{\gamma\subset X^{[n]}} = \mathcal{O}_\gamma(1)^{\oplus(n-1)} \oplus \mathcal{O}_\gamma^{\oplus 2} \oplus \mathcal{O}_\gamma(-1)^{\oplus(n-2)}$;

(ii) *the moduli space $\mathfrak{M}(\beta_\ell - (n-1)\beta_n)$ of all the curves in $X^{[n]}$ homologous to $(\beta_\ell - (n-1)\beta_n)$ is unobstructed, i.e., is smooth with the expected dimension.*

In view of Theorem 5.1 and Proposition 5.2 (ii), we obtain the following global description which is the analogue to Theorem 3.3.

THEOREM 5.3. *Let $n \geq 2$, and ℓ be a line in $X = \mathbb{P}^2$. Then the moduli space $\mathfrak{M}(\beta_\ell - (n-1)\beta_n)$ of curves in $X^{[n]}$ homologous to $(\beta_\ell - (n-1)\beta_n)$ is isomorphic to the Grassmannian bundle* $\mathrm{Grass}(\mathrm{Sym}^n((T_{(\mathbb{P}^2)^*})^*), 2)$ *over* $(\mathbb{P}^2)^*$.

6. The structure of the morphism φ_1

Recall the morphism φ_1 defined in (4.2). From the definition of φ_1, we see that when $n = 2$, $\varphi_1 : X^{[2]} \to \mathrm{Grass}(H^0(X, \mathcal{O}_X(1)), 2) = (\mathbb{P}^2)^*$ is isomorphic to the relative Hilbert scheme $\mathrm{Hilb}^2(\mathbb{P}((T_{(\mathbb{P}^2)^*})^*)/(\mathbb{P}^2)^*)$ over the dual space $(\mathbb{P}^2)^*$. In the rest of this section, we assume that $n \geq 3$.

First of all, we determine the fiber of the morphism $\varphi_1 : X^{[n]} \to \varphi_1(X^{[n]})$. Fix $\xi \in X^{[n]}$. By definition, $\varphi_1(\xi) \in \varphi_1(X^{[n]}) \subset \mathrm{Grass}(H^0(X, \mathcal{O}_X(n-1)), n)$ corresponds to the $n(n-1)/2$-dimensional subspace $H^0(X, I_\xi \otimes \mathcal{O}_X(n-1))$ in $H^0(X, \mathcal{O}_X(n-1))$. So $\eta \in (\varphi_1)^{-1}(\varphi_1(\xi))$ if and only if $H^0(X, I_\eta \otimes \mathcal{O}_X(n-1)) = H^0(X, I_\xi \otimes \mathcal{O}_X(n-1))$. Let $\mathfrak{d} \subset |(n-1)\ell|$ be the linear system corresponding to the subspace $H^0(X, I_\xi \otimes \mathcal{O}_X(n-1))$ in $H^0(X, \mathcal{O}_X(n-1))$. Then, either $\mathfrak{d}$ cuts out a 0-dimensional subscheme ζ with $\zeta \supset \xi$, or $\mathfrak{d}$ has a fixed component C. It turns out that $(\varphi_1)^{-1}(\varphi_1(\xi)) = \{\xi\}$ in the first case. In the second case, we show that C is line in $X = \mathbb{P}^2$ and $(\varphi_1)^{-1}(\varphi_1(\xi)) = C^{[n]}$.

LEMMA 6.1. *Let $n \geq 3$, $X = \mathbb{P}^2$, and W_n be the contraction locus of the morphism $\varphi_1 : X^{[n]} \to \varphi_1(X^{[n]})$ defined by (4.2). Then,*

(i) *a fiber of φ_1 is either a point or the subset $C^{[n]} \subset X^{[n]}$ for some line $C \subset X$;*

(ii) *the restriction $\varphi_1|_{W_n} : W_n \to \varphi_1(W_n)$ is isomorphic to the projective bundle $\mathbb{P}(\mathrm{Sym}^n((T_{(\mathbb{P}^2)^*})^*))$ over $(\mathbb{P}^2)^*$, where $T_{(\mathbb{P}^2)^*}$ is the tangent bundle of $(\mathbb{P}^2)^*$;*

(iii) *$N \cong (\varphi_1|_{W_n})^*M \otimes \mathcal{O}(-1)$ where $N = N_{W_n\subset X^{[n]}}$ is the normal bundle of W_n in $X^{[n]}$, M is some vector bundle over $\varphi_1(W_n) \cong (\mathbb{P}^2)^*$, and $\mathcal{O}(1)$ is the tautological line bundle over the projective bundle $\varphi_1|_{W_n} : W_n \to \varphi_1(W_n)$.*

The proof of Lemma 6.1 (iii) is based on the fact that $N_{C^{[n]}\subset X^{[n]}} \cong \mathcal{O}_{C^{[n]}}^{\oplus 2} \oplus \mathcal{O}_{C^{[n]}}(-1)^{\oplus(n-2)}$ for every line C in $X = \mathbb{P}^2$, which is mentioned in Sect. 5.

Let $n \geq 3$. By Lemma 6.1 (ii), we have $\dim(W_n) = (n+2) \leq \dim(X^{[n]}) - 1$. So $\varphi_1 : X^{[n]} \to \varphi_1(X^{[n]})$ is a birational morphism. Since $K_{X^{[n]}} = -3D_\ell$, $(\beta_\ell -$

$(n-1)\beta_n) \cdot K_{X^{[n]}} = -3 < 0$. This says that $R = \mathbb{R}^+ \cdot (\beta_\ell - (n-1)\beta_n)$ is an extremal ray since $(-K_{X^{[n]}})$ is nef and the Picard number of $X^{[n]}$ is 2. Thus φ_1 is the contraction of R since $((n-1)D_\ell - B_n/2) \cdot (\beta_\ell - (n-1)\beta_n) = 0$. Moreover, when $n \geq 4$, $\dim(W_n) \leq \dim(X^{[n]}) - 2$, so φ_1 is a small contraction.

THEOREM 6.2. *Let $n \geq 4$ and $X = \mathbb{P}^2$. Then, the flip for the small contraction $\varphi_1 : X^{[n]} \to \varphi_1(X^{[n]})$ defined by (4.2) exists.*

The proof of Theorem 6.2 is standard. Let $\tilde{\varphi}_- : \widetilde{X^{[n]}} \to X^{[n]}$ be the blowing-up of $X^{[n]}$ along W_n, and $\widetilde{W_n} \subset \widetilde{X^{[n]}}$ be the exceptional divisor. By Lemma 6.1 (iii), the map $(\varphi_1|_{W_n} \circ \tilde{\varphi}_-|_{\widetilde{W_n}}) : \widetilde{W_n} \to \varphi_1(W_n)$ factors through the natural projection $\mathbb{P}(M^*) \to \varphi_1(W_n)$. In fact, the induced morphism $\tilde{\varphi} : \widetilde{W_n} \to \mathbb{P}(M^*)$ is a $\mathbb{P}^n$-bundle, and the restriction $\mathcal{O}_{\widetilde{X^{[n]}}}(-\widetilde{W_n})|_{\widetilde{W_n}}$ is a tautological line bundle of $\tilde{\varphi} : \widetilde{W_n} \to \mathbb{P}(M^*)$. By the results in [**Nak, F-N**], there exists a morphism $\tilde{\varphi}_+ : \widetilde{X^{[n]}} \to (X^{[n]})_+$ such that $(X^{[n]})_+$ is smooth and contains $\mathbb{P}(M^*)$. Moreover, $\tilde{\varphi}_+$ is the blowing-up of $(X^{[n]})_+$ along $\mathbb{P}(M^*)$, $(\tilde{\varphi}_+)^{-1}(\mathbb{P}(M^*)) = \widetilde{W_n}$, $\tilde{\varphi}_+|_{\widetilde{W_n}} = \tilde{\varphi}$, and there exists a morphism $\varphi_+ : (X^{[n]})_+ \to \varphi_1(X^{[n]})$ forming a commutative diagram:

$$\begin{array}{ccccc} & & \widetilde{X^{[n]}} & & \\ & \swarrow^{\tilde{\varphi}_-} & & \searrow^{\tilde{\varphi}_+} & \\ X^{[n]} & & & & (X^{[n]})_+. \\ & \searrow^{\varphi_1} & & \swarrow^{\varphi_+} & \\ & & \varphi_1(X^{[n]}) & & \end{array}$$

Next, we check that $K_{(X^{[n]})_+}$ is φ_+-ample. So the commutative triangle

$$\begin{array}{ccccc} X^{[n]} & & \overset{\tilde{\varphi}_+ \circ (\tilde{\varphi}_-)^{-1}}{\dashrightarrow} & & (X^{[n]})_+ \\ & \searrow^{\varphi_1} & & \swarrow^{\varphi_+} & \\ & & \varphi_1(X^{[n]}) & & \end{array}$$

satisfies the Flip Conjecture 5-1-10 in [**KMM**], i.e., the flip for φ_1 exists.

Finally, we remark that when $n = 3$, $\varphi_1 : X^{[n]} \to \varphi_1(X^{[n]})$ is the blowing-up of $\varphi_1(X^{[n]})$ along $\varphi_1(W_n) \cong (\mathbb{P}^2)^*$, and W_n is the exceptional locus.

References

[AIK] A. Altman, A. Iarrobino, S. Kleiman, *Irreducibility of the compactified Jacobian*, Real and complex singularities (Proc. 9th Nordic Summer School, Oslo, 1976), 1-12. Sijthoff and Noordhoff , Groningen (1977).

[Bea] A. Beauville, *Variétés Kähleriennes dont la premiére classe de Chern est nulle*, J. Diff. Geom. **18** (1983), 755-782.

[BSG] M. Beltrametti, A. Sommese, *Zero cycles and kth order embeddings of smooth projective surfaces. With the Appendix "Identification of very ample line bundles on $S^{[r]}$" by L. Göttsche*, Sympos. Math., XXXII, Problems in the theory of surfaces and their classification (Cortona, 1988), 33-48, Academic Press, London (1991).

[C-G] F. Catanese, L. Göttsche, *d-very-ample line bundles and embeddings of Hilbert schemes of 0-cycles*, Manuscripta Math. **68** (1990), 337-341.

[E-L] G. Ellingsrud, M. Lehn, *Irreducibility of the punctual quotient scheme of a surface*, Ark. Mat. **37** (1999), 245-254.

[Fo2] J. Fogarty, *Algebraic families on an algebraic surface. II: The Picard scheme of the punctual Hilbert scheme*, Amer. J. Math. **95** (1973), 660-687.

[F-N] A. Fujiki, S. Nakano, *Supplement to "On the inverse of monoidal transformation"*, Publ. RIMS Kyoto University **7** (1971), 637-644.

[Got] L. Göttsche, *Hilbert schemes of zero-dimensional subschemes of smooth varieties*, Lecture Notes in Mathematics **1572**, Springer-Verlag , Berlin (1994).

[Gro] I. Grojnowski, *Instantons and affine algebras I: the Hilbert scheme and vertex operators*, Math. Res. Lett. **3** (1996), 275-291.

[Iar] A. Iarrobino, *Punctual Hilbert schemes*, Mem. of A.M.S. **188** (1977).

[KMM] Y. Kawamata, K. Matsuda, K. Matsuki, *Introduction to the minimal model problem*, Algebraic Geometry, Sendai Adv. Stud. Pure Math. **10**, 283–360, Kinokuniya and Amsterdam North-Holland, Tokyo (1987).

[Leh] M. Lehn, *Chern classes of tautological sheaves on Hilbert schemes of points on surfaces*, Invent. Math. **136** (1999), 157-207.

[L-Q] W.-P. Li, Z. Qin, *On 1-pointed Gromov-Witten invariants of the Hilbert schemes of points on surfaces*, Proceedings of 8th Gökova Geometry-Topology Conference (2001), Turkish J. Math. **26** (2002), 53-68.

[Mum] D. Mumford, *Lectures on curves on an algebraic surface.* Princeton University Press, Princeton, N.J., 1966.

[Na1] H. Nakajima, *Heisenberg algebra and Hilbert schemes of points on projective surfaces*, Ann. Math. **145** (1997), 379-388.

[Nak] S. Nakano, *On the inverse of monoidal transformation*, Publ. RIMS Kyoto University **6** (1970), 483-502.

[Sim] C.T. Simpson, *Moduli of representations of the fundamental group of a smooth projective variety I*, Publ. Math. IHES **79** (1994), 47-129.

Department of Mathematics, HKUST, Clear Water Bay, Kowloon, Hong Kong
E-mail address: `mawpli@uxmail.ust.hk`

Department of Mathematics, University of Missouri, Columbia, MO 65211, USA
E-mail address: `zq@math.missouri.edu`

Department of Mathematics, University of Missouri, Columbia, MO 65211, USA
E-mail address: `qi@math.missouri.edu`

Contemporary Mathematics
Volume **322**, 2003

Limiting linear subspaces on non-reduced schemes

Xian Wu

ABSTRACT. Let X be a hypersurface of degree d in $\mathbb{P}^n$ and F_X be the scheme of $\mathbb{P}^r$'s contained in X. When one deforms a generic X into a degenerate one, there is a natural subscheme F_{lim} of F_X which consists of the limiting $\mathbb{P}^r$'s in X with respect to the deformation. One can often learn interesting and useful information about generic F_X from F_{lim}, or vice versa, by taking various degenerations. In this paper, we will study F_{lim} when X is deformed into the m-fold of a generic hypersurface $L = X_l$ of degree l in $\mathbb{P}^n$, where $ml = d$. Our main tool is the refined intersection product of Fulton-MacPherson. There is a nice relation between the Segre class of F_X and the Segre class of F_L which leads to an identity for the class of the limiting $\mathbb{P}^r$'s in the non-reduced hypersurface X. In fact, such an identity actually holds for arbitrary vector bundles.

0. Introduction

Let X be a hypersurface of degree d in $\mathbb{P}^n$ and F_X be the scheme of $\mathbb{P}^r$'s in X. If one deforms a generic X into a special form, the dimension of F_X will jump. For example, consider the case of 27 lines in a generic cubic surface. If we degenerate the surface into the union of a plane and a quadric, then there are infinitely many lines in the union. For a generic deformation, one can identify 27 limiting lines as follows. The intersection of the plane, the quadric, and a cubic (which depends on the deformation) consists of six points. A line in the plane is a limiting line if and only if it passes through any two of those six points. That gives us $\binom{6}{2} = 15$ limiting lines in the plane. On the other hand, a line in the quadric is a limit if and only if it passes through any one of those six points. Since there are exactly two lines in a generic quadric that pass through any given point on the quadric, we see that there are 12 limiting lines in the quadric.

In general, there is a naturally defined subscheme F_{lim} of F_X which consists of the limiting $\mathbb{P}^r$'s in X with respect to the deformation. Using the fundamental construction of Fulton-MacPherson, one sees that the class of F_{lim} can be represented as the refined intersection product $\mathcal{R}_X$ (Theorem 1.3, [W2]). By definition,

$$\mathcal{R}_X = \{c(\mathrm{Sym}^d U^*) \cap s(F_X, G)\}_k, \tag{0.1}$$

where G is the Grassmannian of $\mathbb{P}^r$'s in $\mathbb{P}^n$, U is the universal subbundle on G, $c(\operatorname{Sym}^d U^*)$ is the Chern class of $\operatorname{Sym}^d U^*$, $s(F_X, G)$ is the Segre class of F_X in G, and k is the expected dimension.

A natural degeneration is to take X as the d-fold of a hyperplane $\mathbb{P}^{n-1}$ in $\mathbb{P}^n$. More generally, we will take $X = X_l^m$ as the m-fold of a generic hypersurface $L = X_l$ of degree l in $\mathbb{P}^n$, where $ml = d$. We will hence study the non-reduced F_X with a smooth center F_L. From (0.1), to compute $\mathcal{R}_X$ and hence $F_{\lim}$, it will be enough to know the Segre class $s(F_X, G)$. As it turns out, $s(F_X, G)$ is nicely related to $s(F_L, G)$ (which is well-known and easy to compute) as stated below. To simplify the notation, we will identify $\mathcal{R}_X$ with its image in the Chow ring of G. Furthermore, we will also identify an element in the Chow ring $A_*(G)$ with its dual in $A^*(G)$ when its meaning is clear from the context.

PROPOSITION 1.4 AND COROLLARY 1.7. *Let X and L be the hypersurfaces in $\mathbb{P}^n$ defined above. Then*

$$s(F_X, G) = \operatorname{Adams}(m, s(\operatorname{Sym}^l U^*)) \cap m^{r_l}[F_L] = \operatorname{Adams}(m, s(F_L, G)),$$

where $r_l = \binom{r+l}{l}$ and Adams$(*, *)$ *is the Adams operator of K-theory. Therefore,*

$$\mathcal{R}_X = \operatorname{Adams}(m, c(\operatorname{Sym}^l U^*))_{r_l} c_h(\operatorname{Sym}^d U^* - \operatorname{Adams}(m, c(\operatorname{Sym}^l U^*))), \tag{0.2}$$

where

$$h = \operatorname{rank}(\operatorname{Sym}^d U^* - \operatorname{Adams}(m, c(\operatorname{Sym}^l U^*))) = \binom{r+d}{d} - r_l$$

The above formulas provide a way to compute $\mathcal{R}_X$ using standard techniques of intersection theory. For example, in the case of quintic threefold $X_5 = X_1^5$, one can conclude that there are 115 limiting lines, each with multiplicity 25 in the 5-fold of $L = X_1 = \mathbb{P}^3$. Notice that $115 \times 25 = 2875$ as expected. More examples will be given in Section 2.

In general, since the class of the limiting $\mathbb{P}^r$'s in X must be equal to the class of $\mathbb{P}^r$'s in a generic hypersurface of degree d, $\mathcal{R}_X$ is also equal to the top Chern class of $\operatorname{Sym}^d U^*$. Therefore, formula (0.2) leads to a family of identities involving the characteristic classes of the bundle $\operatorname{Sym}^l U^*$ (Corollary 1.10). As it turns out, these identities actually hold for arbitrary vector bundles as stated below.

PROPOSITION 1.11. *Let m, l, and $d = ml$ be positive integers. The following identity holds for any vector bundle V of rank $r+1$ over a manifold:*

$$c_{\text{top}}(\operatorname{Sym}^d V) = \operatorname{Adams}(m, c(\operatorname{Sym}^l V))_{r_l} c_h(\operatorname{Sym}^d V - \operatorname{Adams}(m, c(\operatorname{Sym}^l V))),$$

where r_l is the rank of the bundle $\operatorname{Sym}^l V$, *and h is the rank of the virtual bundle* $\operatorname{Sym}^d V - \operatorname{Adams}(m, c(\operatorname{Sym}^l V))$ *given in Proposition 1.4.*

Since $\operatorname{Adams}(m, c(\operatorname{Sym}^l V))_{r_l} = m^{r_l} c_{r_l}(\operatorname{Sym}^l V)$, an immediate consequence of Proposition 1.11 is that the class of $\mathbb{P}^r$'s in a generic hypersurface of degree d is divisible by m^{r_l} for any $ml = d$. It will be interesting to see other applications of the above identity.

Our results can also be used to study limits for other types of degenerations. For example, consider the case that a quintic threefold is deformed into the union of a hyperplane X_1 and a double quadric X_2^2. Using Proposition 1.4 and the refined residual intersection decomposition [W3], one can conclude that there are 635 and

2240 limiting lines in X_1 and X_2^2, respectively. More such examples are given in Section 2.

Finally, we want to point out that the results here can be formally generalized to study other limiting problems. For example, although actual computation will be much more complicated, one should be able to generalize formulas from Proposition 1.4 to extend a study of limiting rational curves in [W4] to non-reduced cases.

I would like to thank the referee for his/her comments and suggestions.

1. Refined intersection products and limiting linear subspaces

Let us first recall some definitions and facts from [W2] which will be needed for the present work.

For any hypersurface X (which can be singular or non-reduced) of degree d in $\mathbb{P}^n$, we will use F_X to denote the scheme of $\mathbb{P}^r$'s in X. As a set, F_X consists of all $\mathbb{P}^r$'s which are contained in X. To see the scheme structure of F_X, let $G = G(r+1, n+1)$ be the Grassmannian of $\mathbb{P}^r$'s in $\mathbb{P}^n$ and $\operatorname{Sym}^d U^*$ be the d-th symmetric power of the dual of U, where U is the universal subbundle on G. Then F_X can be defined as the zero-scheme of s_X, where s_X is the section of $\operatorname{Sym}^d U^*$ induced by the equation that defines X. Next, we consider the fiber square

$$\begin{array}{ccc} F_X & \longrightarrow & G \\ \downarrow & & \downarrow{\scriptstyle s_X} \\ G & \xrightarrow{\ i\ } & \operatorname{Sym}^d U^* \end{array} \tag{1.1}$$

where i is the zero section imbedding. The fundamental construction of Fulton-MacPherson defines the refined intersection product, called $\mathcal{R}_X$ in [W2], from the fiber square above. It is well-known that this class can be expressed by the following formula (see [Chapter 6, F]):

$$\mathcal{R}_X = i^!([G]) = \{c(\operatorname{Sym}^d U^*) \cap s(F_X, G)\}_k, \tag{1.2}$$

where $c(\operatorname{Sym}^d U^*)$ is the Chern class of $\operatorname{Sym}^d U^*$, $s(F_X, G)$ is the Segre class of F_X in G and

$$k = \dim(G) - \operatorname{rank}(\operatorname{Sym}^d U^*) = (r+1)(n-r) - \binom{r+d}{d}$$

is the expected dimension of F_X. Recall next that if we deform X from a generic hypersurface to a special one, then the dimension of F_X will jump. In this case, there is a proper subscheme F_{lim} of F_X which consists of the limiting $\mathbb{P}^r$'s in X (see [W2] or [Chapter 11, F] for more precise definitions). Although $\mathcal{R}_X$ is defined pointwise from data on X and is totally independent of deformations, the theory of dynamic intersections (see [W2] or [Chapter 11, F]) implies that the subscheme F_{lim} of the limiting $\mathbb{P}^r$'s in X is represented by $\mathcal{R}_X$. For the convenience of readers, we recall some facts about $\mathcal{R}_X$ below.

LEMMA 1.3 (see [Theorem 1.3, W2] or [F]). *Let $\mathcal{R}_X$ be the refined intersection*

(2) *Geometrically, it is equal to the class of* F_X *if* X *is generic and otherwise is equal to the class of the subscheme* F_{lim} *of the limiting* $\mathbb{P}^r$*'s in* X *with respect to a generic deformation.*
(3) *Finding the local distribution of the limiting* $\mathbb{P}^r$*'s is equivalent to finding the canonical decomposition of* $\mathcal{R}_X$ *in terms of its distinguished varieties.*

A difficulty in using the construction given above is that $\mathcal{R}_X$ is in general not easy to compute. In [W1] and [W2], we considered degenerations that deform X into the union of two generic hypersurfaces of lower degrees. The method of infinitesimal deformations is used there to characterize and to compute F_{lim} and $\mathcal{R}_X$.

The basic degeneration that we will consider here, as indicated in the introduction, is to deform a generic X into a multiple of a lower degree hypersurface. To be more precise, we will deform a generic hypersurface of degree d into the m-fold of a generic hypersurface X_l of degree l with $d = ml$. To simplify the notation, we will identify a hypersurface with its equation. Hence, we have $X = L^m$, where $L = X_l$. In this case, F_X is equal to F_L as a set but carries a non-reduced scheme structure. The infinitesimal method will be difficult to apply here since the limiting $\mathbb{P}^r$'s will now have high multiplicity and may not deform infinitesimally with X at all. However, since F_X basically has only one component in this case, a direct computation of F_{lim} based on formula (1.2) and Lemma 1.3 becomes possible. In fact, the following proposition tells us how this can be done.

PROPOSITION 1.4. *Let* X *be the* m*-fold of a generic hypersurface* L *of degree* l *in* $\mathbb{P}^n$*. Then*

$$s(F_X, G) = \mathrm{Adams}(m, s(\mathrm{Sym}^l U^*)) \cap m^{r_l}[F_L] = \mathrm{Adams}(m, s(F_L, G)), \tag{1.5}$$

where $r_l = \binom{r+l}{l}$ *is the rank of* $\mathrm{Sym}^l U^*$*,* $\mathrm{Adams}(*,*)$ *is the Adams operator of* K*-theory,* $[F_L]$ *is the class of* F_L*, and* $s(\mathrm{Sym}^l U^*)$ *is the Segre class of* $\mathrm{Sym}^l U^*$*.*

REMARK. We may consider the last equality in (1.5) as a definition. Notice that (see proof of Corollary 1.7)

$$\begin{aligned} &s(F_L, G) = s(\mathrm{Sym}^l U^*) \cap [F_L], \\ &\mathrm{Adams}(m, c(\mathrm{Sym}^l U^*))_{\mathrm{top}} = m^{r_l} c_{\mathrm{top}}(\mathrm{Sym}^l U^*) = m^{r_l}[F_L]. \end{aligned} \tag{1.6}$$

In particular, substituting (1.5) into (1.2), we get the following formula for $\mathcal{R}_X$ which can be easily computed using standard techniques in intersection theory.

COROLLARY 1.7. *Let* X *be the hypersurface given in Proposition 1.4 and* $d = ml$*. Then*

$$\mathcal{R}_X = m^{r_l} c_{r_l}(\mathrm{Sym}^l U^*) c_h(\mathrm{Sym}^d U^* - \mathrm{Adams}(m, c(\mathrm{Sym}^l U^*))), \tag{1.8}$$

where

$$r_l = \mathrm{rank}(\mathrm{Sym}^l U^*) = \binom{r+l}{l}$$

and

$$h = \mathrm{rank}(\mathrm{Sym}^d U^*) - \mathrm{rank}(\mathrm{Sym}^l U^*) = \binom{r+d}{d} - r_l.$$

In particular, the class of the limiting $\mathbb{P}^r$*'s is divisible by* m^{r_l} *for any* $ml = d$*.*

REMARK. Notice that, for any given d, we can always take $l = 1$ and $m = d$. We hence have that the class of $\mathbb{P}^r$'s in a generic hypersurface of degree d is divisible by d^{r+1}.

PROOF OF COROLLARY 1.7. Since F_L is smooth, the Segre class of F_L in G is equal to the Segre class of the normal bundle of F_L in G, that is, the formal inverse of the Chern class of the normal bundle of F_L in G. On the other hand, since F_L is the zero-scheme of a regular section of $\mathrm{Sym}^l U^*$, the normal bundle of F_L in G is just the restriction of $\mathrm{Sym}^l U^*$ to F_L. That together with (1.2), (1.5), and the formula

$$\mathrm{Adams}(l, A)_i = l^i A_i$$

give the formula (1.8) for $\mathcal{R}_X$ (more precisely, the image of $\mathcal{R}_X$ in $A^*(G)$). □

PROOF OF PROPOSITION 1.4. To simplify the notation, we will take $l = 1$ and $m = d$. Let s_L be the regular section of U^* induced by L. Since we are only interested in computing the characteristic classes, we can apply the splitting principle and may assume that

$$U^* = \bigoplus_{i=1}^{r+1} L_i$$

is the direct sum of $r + 1$ line bundles. Next, assume that

$$s_L = \bigoplus_{i=1}^{r+1} s_i,$$

where s_i is a regular section of L_i. Let Z_i be the zero-scheme of s_i, which is a hypersurface in G. Therefore,

$$F_L = \bigcap_{i=1}^{r+1} Z_i.$$

Since X is the d-fold of L, s_L^d induces a section of $\mathrm{Sym}^d U^*$ and F_X is equal to the zero-scheme of s_L^d. By symmetry, we may assume that F_X is the set-theoretic intersection of all Z_i's with each Z_i having a multiplicity of d along F_X. Notice that

$$\begin{aligned} s(F_L, G) &= s(U^*) \\ &= c^{-1}(U^*) \\ &= c^{-1}(\bigoplus_{i=1}^{r+1} L_i) \\ &= (\prod_{i=1}^{r+1}(1 + Z_i))^{-1} \qquad (1.9) \\ &= (\sum_{j=0}^{r+1} \sum_{i_1, i_2, \ldots, i_j} Z_{i_1} Z_{i_2} \ldots Z_{i_j})^{-1} \\ &= (\sum^{r+1} D \,)^{-1} \end{aligned}$$

We now blow up G along F_L and use b to denote the blow-up. We will also use the same notation b to denote the restriction of the blow up to subschemes. Using the birational invariant property of Segre classes and (1.9), we then have

$$\begin{aligned}
s(F_X, G) &= b_*(s(b^*(F_X), b^*G)) \\
&= b_*(s(\bigcap_{i=1}^{r+1} db^*(Z_i), b^*G)) \\
&= b_*(\prod_{i=1}^{r+1} s(db^*(Z_i), b^*G)) \\
&= b_*(\prod_{i=1}^{r+1}(1 + db^*(Z_i))^{-1}) \\
&= b_*(\sum_{j=0}^{r+1} d^j \sum_{i_1, i_2, \ldots, i_j} b^*(Z_{i_1})b^*(Z_{i_2})\ldots b^*(Z_{i_j}))^{-1} \\
&= (\sum_{j=0}^{r+1} d^j \sum_{i_1, i_2, \ldots, i_j} Z_{i_1} Z_{i_2} \ldots Z_{i_j})^{-1} \\
&= (\sum_{j=0}^{r+1} d^j P_j)^{-1} \\
&= \mathrm{Adams}(d, s(F_L, G)).
\end{aligned}$$

This completes our proof of Proposition 1.4. □

By Lemma 1.3, the image of $\mathcal{R}_X$ in $A_*(G)$ is always equal to the top Chern class of $\mathrm{Sym}^d U^*$. We hence have the following family of identities.

COROLLARY 1.10. *The following identity holds for any positive integers l, m, and $d = ml$.*

$$c_{\mathrm{top}}(\mathrm{Sym}^d U^*) = \mathrm{Adams}(m, c(\mathrm{Sym}^l U^*))_{r_l} c_h(\mathrm{Sym}^d U^* - \mathrm{Adams}(m, c(\mathrm{Sym}^l U^*))),$$

where r_l and h are same as given in Corollary 1.7.

In fact, those identities actually hold for any vector bundle as stated below.

PROPOSITION 1.11. *Let V be a vector bundle of rank $r+1$ over a manifold. The following identity holds for any positive integers l, m, and $d = ml$:*

$$c_{\mathrm{top}}(\mathrm{Sym}^d V) = c_{r_l}(\mathrm{Sym}^l V(m)) c_h(\mathrm{Sym}^d V - \mathrm{Sym}^l V(m)), \tag{1.12}$$

where

$$r_l = \mathrm{rank}(\mathrm{Sym}^l V) = \binom{r+l}{l},$$

$$h = \mathrm{rank}(\mathrm{Sym}^d V - \mathrm{Sym}^l V(m)) = \binom{r+d}{d} - r_l,$$

and $\mathrm{Sym}^l V(m)$ is a vector bundle with

$$c_i(\mathrm{Sym}^l V(m)) = m^i c_i(\mathrm{Sym}^l V).$$

In other words,

$$c(\operatorname{Sym}^l V(m)) = \operatorname{Adams}(m, c(\operatorname{Sym}^l V)).$$

In particular, $c_{\text{top}}(\operatorname{Sym}^d V)$ *is divisible by* m^{r_l} *for any* $ml = d$.

PROOF. From the proof of Proposition 1.4 and Corollary 1.7, identity (1.12) holds if V is generated by its sections. In general, taking an ample line bundle H and replacing V by $V \otimes H^t$, we then see that the identity holds for t sufficiently large. Since it can be considered as a polynomial in t, it must hold for all t. In particular, it holds for $t = 0$. □

REMARK. The above identity can also be proved by elementary methods. In fact, it is equivalent to an identity of symmetric polynomials in $r+1$ variables by the splitting principle. We verified this polynomial identity in the case of $l = 1$ and $m = d$ by combinatorial methods, but the argument runs to few pages. It would be interesting to see if it has some meaning that comes directly from algebra.

2. A few examples

Both sides of the formula in Corollary 1.10 can be computed by using standard techniques in intersection theory (with the help of computers). We have computed many examples using the Maple package Schubert written by Katz and Strømme [KS]. Besides the example of lines in a generic quintic threefold given earlier, we give three more examples below. Standard notations of Schubert calculus (see [GH]) are used.

EXAMPLE 2.1. (Degeneration of hypersurfaces X_7 of degree 7 in $\mathbb{P}^5$ to the 7-fold X_1^7 of $\mathbb{P}^4$ and their lines.)

It is easy to compute that in this case

$$\operatorname{Adams}(7, c(U^*))_2 = 49\sigma_{1,1}$$

and

$$\sigma_{1,1} \cdot c_6(\operatorname{Sym}^7 U^* - \operatorname{Adams}(7, c(U^*))) = 14,245.$$

On the other hand, our computation shows that the number of lines in a generic hypersurface of degree 7 in $\mathbb{P}^5$, which is given by the degree of $c_8(\operatorname{Sym}^7 U^*)$, is equal to

$$698,005 = 49 \cdot 14,245$$

as expected.

EXAMPLE 2.2. (Degeneration of hypersurfaces X_4 in $\mathbb{P}^7$ to X_l^m and their $\mathbb{P}^2$'s.)

There are two degenerations of a generic X_4 into X_l^m.

Case 1: $X_4 \to X_1^4$. We have in this case that

$$\operatorname{Adams}(4, c(U^*))_3 = 4^3\sigma_{1,1,1}$$

and

$$c_{12}(\operatorname{Sym}^4 U^* - \operatorname{Adams}(4, c(U^*))) = 51,520\sigma_{4,4,4}.$$

Notice that

$$64\sigma \quad 51,520\sigma \quad = 3,297,280\sigma$$

Case 2: $X_4 \to X_2^2$. Similarly, we have that

$$\text{Adams}(2, c(\text{Sym}^2 U^*))_6 = 2^6 c_6(\text{Sym}^2 U^*)$$

and

$$2^6 c_6(\text{Sym}^2 U^*) c_9(\text{Sym}^4 U^* - \text{Adams}(2, c(\text{Sym}^2 U^*))) = 3,297,280$$

as expected.

As another application, we will use our formulas to compute limits for other types of degenerations. These will be more interesting as the degenerations will contain two components. Mainly, we will consider degenerations of a generic hypersurface X_d of degree d in $\mathbb{P}^r$ to the union of the m-fold X_k^m of a generic hypersurface X_k of degree k and the n-fold X_l^n of a generic hypersurface X_l of degree l with $mk + nl = d$. The main tool needed is the refined residual intersection decomposition as given in [W3]. It decomposes $\mathcal{R}_X = [F_{\text{lim}}(X)]$ into the sum of two classes $[F_{\text{lim}}(X_k^m)]$ and $[F_{\text{lim}}(X_l^n)]$. When either $n > 1$ or $m > 1$ (non-reduced cases), the formulas from this work are needed to compute $[F_{\text{lim}}(X_k^m)]$ or $[F_{\text{lim}}(X_l^n)]$. For details and notation, see [W3].

EXAMPLE 2.3. (Degenerations of quintic threefolds and their lines.)
There are seven different degenerations of a generic X_5 into $X_k^m \cup X_l^n$ and the five of them that contain non-reduced components are listed below.

Case 1: $X_5 \to X_1^4 \cup X_1$.

$$[F_{\text{lim}}(X_1^4)] = \mathbb{R}_{F_{X_1^4}} = \mathbb{M}_{F_{X_1^4}} + \mathbb{A}_{F_{X_1^4}} = 2400 + 320 = 2720;$$

$$[F_{\text{lim}}(X_1)] = \mathbb{R}_{F_{X_1}} = \mathbb{M}_{F_{X_1}} + \mathbb{A}_{F_{X_1}} = 1275 - 1120 = 155.$$

Case 2: $X_5 \to X_1^3 \cup X_2$.

$$[F_{\text{lim}}(X_1^3)] = \mathbb{R}_{F_{X_1^3}} = \mathbb{M}_{F_{X_1^3}} + \mathbb{A}_{F_{X_1^3}} = 3195 - 540 = 2655;$$

$$[F_{\text{lim}}(X_2)] = \mathbb{R}_{F_{X_2}} = \mathbb{M}_{F_{X_2}} + \mathbb{A}_{F_{X_2}} = 1300 - 1080 = 220.$$

Case 3: $X_5 \to X_1^3 \cup X_1^2$.

$$[F_{\text{lim}}(X_1^3)] = \mathbb{R}_{F_{X_1^3}} = \mathbb{M}_{F_{X_1^3}} + \mathbb{A}_{F_{X_1^3}} = 3195 - 1080 = 2115;$$

$$[F_{\text{lim}}(X_1^2)] = \mathbb{R}_{F_{X_1^2}} = \mathbb{M}_{F_{X_1^2}} + \mathbb{A}_{F_{X_1^2}} = 2920 - 2160 = 760.$$

Case 4: $X_5 \to X_1^2 \cup X_3$.

$$[F_{\text{lim}}(X_1^2)] = \mathbb{R}_{F_{X_1^2}} = \mathbb{M}_{F_{X_1^2}} + \mathbb{A}_{F_{X_1^2}} = 2920 - 540 = 2380;$$

$$[F_{\text{lim}}(X_3)] = \mathbb{R}_{F_{X_3}} = \mathbb{M}_{F_{X_3}} + \mathbb{A}_{F_{X_3}} = 1575 - 1080 = 495.$$

Case 5: $X_5 \to X_2^2 \cup X_1$.

$$[F_{\text{lim}}(X_2^2)] = \mathbb{R}_{F_{X_2^2}} = \mathbb{M}_{F_{X_2^2}} + \mathbb{A}_{F_{X_2^2}} = 2880 - 640 = 2240;$$

$$[F_{\text{lim}}(X_1)] = \mathbb{R}_{F_{X_1}} = \mathbb{M}_{F_{X_1}} + \mathbb{A}_{F_{X_1}} = 1275 - 640 = 635.$$

Notice that in each of the cases above we have that

$$[F_{\text{lim}}(X_k^m)] + [F_{\text{lim}}(X_l^n)] = 2875 = [F_{X_5}]$$

as expected.

EXAMPLE 2.4. (Degeneration of cubic hypersurface X_3 in $\mathbb{P}^8$ and their $\mathbb{P}^3$'s.) The number of $\mathbb{P}^3$'s in a generic X_3 in $\mathbb{P}^8$ is $321,489$. There are two degenerations of a generic X_3 into $X_k^m \cup X_l^n$.

Case 1: $X_3 \to X_1^2 \cup X_1$.

$$[F_{\lim}(X_1^2)] = \mathbb{R}_{F_{X_1^2}} = \mathbb{M}_{F_{X_1^2}} + \mathbb{A}_{F_{X_1^2}} = -50,173,344 + 50,481,792 = 308,448,$$

$$[F_{\lim}(X_1)] = \mathbb{R}_{F_{X_1}} = \mathbb{M}_{F_{X_1}} + \mathbb{A}_{F_{X_1}} = 410,385 - 397,344 = 13,041.$$

So we see that 308,448 of 321,489 $\mathbb{P}^3$'s went to X_1^2 and the rest went to X_1.

Case 2: $X_3 \to X_2 \cup X_1$,

$$[F_{\lim}(X_2)] = \mathbb{R}_{F_{X_2}} = \mathbb{M}_{F_{X_2}} + \mathbb{A}_{F_{X_2}} = 168,528 - 168,528 = 0;$$

$$[F_{\lim}(X_1)] = \mathbb{R}_{F_{X_1}} = \mathbb{M}_{F_{X_1}} + \mathbb{A}_{F_{X_1}} = 410,385 - 88,896 = 321,489.$$

The degeneration in Case 2 does not contain non-reduced components. We listed it here as it surprisingly shows that all limiting $\mathbb{P}^3$'s go into X_1. This implies that U^* is not a positive bundle although it is generated by its sections.

Many other examples have been computed but we chose to list the those above as they all have the maximal codimension so that we will get numbers instead of classes.

References

[F] W. Fulton, *Intersection theory*, Springer-Verlag, New York, 1984.

[GH] P. Griffiths and J. Harris, *Principles of algebraic geometry*, John Wiley and Sons, New York, 1978.

[KS] S. Katz and S. Strømme, *Schubert: a maple package for intersection theory.*

[W1] X. Wu, *Chern classes and degenerations of hypersurfaces and their lines*, Duke Math. Jour., **67**, No.3, (1992), 633-652.

[W2] X. Wu, *Refined intersection products and limiting linear subspaces of hypersurfaces*, Math. Ann., **299**, (1994), 415-433.

[W3] X. Wu, *Residual intersections and some applications*, Duke Math. Jour., **75**, No.3, (1994), 733-758.

[W4] X. Wu, *On limiting rational curves*, Manu. Math., **85**, (1994), 243-253.

DEPARTMENT OF MATHEMATICS, UNIVERSITY OF SOUTH CAROLINA, COLUMBIA, SC 29208,
E-mail address: `wux@math.sc.edu`

Contemporary Mathematics
Volume **322**, 2003

GEOMETRY OF CANONICAL COVERS OF VARIETIES OF MINIMAL DEGREE WITH APPLICATIONS TO CALABI-YAU THREEFOLDS

BANGERE P. PURNAPRAJNA

ABSTRACT. In this article we deal with surfaces and higher dimensional varieties of general type whose canonical linear series gives a morphism to a variety of minimal degree. We show that these varieties play an important role in various contexts. The results dealt in this article include generators of the canonical ring, degree and structure of the canonical morphism among other things. We also show that these results have applications to linear series on Calabi-Yau threefolds.

INTRODUCTION

In this article we will deal with results on varieties of general type with special emphasis on the case of an algebraic surface. These include results on the degrees of the generators of the canonical ring of a variety of general type, the study of the structure of the canonical morphism, determining its possible degrees and constructing new examples of surfaces of general type. We also give some evidence to a conjecture regarding projective normality of bi-canonical maps of a surface of general type. All of these have applications to Calabi-Yau threefolds.

For the nature of questions that this article deals with, surfaces of general type fall nicely into following two types:

I. Surfaces of general type X with base-point free canonical bundle.

One can further subdivide this into:

(a) Surfaces whose canonical morphism maps X onto a surface of minimal degree.

(b) Surfaces whose canonical morphism does not map X onto a surface of minimal degree.

We will call these surfaces Type I(a) and Type I(b) respectively.

II. Surfaces of general type whose canonical bundle is not base-point free.

By a surface of minimal degree we mean a non degenerate surface in projective space whose degree is equal to its codimension plus 1. The surfaces of minimal degree are classically known: they are (linear) $\mathbf{P}^2$, the Veronese surface in $\mathbf{P}^5$, smooth rational scrolls or cones over a rational normal curve (see [EH]).

This article deals with surfaces of Type I(a) and its higher dimensional analogues. These surfaces play an important role in a variety of contexts. The varieties of minimal degree have a ubiquitous presence in the study of geometry of varieties of general type, especially in the case of an algebraic surface. These surfaces play the role of a building block on many occasions. As a motivation to study the varieties of Type I(a), we mention a sample of circumstances where they come up:

(1) Analogy with curves: Let C be a curve of genus $g \geq 2$, then its canonical bundle K_C is base-point free and ample. If C is non-hyperelliptic, then the canonical morphism gives an embedding. If C is hyperelliptic, the canonical morphism maps C two to one onto a rational normal curve of degree $g-1$, which is a variety of minimal degree. The surfaces of Type I(a) can be thought of as two-dimensional (partial) analogues of hyperelliptic curves. Needless to say, the situation is far more complex in the case of an algebraic surface as higher degree covers are involved.

(2) The surfaces of Type I(a) play an important role in the classification of surfaces of general type with small K_X^2 in the pioneering work of Horikawa ([H1], [H2], [H3] and [H4]). He makes a detailed study of double covers and studies their degeneration and their moduli.

(3) Low degree canonical morphism has stimulated interest among geometers. Konno [Ko], Mendes-Lopez, and Pardini [MP] have studied degree three canonical maps. Surfaces of Type I(a) play an important role here. Konno studies surfaces of general type with $K_X^2 = 3p_g - 6$. If the image is a surface of minimal degree, Konno shows that $p_g \leq 5$. In [T], Tan has constructed some low degree covers of $\mathbf{P}^2$.

(4) To determine the possible degrees of the canonical morphism and to construct new examples of surfaces of general type has been of interest to many geometers. Various authors including Beauville [Be], Hirzebruch [Hi], Gang [Ga], Konno [Ko], Persson [P], Tan [T] among others have studied this topic. The Type I(a) surfaces occur frequently in this context.

(5) The study of surfaces of Type I(a) has a direct bearing on the study of linear series on some threefolds, most importantly Calabi-Yau threefolds as the results in [OP] and author's result with F. J. Gallego in [GP4] show. Results proved there included very ampleness results, which are analogues of Lefschetz theorem for Abelian varieties and St. Donat's theorem for K3 surfaces, for Calabi-Yau threefolds. Such an analogue is proven in [GP4]. They also appear in the work of Beltramatti and Szemberg ([BS]) in their study of k-jet ampleness on Calabi-Yau threefolds. Surfaces of Type I(a) played an important role in these results. Also, one of the important sources of examples of Calabi-Yau threefolds is to take covers of threefolds of minimal degree. The threefolds of minimal degree play an important role in all of these topics.

(6) To determine the degrees of the generators of the canonical ring of a variety of general type has been an object of intense study. In this context, the surfaces of Type I(a) play an important role. The results of Ciliberto[Ci], Catanese [Ca1], Green [G] and Reid [R] provide good illustration in this regard.

The above mentioned motivations lead us to some interesting questions. These questions have caught quite an amount of attention over the years. We will first concentrate on the two dimensional analogue of the curve case as mentioned in (1) above.

In the case of an algebraic curve, the degree d of the canonical morphism is bounded by 2. It is known by the results in [Be] that if the image Y of the canonical morphism of a surface of general type has $p_g(Y) = 0$, then the degree of the morphism is bounded. The surfaces of minimal degree form an important subclass of surfaces with $p_g = 0$.

Let X be a surface of general type with base-point free canonical bundle. Let $\pi : X \longrightarrow Y$ be the morphism induced by the canonical linear series K_X. Assume that Y is a surface of minimal degree. In other words X is of Type I(a). As the results in [Be] and [Ga] suggest, if $p_g(X)$ is large enough, the degree of the canonical morphism is bounded. Can one get more precise information about the degree of π? This leads us to:

Question 1: What integers can occur as degree of the canonical morphism, if the image of the canonical morphism is a surface of minimal degree?

To construct new examples of surfaces of general type has aroused considerable interest among geometers. Low degree canonical covers have sparked special interest ([Be], [H1], [H2], [H3], [H4], [T], [Ko], [MP]).

Question 2: Can one classify low degree covers for surfaces of Type I(a)? Can one construct new examples of low degree covers with arbitrarily large p_g and other invariants like c_1^2, q?

Another theme of interest is to study the structure of the canonical morphism.

The results of Konno [Ko] deals with triple covers for Type I(a), he shows that p_g is bounded by 5 in this case. This is not the case for covers of even degree as we show in [GP1]. The above question is interesting in the light of these results.

Another object of intense study has been the canonical ring of a surface Of general type ([Bo], [Ci], [Ca 1,2], [G], [R] among others). In the context of the present article, we are particularly interested in the result of Green and Ciliberto that bounds the degrees of the generators of the canonical ring when the surface of general type is of Type I(b). More precisely they prove (independently) the following beautiful theorem:

Let X be a regular surface of general type with base-point free canonical divisor. Assume that the canonical morphism π satisfies the following conditions:
(1) π does not map X generically $2:1$ onto the projective plane.
(2) $\pi(X)$ is not a surface of minimal degree (other than linear $\mathbf{P}^2$).
That is, X is not a surface of Type I(a).
Then the canonical ring of X is generated in degree less than or equal to 2.

Note that for the case of a hyperelliptic curve of genus $g \geq 2$, the generators of the canonical ring are classically known. The following question is natural for surfaces of Type I(a):

Note that result of Ciliberto and Green answers this question for surfaces of Type I(b).

For higher dimensional varieties of general type, Green in [G] showed that the degrees of the generators of the canonical ring are bounded by the dimension of the variety of general type X whenever its canonical morphism *does not* map X onto a variety of minimal degree. One can also ask for higher dimensional analogues of Question 3. We will first survey recent results in [GP1] and [GP2] that answer some of these questions. We will also put them in context with earlier results by other authors.

Acknowledgements: We thank Francisco Gallego for his invaluable help during the preparation of this article. We also thank N. Mohan Kumar for helpful discussions and suggestions. Thanks also to Pete Vermiere and A. Prabhakar Rao for helpful suggestions and Zhenbo Qin for his encouragement. The author was partially supported by the General Research Fund of the University of Kansas at Lawrence.

1. Preliminaries

In this section we will recall some known facts about the push forward of the structure sheaf of a variety by a flat, finite morphism. We summarize these facts below and refer for the proof to Section 2 in [HM].

Let X and Y be algebraic varieties. Let $\pi : X \longrightarrow Y$ be a finite, flat morphism of degree n. We have the following facts:

(1.1) The sheaf $\pi_*\mathcal{O}_X$ is a rank n, locally free sheaf on Y of algebras over $\mathcal{O}_Y$.

(1.2) There exists a map

$$\frac{1}{n}\mathrm{tr} : \pi_*\mathcal{O}_X \longrightarrow \mathcal{O}_Y$$

of sheaves of $\mathcal{O}_Y$-modules defined locally as follows: Given $\alpha \in \pi_*\mathcal{O}_X$ we consider the homomorphism of $\mathcal{O}_Y$-modules

$$\pi_*\mathcal{O}_X \xrightarrow{\cdot\alpha} \pi_*\mathcal{O}_X$$

induced by multiplication by α. Then we define $\frac{1}{n}\mathrm{tr}(\alpha)$ as the trace of such homomorphism divided by n.

(1.3) The map $\frac{1}{n}\mathrm{tr}$ is surjective, in fact, the map $\mathcal{O}_Y \hookrightarrow \pi_*\mathcal{O}_X$ induced by π is a section of $\frac{1}{n}\mathrm{tr}$. Therefore the sequence

$$0 \longrightarrow E \longrightarrow \pi_*\mathcal{O}_X \xrightarrow{\frac{1}{n}\mathrm{tr}} \mathcal{O}_Y \longrightarrow 0$$

splits. E is the kernel of $\frac{1}{n}\mathrm{tr}$ and locally consists of the trace 0 elements of $\pi_*\mathcal{O}_X$. We will call E the *trace-zero module* of π.

(1.4) $\pi_*(\mathcal{O}_X)$ is a sheaf of $\mathcal{O}_Y$-algebras, therefore it has a multiplicative structure. Its multiplication map is an $\mathcal{O}_Y$-linear map

$$[\mathcal{O}_Y \oplus E] \otimes [\mathcal{O}_Y \oplus E] \longrightarrow \mathcal{O}_Y \oplus E$$

made of four components. The first component

$$\mathcal{O}_Y \otimes \mathcal{O}_Y \longrightarrow \mathcal{O}_Y \oplus E$$

is given by the multiplication in $\mathcal{O}_Y$ and therefore goes to $\mathcal{O}_Y$. The components

$$\mathcal{O}_Y \otimes E \longrightarrow \mathcal{O}_Y \oplus E$$
$$E \otimes \mathcal{O}_Y \longrightarrow \mathcal{O}_Y \oplus E$$

are given by the left and right module structure of E over $\mathcal{O}_Y$ and therefore go to E. Finally there is a fourth component

$$E \otimes E \longrightarrow \mathcal{O}_Y \oplus E$$

which factors through

$$S^2 E \longrightarrow \mathcal{O}_Y \oplus E\ ,$$

for multiplication in $\pi_* \mathcal{O}_Y$ is commutative.

Conventions and Notations:

Unless otherwise stated all surfaces are assumed to be smooth.

Suppose X is a variety, L is a line bundle on X. The following notation and conventions are followed:

$H^1(X, L)$ will be simply denoted by $H^1(L)$ in a given context.

K_X = canonical bundle of X.

Regularity of $X = q(X) = dim H^1(\mathcal{O}_X)$.

$p_g(X)$ or $p_g = dim H^0(K_X)$

By a canonical cover X of a variety Y, we mean that Y is the image of X under the canonical morphism of X. Unless otherwise stated, all canonical morphisms are assumed induced by the complete linear series $|K_X|$.

Throughout this article, we work over an *algebraically closed field of characteristic zero.*

2. Ring generation for covers of surfaces of minimal degree

In this section we review some recent results of the author with F. J. Gallego regarding the generators of the canonical ring of surfaces of general type whose canonical morphism maps it to a surface of minimal degree (i.e., surfaces of Type I(a)). We determine the precise degrees of the generators of the canonical ring. These results complement the results of Ciliberto and Green on the generators of the canonical ring of surfaces of general type. We obtain the following:

Theorem 2.1 (cf. [GP1], Theorem 2.1). *Let S be a regular surface of general type with at worst canonical singularities and such that its canonical bundle K_S is base-point free. Let π be the canonical morphism of S. Let n be the degree of π*

2) if $n \neq 2$ or $r \neq 1$, the canonical ring of S is generated by its part of degree 1, $r(n-2)$ generators in degree 2 and $r-1$ generators in degree 3.

This answers a stronger version of Question 3 as we not only determine the degrees of the generators but also the number of generators in each degree bigger than one. So the generators of the canonical ring of surfaces of Type I(a) does depend on the degree of the morphism as well as on the degree of the variety of minimal degree. It also shows that the degrees of the generators are bounded by 3 and are required as the examples constructed in [GP1] show. The result also shows that the surfaces of Type I(a) behave quite differently from the surfaces of Type I(b) considered by Ciliberto and Green.

Combining the results of Ciliberto, Green and ours, one obtains the following characterization of surfaces of general type of Type I whose canonical ring is generated in degree less than or equal to two:

Corollary 2.2 (cf. [GP1], Corollary 2.8). *Let S be a smooth regular surface of general type and such that K_S is globally generated. Let π be the canonical morphism of S. The canonical ring of S is generated in degree less than or equal to 2 if and only if*

(1) *π does not map S generically $2:1$ onto $\mathbf{P}^2$; and*
(2) *$\pi(S)$ is not a surface of minimal degree other than linear $\mathbf{P}^2$.*

We now turn to higher dimensional analogues of the above results on surfaces. In [G] and [Ca1], Green and Catanese have proved results on the generators of the canonical ring of a *pluriregular* variety of general type. By a *pluriregular* variety we mean a variety for which all intermediate cohomology of the structure sheaf vanish. In [G], Green proved that the canonical ring of a pluriregular variety of general type of dimension n is generated in degree less than or equal to n if its canonical morphism *does not* map X onto a variety of minimal degree. Here is yet another instance where varieties of minimal degree come into the picture.

Before we state the results on higher dimensional varieties of general type, it might be helpful to give some motivation. To do this, we will recall the well known classical result regarding canonical ring of a hyperelliptic curve C of genus g; note that its canonical bundle maps C to a rational normal curve, a curve of minimal degree. In other words C is a curve of Type I(a); its canonical ring is generated in degree 1 and requires $g-2$ generators of degree 2 if $g \geq 3$. If $g=2$, the canonical ring is generated in degree 1 and requires one generator in degree 3. We have seen that for the case of an algebraic surface, the degrees of the generators of the canonical ring of a surface of Type I(a) are bounded. Note also that, Theorem 2.1 showed that the converse to the result of Ciliberto and Green holds.

If one looks at the pattern in the numbers involved in the degrees of the generators in dimension one and two, one might be tempted to guess that this pattern might hold for higher dimensions. In fact we found that this is false. In [GP2], we took up a detailed study, among other things, of the canonical ring of a variety of general type of arbitrary dimension which are triple canonical covers of a variety of minimal degree. We found that there is no converse to Green's result. What we mean here is a result saying that the canonical ring of a pluriregular variety of general type X is generated in degree less than or equal to n *if and only if* the canonical morphism does not map X onto a variety of minimal degree, is false in dimension greater than 2. The reason the converse fails is because for these higher dimensional varieties of general type, the canonical ring is generated in degree much lower than the Green's bound, in fact in roughly degree half of Green's bound.

We will now state the results that illustrate what we have been saying above:

Theorem 2.3 (cf. [GP2], Theorem 1). *Let X be a smooth, pluriregular variety of general type of dimension $m \geq 2$ and let $Y \subset \mathbf{P}^{r+m-1}$ be an m-fold of minimal degree r. Assume that K_X is base-point free and let $\pi : X \longrightarrow Y$ be the canonical morphism of X. Assume that π is a generically finite morphism of degree 3. Then*

1) *if m is odd, the canonical ring of X is generated by its part of degree 1 and $\frac{r}{2}$ linearly independent generators in degree $\frac{m+1}{2}$;*
2) *if m is even, the canonical ring of X is generated by its elements of degree 1, r linearly independent generators in degree $\frac{m+2}{2}$ and $r-1$ linearly independent generators in degree $m+1$.*

The questions on very ampleness and projective normality of pluricanonical bundles have a long history ([K], [Bo], [Ca1], [Ci]. See also [GP5] for recent and more general results on this topic and higher syzygies). The above result, apart from its intrinsic interest, has applications in this direction. The degrees of the generators of the canonical ring are related to the degrees of the generators of the pluricanonical rings and therefore, with the very ampleness and projective normality associated to pluricanonical morphisms of the variety X. As a corollary of Theorem 2.3 above we give a necessary and sufficient condition for multiples of the canonical bundle on X to give projectively normal embeddings. We now state the precise result for odd dimensional variety of general type X:

Theorem 2.4 (cf. [GP2], Theorem 2.9). *Let X be a smooth, pluriregular variety of general type of odd dimension $m \geq 3$ and let $Y \subset \mathbf{P}^{r+m-1}$ be an m-fold of minimal degree r. Assume that K_X is ample and base-point free and let $\pi : X \longrightarrow Y$ be the canonical morphism of X. Assume that π has degree 3. Then the line bundle $K_X^{\otimes n}$ satisfies property N_0 if and only if $n \geq \frac{m+1}{2}$.*

In the even dimensional case using Theorem 2.3 we obtain a sufficient condition for a multiple of K_X to induce a projectively normal embedding.

Theorem 2.5 (cf. [GP2], Theorem 2.13). *Let X be a smooth, pluriregular variety of general type of even dimension m with ample and base-point free canonical bundle. Assume that the canonical morphism has degree 3 and maps X onto a variety Y of minimal degree r.*

1) *If $r = 1$, then the line bundle $K_X^{\otimes n}$ satisfies property N_0 if and only if $n \geq \frac{m+2}{2}$.*
2) *If $r > 1$, then the line bundle $K_X^{\otimes n}$ fails to satisfy property N_0 if $n \leq \frac{m}{2}$ and satisfies property N_0 if $n \geq m+1$.*

The following question regarding projective normality in the even dimensional case remains open *variety Y of minimal degree $r > 1$. Does $K_X^{\otimes n}$ satisfy property N_0 if $\frac{m+2}{2} \leq n \leq m$?*

The proofs of all these theorems involves, among other things, reducing the problem to curves lying on varieties of general type and dealing with certain multiplication maps on curves. The reason why the above question cannot be addressed by the methods developed in [GP2] is because some multiplication maps of global sections on curves were not surjective. We refer the reader to Theorem 2.11 and Theorem 2.13 in [GP2] for details.

3. Techniques and Method of Proof

We now illustrate the techniques involved in the proofs of theorems on ring generation. We will describe in some detail the proof for a quadruple cover. The proof given in this article though different from that in [GP1,2] does capture some of the spirit of the proofs given there. Also the examples dealt here serve as a positive test case for a conjecture on projective normality of bi-canonical maps on surfaces. The philosophy is the following: a variety possessing a finite morphism is endowed with an extra structure which provides more information about the geometry of the variety and the structure of its coordinate rings.

Let $\varphi : X \to Y$ be a flat, finite morphism. We have $\varphi_* \mathcal{O}_X = \mathcal{O}_Y \oplus E$, where E is trace-zero module as noted in the Section 1. On some occasions the structure of $\varphi_* \mathcal{O}_X$ as $\mathcal{O}_Y$ module might be relatively simple. Note that E might globally split completely into line bundles as when φ is a cyclic cover or Y is $\mathbf{P}^1$. In these cases one might be able to give a very detailed description of the multiplicative structure of $\varphi_* \mathcal{O}_X$ as $\mathcal{O}_Y$ algebra, i.e., one might be able to describe locally how elements of direct summands of $\varphi_* \mathcal{O}_X$ multiply among each other.

Let φ denote the canonical morphism. The relation of the algebra structure of $\varphi_* \mathcal{O}_X$ with the ring structure of $R = \bigoplus_{l=0}^{\infty} H^0(K_X^{\otimes l})$ is the following: $\bigoplus_{l=0}^{\infty} H^0(K_X^{\otimes l})$ is a graded ring and to know how its multiplication works we need to know how elements of its graded pieces multiply among each other. The splitting of $\varphi_* \mathcal{O}_X$ as module over $\mathcal{O}_Y$ induces a decomposition of the graded pieces of the ring as direct sum of some vector spaces. The multiplication between elements of those vector spaces is controlled by the multiplication in $\varphi_* \mathcal{O}_X$ in some situations. In general all of these are hard to determine. In [GP1] we are able to get enough information to make things work. We first reduce the problem of finding the ring generators of the canonical ring on a surface X to a similar question on curve $C \in |K_X|$ lying on the surface of general type X.

In the context at hand, $\varphi : X \longrightarrow Y$ represents the canonical morphism from X, a surface of general type, to a surface of minimal degree Y. Note that this may be only generically finite. Since Y is the image of X under the canonical morphism, $C \in |K_X|$ comes equipped with a morphism π to $\mathbf{P}^1$ lying on Y. This morphism will turn out to be finite and flat. We use this morphism to find the ring generation on the curve C by pushing down to $\mathbf{P}^1$ the relevant multiplication maps of global sections of vector bundles and then lift this information back to X. Things get technical. But we will illustrate Theorem 2.1 in an example, where the vector bundle E already splits on the surface of minimal degree Y. So in our demonstration example, there is no need to go to curves unlike in the proof of the general case in [GP1]. But even in this example, there are some subtleties that will give us some flavor of the proof for the general case.

We also prove here that the following question on linear series has a positive answer for these examples.

Question 3.1:. *Let S be a regular surface of general type with base point free canonical bundle K_S. Then $K_S^{\otimes n}$ for $n \geq 2$ embeds S as a projectively normal variety if and only if there is a smooth non-hyperelliptic curve C in the linear system $|K_S|$.*

To illustrate these ideas we look at the following Example 3.4 in [GP1], the proof given here works for other examples in [GP1] as well: Let $\mathbf{Y} = \mathbf{P}^1 \times \mathbf{P}^1$. Let f be a fiber of one of the fibrations of $\mathbf{P}^1$ and let f' be a fiber of the other fibration. Let Y be embedded in $\mathbf{P}^3$ by $|f + f'|$. Let $a_1 = 1$,$a_2 = 2$,$b_1 = 2$ and $b_2 = 1$ (*). For $i = 1, 2$, let D_i be a smooth divisor linearly equivalent to $2(a_i f + b_i f')$ such that D_1 and D_2 intersect at $D_1 \cdot D_2$ distinct points. This can be arranged since $2(a_1 f + b_1 f')$ and $2(a_2 f + b_2 f')$ are very ample. Let $X' \xrightarrow{\varphi_1} \mathbf{Y}$ be the double cover of $\mathbf{Y}$ ramified along D_1. Let D_2' be the inverse image in X' of D_2 by φ_1. Since D_2 is smooth and meets D_1 at distinct points, D_2' is also smooth. Let $X \xrightarrow{\varphi_2} X'$ be the double cover of X' ramified along D_2'. Since X' and D_2' are both smooth, so is X. Let us call $\varphi = \varphi_1 \circ \varphi_2$. It can be checked that X is a regular surface of general type, that $K_X = \varphi^* \mathcal{O}_{\mathbf{Y}}(1)$ and that φ is induced by the complete canonical series of X.

Since the values of a_1, a_2, b_1 and b_2 are as in (*) above, one can also prove that

$$\varphi_* \mathcal{O}_X = \mathcal{O}_{\mathbf{Y}} \oplus \mathcal{O}_{\mathbf{Y}}(-f - 2f') \oplus \mathcal{O}_{\mathbf{Y}}(-2f - f') \oplus \mathcal{O}_{\mathbf{Y}}(-3f - 3f') \ .$$

Interested reader can look up the details of all these in Example 3.4 of [GP1].

Let us denote $L_1 = \mathcal{O}_{\mathbf{Y}}(-f-2f')$, $L_2 = \mathcal{O}_{\mathbf{Y}}(-2f-f')$ and $L_3 = \mathcal{O}_{\mathbf{Y}}(-3f-3f')$. Call $E = L_1 \oplus L_2 \oplus L_3$. Note that the cover X above is a special case of a bidouble cover and by the very nature of it, we have:

$$(**) L_i \otimes L_i \text{ goes into } \mathcal{O}_{\mathbf{Y}}$$
$$\text{and the map}$$
$$L_1 \otimes L_2 \longrightarrow L_3$$
$$\text{is surjective, in fact, it is an isomorphism.}$$

We also have,

$$K_X = \varphi^* \mathcal{O}_{\mathbf{Y}}(1) = \varphi^* \mathcal{O}_{\mathbf{Y}}(f + f') \ .$$

Since the morphism φ is finite $H^0(K_X^{\otimes l}) = H^0(\varphi_*(K_X^{\otimes l}))$. Let $R_l = H^0(K_X^{\otimes l})$. So by projection formula we have the following:

We will demonstrate Theorem 2.1 in this special case. Let $\beta(s,t)$ be the multiplication map

$$H^0(K_X^{\otimes s}) \otimes H^0(K_X^{\otimes t}) \longrightarrow H^0(K_X^{\otimes s+t}) \text{ for all } s,t > 0 \ .$$

Note that codimension of $H^0(K_X) \otimes H^0(K_X) \longrightarrow H^0(K_X^{\otimes 2})$ is the same as the codimension of $R_1 \otimes R_1 \longrightarrow R_2$. But $R_2 = H^0(\mathcal{O}_{\mathbf{Y}}(2f+2f')) \oplus H^0(\mathcal{O}_{\mathbf{Y}}(f)) \oplus H^0(\mathcal{O}_{\mathbf{Y}}(f'))$. Therefore an element of R_l, i.e., a global section of $H^0(K_X^{\otimes l})$ is a sum of 4 components, one in each piece of the above decomposition of R_l. On the other hand, the product of an element of R_l belonging to one of the blocks with an element of R_1 is determined by the ring structure of $\mathcal{O}_{\mathbf{Y}}$ and by the module structure of E. More precisely, the restriction of $\beta(l,1)$ to $H^0(\mathcal{O}_{\mathbf{Y}}(l)) \otimes H^0(\mathcal{O}_{\mathbf{Y}}(1))$ maps, in fact isomorphically, onto $H^0(\mathcal{O}_{\mathbf{Y}}(l+1))$. The restriction of $\beta(l,1)$ to each of the blocks $H^0(L_i \otimes \mathcal{O}_{\mathbf{Y}}(l)) \otimes H^0(\mathcal{O}_{\mathbf{Y}}(1))$ maps to the corresponding $H^0(L_i \otimes \mathcal{O}_{\mathbf{Y}}(l+1))$. This restriction is 0 if $L_i \otimes \mathcal{O}_{\mathbf{Y}}(l)$ has no sections and an isomorphism otherwise.

In the light of the explanation in the preceding paragraph and the fact that

$$R_1 = H^0(K_X) = H^0(\mathcal{O}_{\mathbf{Y}}(f+f')) \ .$$

$$R_2 = H^0(K_X^{\otimes 2}) = H^0(\mathcal{O}_{\mathbf{Y}}(2f+2f')) \oplus H^0(\mathcal{O}_{\mathbf{Y}}(f)) \oplus H^0(\mathcal{O}_{\mathbf{Y}}(f')) \ .$$

we deduce that the codimension of $\beta(1,1)$, which is nothing but the codimension of $R_1 \otimes R_1 \longrightarrow R_2$ is 4, codimension of $\beta(2,1)$, i.e., codimension of $R_2 \otimes R_1 \longrightarrow R_3$ is 1. For $l \geq 3$, it can be checked that $\beta(3,1)$ surjects.

Since in the case of this example $r = 2$, $n = 4$; we have that $\bigoplus_{l=0}^{\infty} H^0(K_X^{\otimes l})$ is generated in degree 1, needs $r(n-2) = 4$ generators in degree 2 and $r - 1 = 1$ generator in degree 3. Thus this example gives a good illustration of Theorem 2.1.

Now we will show that $K_X^{\otimes n}$ embeds X as a projectively normal variety for $n \geq 2$. To prove this it is enough to show that $\beta(2,2)$ surjects. That is, codimension of $R_2 \otimes R_2 \longrightarrow R_4$ is zero. Note that $R_4 = H^0(\mathcal{O}_{\mathbf{Y}}(4f+4f')) \oplus H^0(\mathcal{O}_{\mathbf{Y}}(3f+2f')) \oplus H^0(\mathcal{O}_{\mathbf{Y}}(2f+3f')) \oplus H^0(\mathcal{O}_{\mathbf{Y}}(f+f'))$. The fact that first three summands are captured by $\beta(2,2)$ is obvious. The only thing to check is that the last summand of R_4 is captured by $\beta(2,2)$. But this follows because of observation (**) made in the previous page. Hence $K_X^{\otimes l}$ satisfies N_0 for all $n \geq 2$. Looking back on the proof one realizes that, it is the $\mathcal{O}_Y$ algebra structure on $\varphi_* \mathcal{O}_X$ that is playing the crucial role in the multiplication maps above.

In general it is more difficult as $f_* \mathcal{O}_X$ does not split, so we do not know the algebra structure so well. In fact, we deal with a large class of surfaces at once, so many different algebra structures are possible. In addition to this there can be singular targets, non-generically finite morphims among other difficulties. The methods developed in [GP1] applies to all these settings. For this we need to go to curves. The lifting of information from curves to surfaces needs work and we refer the reader to the proofs of Theorem 2.1, Proposition 2.2 in [GP1] and Lemma 2.2, Lemma 2.3 and Proposition 2.4 in [GP2] for details.

4. On the degree and structure of the canonical morphism

In this section we deal with results on the structure of the canonical Morphism $X \xrightarrow{\pi} Y$. This study also tells about the structure of X and Y. These results answer Question 1 and part of Question 2 mentioned in the introduction. Other than of intrinsic value, answers to these questions are helpful in finding new examples. In a recent work [GP3], F.J. Gallego and I have classified quadruple Galois covers thus answering part of Question 2. To determine the bound on the degree of the canonical morphism of a *minimal* surface of general type has been done in ([Be] and [Ga]). We give much more precise answers when the image of the canonical morphism is a variety of minimal degree. As a consequence of the results that we will mention in this section, it follows that many a priori possible examples of surfaces of general type cannot exist. We will show in the next section, these results have applications for Calabi-Yau threefolds.

We will briefly review the two results that fit into the present context regarding the bound on the degree of the canonical morphism. In [Be], Beauville proved the following: Let $\pi : X \longrightarrow Y$ be the canonical map from a minimal surface of general type X to a surface Y with $p_g(Y) = 0$, then the degree of the canonical map $d \leq 9$ if $\chi(\mathcal{O}_X) \geq 31$. In addition if Y is not ruled, then $d \leq 4$. In [Ga], Gang showed that if $p_g(X) \geq 132$, then $d \leq 8$. The proofs of these results crucially hinge on Miyaoka's famous inequality [Mi] (See also Yau [Y].)

In [GP1], we determine which positive integers occur as degree of the canonical morphism π when the image under it is a variety of minimal degree Y. This solves partly Question 1 raised in the introduction.

One would hope that a natural way to construct examples of canonical covers would be to take cyclic covers of surfaces of minimal degree. Unfortunately this is not possible. In [GP1], we proved the following theorem which rules out the existence of cyclic canonical covers.

Theorem 4.1 (cf. [GP1], Proposition 3.1). *Let X be a surface of general type with at worst canonical singularities and with base-point free canonical bundle. Assume that the complete canonical series of X restricts to a complete linear series on a general hyperplane section (e.g., if X is regular). Let $\pi : X \longrightarrow Y$ be the canonical morphism to a surface of minimal degree. Let n be the degree of π. Let U be a smooth open set of Y whose complement has codimension 2 and let L be a line bundle on U. Assume that*

$$(\pi_* \mathcal{O}_X)|_U = \mathcal{O}_U \oplus L^{-1} \oplus \cdots \oplus L^{\otimes 1-n} .$$

Then $n = 2$ or 3.

An immediate corollary of the above result is the following:

Corollary 4.2 (cf. [GP1], Corollary 3.2). *Let X be a regular surface of general type with at worst canonical singularities and with base-point free canonical bundle. Let Y be the image of X by its canonical morphism $X \xrightarrow{\pi} Y$. If Y is a surface of minimal degree and π is a cyclic cover, then the degree of π is 2 or 3.*

a more general theorem for varieties of general type of arbitrary dimension from which the preceding statement can be deduced as an easy corollary. The result below is more general and holds for targets that are not necessarily rational scrolls but all scrolls. We give a proof for the case of an algebraic surface to show the flavor for higher dimensions.

Theorem 4.3 (cf. [GP2], Theorem 3.2). *Let X be a smooth variety of general type of dimension m with base-point free canonical bundle X. Let W be a linear subseries of $|K_X|$ without base points and let $X \xrightarrow{\pi} Y$ be the morphism induced by W. Assume that π is generically finite of degree n and that Y is a scroll. Then either n is even or m is odd. In particular, if $n = 3$, m is odd.*

Proof (for the case of an algebraic surface): Let X be an algebraic surface of general type. Let f be a fiber of Y and let C be a section of Y. Let $-d = C^2$. Since Y is a scroll, its hyperplane section is linearly equivalent to $C+mf$, for some integer m. Then $K_X = \pi^*(C + mf)$. Then deg $\pi = (\pi^* f) \cdot (\pi^* C) = (\pi^* f) \cdot (K_X + \pi^* f)$, which is an even number by adjunction. □

The following theorem says, among other things that a variety of minimal degree Y with degree an odd integer, cannot occur as the image of a canonical morphism $\pi : X \longrightarrow Y$ of degree 3 if the dimension of X is odd.

Theorem 4.4 (cf. [GP2], Theorem 3.1). *Let X be a smooth variety of general type of dimension $m \geq 2$ and let $Y \subset \mathbf{P}^{r+m-1}$ be an m-fold of minimal degree. Assume that K_X is base-point-free and let W be a sublinear series of $|K_X|$ without base points. Let $\pi : X \longrightarrow Y$ be the morphism induced by W. Assume that π is generically finite of degree 3. Then*
1) If m is odd, then the degree r of Y is even (in particular Y is not linear $\mathbf{P}^m$)
2) If $h^1(\mathcal{O}_X) = 0$, then π is induced by the complete canonical series of X.

We will now state a structure theorem concerning triple canonical covers of varieties of minimal degree assuming that the canonical morphism is flat. It turns out that in this case the variety of general type is always *quasi-cyclic* (i.e., it behaves like a cyclic cover but may not necessarily come from a group action) and pluriregular. The theorem below lists also all possible targets and as examples in Proposition 3.4 of [GP2] show all these possibilities exist.

Theorem 4.5 (cf. [GP2], Theorem 3.3). *Let X be a smooth variety of general type of dimension $m \geq 2$. Assume that K_X is ample and base-point free. Let W be a linear subseries of $|K_X|$ without base points. Let π be the morphism induced by W. Assume that π is flat and has degree 3. Let $Y = \pi(X)$ be a variety of minimal degree r. Then Y is one of the following:*

1) $\mathbf{P}^m$ with $m \geq 2$ even.
2) A smooth rational normal scroll of odd dimension $m \geq 3$ and even degree r. In this case X is fibered over $\mathbf{P}^1$, with general fiber a smooth variety of general type of dimension $m-1$. The restriction of π to the general fiber is finite and flat, induced by the complete canonical series, and maps onto $\mathbf{P}^{m-1}$, fitting therefore in case 1) above.
3) A smooth quadric hypersurface of odd dimension $m \geq 3$.

In addition, π is in fact induced by the complete canonical linear series of X; $\pi_*\mathcal{O}_X = \mathcal{O}_Y \oplus L^{-1} \oplus L^{-2}$ for a line bundle L such that L^{-2} is not effective; $K_Y = L^{-2}(1)$ and $H^1(\mathcal{O}_X) = \cdots = H^{m+1}(\mathcal{O}_X) = 0$.

The idea of the proof is as follows: since we are working over an algebraically closed field of characteristic zero, $\pi_*\mathcal{O}_X = \mathcal{O}_Y \oplus E$. We first show that the vector bundle E splits. To prove the necessary vanishing to show that X is indeed pluriregular needs some work. The crucial idea is to show that $H^1(K_X(-G)) = 0$, where G is the inverse image of a general fiber F of Y. The vanishing, after some work follows from pushing down via π the bundle $K_X(-G)$ to Y. For details we refer the reader to Theorem 3.3 in [GP2].

Before we turn our attention to new examples of regular surface of general type constructed in [GP1], we review some of the previously known results which are related to the present situation.

As noted before, the results of [Be] bound the degree of the canonical morphism. To construct new examples of surfaces of general type with prescribed numerical invariants like p_g, q, K_X^2 is quite important. Among other things, the examples give a better understanding of the geography of surfaces of general type. Low degree covers have especially attracted attention and one of the chief sources of these examples has been to take covers of surfaces of minimal degree. These appear in the works of various authors. The cases when π is generically finite morphism and has degree 2 and 3 have been completely studied by Horikawa and Konno (see [H1, Theorem 1.6] [H2, Theorem 2.3.I], [H3, Theorem 4.1] and [Ko, Lemma 2.2 and Theorem 2.3]; see also Mendes Lopes and Pardini, [MP]). As it turns out there exist generically double covers of linear $\mathbf{P}^2$, the Veronese surface, smooth rational normal scrolls $S(a,b)$ with $b \leq 4$ and cones over rational normal curves of degree 2, 3 and 4 and generically triple covers of $\mathbf{P}^2$ (in particular cyclic triple covers of $\mathbf{P}^2$ ramified along a sextic with suitable singularities) and of the cones over rational normal curves of degree 2 and 3. Horikawa (see [H4, Theorem 2.1]) also describes all generically finite quadruple covers $X \xrightarrow{\pi} Y$, where X is smooth, minimal regular surface, π is the canonical morphism of X and Y is linear $\mathbf{P}^2$.

The examples of Horikawa and Konno just reviewed are examples of covers of degree d less than or equal to 3 of surfaces of minimal degree and quadruple covers of $\mathbf{P}^2$. In [T], Tan has constructed a regular surface of general type X whose canonical morphism is a Galois covering over $\mathbf{P}^2$ with group $Z_3 \oplus Z_3$ and $p_g(X) = 3$. In [Be], Beauville has constructed surfaces with arbitrarily large $\chi(\mathcal{O}_X)$ for each $d \leq 8$. However these surfaces have positive irregularity. In [P], Persson has given an example with $d = 16$.

In [GP1] we construct three new sets of examples of regular surfaces of general type which are quadruple covers of surfaces of minimal degree under the canonical morphism. These examples are 4:1 covers of smooth rational normal scrolls isomorphic to the Hirzebruch surfaces $\mathbf{F}_0$ and $\mathbf{F}_1$ and of quadric cones in $\mathbf{P}^3$. In fact, there is more, for each integer $m \geq 4$, we construct 4:1 cover X such that $p_g(X) = m$. This is in sharp contrast to the results in [Ko] on triple canonical

We will describe briefly one of the examples that we have constructed in [GP1]. For details we refer the reader to Example 3.4, Example 3.5 and Example 3.7 in [GP1].

Example 4.6 (cf. [GP1], Example 3.4).
We construct finite quadruple covers $X \xrightarrow{\varphi} Y$, where X is a smooth minimal regular surface of general type, φ is the canonical morphism of X and Y is a smooth rational scroll $S(m,m)$, $m \geq 1$.

In the article [GP1] there are two sets of examples that are contained under the title of Example 3.4. We will briefly describe one of them below. Let f be a fiber of one of the fibrations of $\mathbf{P}^1$ and let f' be a fiber of the other fibration. Then Y is $\mathbf{P}^1 \times \mathbf{P}^1$ and it is embedded in $\mathbf{P}^{2m+1}$ by $|f + mf'|$. The way one constructs the example is to first take a double cover of F_0 ramified along smooth D_1 linearly equivalent to $2(a_1 f + b_1 f')$. Call this cover X' and the map φ_1. Now take a double cover of X' ramified along the inverse image of the divisor D_2, where D_2 is linearly equivalent to $2(a_2 f + b_2 f')$. Call this cover X and the map φ_2. Here $a_1 = 1$, $a_2 = 2$, $b_1 = m + 1$ and $b_2 = 1$, with $m \geq 1$. Let $\varphi = \varphi_1 \circ \varphi_2$. Since D_i for $i = 1, 2$ are very ample, it can be arranged that $D_1 \cdot D_2$ meet in distinct points. So the 4:1 cover $X \xrightarrow{\varphi} Y$, is smooth. Ofcourse it remains to be shown that $K_X = \varphi^* \mathcal{O}_Y(1)$ and that X is regular with arbitrarily large p_g. For all these details we refer the reader to [GP1]. Note that some of these details were mentioned for the particular case $m = 1$ after Question 3.1.

5. Applications to Calabi-Yau threefolds

The results on ring generation and structure of the canonical morphism in the previous sections have ramifications for Calabi-Yau threefolds. A Calabi-Yau threefold X is a three dimensional projective variety with trivial canonical bundle and $H^1(\mathcal{O}_X) = 0$. This can be viewed as a three dimensional analogue of a K3 surface. In [GP4] we had proved results on linear series on polarized Calabi-Yau threefolds (X, B) whenever the linear series $|B|$ did not map X onto a threefold of minimal degree. The important case when $|B|$ did map X onto a variety of minimal degree was taken up in [GP1]. The techniques of handling these two cases are very different.

We start with the Calabi-Yau threefold analog of Theorem 2.1:

Theorem 5.1 (cf. [GP1], Theorem 4.1). *Let X be a Calabi-Yau threefold with at worst canonical singularities and let B be a big and base-point free line bundle on X. Let π be the morphism induced by $|B|$. Let n be the degree of π and assume that the image of π is a variety of minimal degree r. Then*
1) if $n = 2$ and $r = 1$ (i.e., if π is generically $2:1$ onto $\mathbf{P}^3$), the canonical ring of X is generated by its part of degree 1 and one generator in degree 4;
2) if $n \neq 2$ or $r \neq 1$, the canonical ring of X is generated by its part of degree 1, $r(n-2)$ generators in degree 2 and $r - 1$ generators in degree 3.

This result complements the results obtained by the author and F. J. Gallego in [GP4]. We also get a characterization of those polarized Calabi-Yau threefolds (X, B) such that $\bigoplus_{l=0}^{\infty} H^0(B^{\otimes l})$ is generated in degree less than or equal to 2. Combining Theorem 5.1 above with the result Theorem 1.7 in [GP4], we obtain:

Corollary 5.2 (cf. [GP2], Corollary 4.2). *Let X be a Calabi-Yau threefold with at worst canonical singularities. Let B be a big and base-point free line bundle on X and let π be the morphism induced by $|B|$. Then $\bigoplus_{l=0}^{\infty} H^0(B^{\otimes l})$ is generated in degree less than or equal to 2 if and only if*

(1) *π does not map X generically $2:1$ onto $\mathbf{P}^3$; and*
(2) *$\pi(X)$ is not a threefold of minimal degree other than $\mathbf{P}^3$.*

In [GP4], we made the following conjecture regarding linear series on a Calabi-Yau threefold X:

Conjecture 5.3 (cf. [GP4], Conjecture 1.9). *Let X be a Calabi-Yau threefold and let B be an ample and base-point free line bundle. Then $B^{\otimes 2}$ satisfies property N_0 if and only if there is a smooth non-hyperelliptic curve C in $|B \otimes \mathcal{O}_S|$, for some $S \in |B|$.*

The conjecture would also give a characterization of when $B^{\otimes 2}$ is very ample. This question of when $B^{\otimes 2}$ is very ample is also open.

The conjecture is known to be true if the variety of minimal degree happens to be $\mathbf{P}^3$ (see Theorem 1.7 in [GP4] or [GP6].) The very ampleness and projective normality of $B^{\otimes n}$ for all $n \geq 2$ was proved in [GP4] whenever the linear system $|B|$ didn't map X onto a variety of minimal degree. The way it was done was to prove first that the $\bigoplus_{l=0}^{\infty} H^0(B^{\otimes l})$ is generated in degree less than or equal to two. Then we proceeded by showing that the map

$$H^0(B^{\otimes l}) \otimes H^0(B) \longrightarrow H^0(B^{\otimes l+1})$$

surjects for all $l \geq 3$. Theorem 5.1 says that the Conjecture 5.3 is not amenable to such tricks and that it needs subtler methods!

Constructing new examples of Calabi-Yau threefolds have evoked enthusiasm among geometers in recent years. One of the important sources for these examples is to take covers of varieties of minimal degree, most simple among them being $\mathbf{P}^3$. A natural question to ask is: does there exist Calabi-Yau cyclic covers of $\mathbf{P}^3$ or in general of threefolds of minimal degree given by a complete linear series?

The results on the structure of the canonical morphism of a surface of general type in the preceding section have applications for polarized Calabi-Yau threefolds. As we will see results in Section 4 prevent many a priori natural examples of Calabi-Yau threefolds from existing. It also shows that examples of prime degree Calabi-Yau covers of threefolds of minimal degree induced by complete linear series cannot come from group actions.

We will now state a corollary of Theorem 4.1 which rules out the existence of higher degree cyclic covers.

Corollary 5.4 (cf. [GP1], Corollary 4.4). *Let X be a Calabi-Yau threefold with at worst canonical singularities. Let B be a base-point free line bundle. Let $X \xrightarrow{\pi} Y$ be the morphism induced by the complete linear series $|B|$. Let n be the degree of π. Let U be a smooth open set of Y whose complement has codimension*

If Y is a variety of minimal degree, then the degree $n = 2$ or $n = 3$. In particular, if π is a cyclic cover, the degree of π is 2 or 3.

Note that there may exist covers that may not come from a cyclic group action and still behave like cyclic covers. We will call such covers *quasi-cyclic*, that is the morphism π satisfying (*) of the above result.

We now state a corollary of Theorem 4.3 which excludes odd degree Calabi-Yau covers of smooth scrolls.

Corollary 5.5 (cf. [GP1], Corollary 4.6). *Let X be a Calabi-Yau threefold with at worst canonical singularities. If $X \xrightarrow{\pi} Y$ is a generically finite morphism onto a smooth scroll $Y \subset \mathbf{P}^N$ given by the complete linear series, then the degree of π is even. In particular, there are no generically finite covers of odd degree of smooth rational normal scrolls.*

In [GP4], we had proved a structure theorem for a polarized Calabi-Yau threefold (X, B) where the complete linear series $|B|$ maps X onto a threefold of minimal degree. We had classified all such polarized Calabi-Yau's. This theorem can be sharpened by Corollary 5.4 and Corollary 5.5 to the form presented below:

Theorem 5.6 (cf. [GP1], Proposition 4.7). *Let X be a smooth Calabi-Yau threefold, let π be the morphism induced by the complete linear series of an ample and base-point free line bundle B on X and let n be the degree of π. If $\pi(X)$ is a variety of minimal degree then one of the following occurs:*

1) *$Y = \mathbf{P}^3$ and $n \leq 24$.*
2) *Y is a smooth rational hypersurface in $\mathbf{P}^4$ and $n = 2, 4, 6, 8, 10, 12$ or 14.*
3) *Y is a smooth rational normal scroll of dimension 3 in $\mathbf{P}^5$ and $n = 2, 4, 6, 8, 10$ or 12.*
4) *Y is a smooth rational normal scroll of dimension 3 in $\mathbf{P}^N$, $N \geq 6$, $n = 2$ and X is fibered over $\mathbf{P}^1$ with a smooth K3 surface as a general fiber. The restriction of B to the general fiber of X is hyperelliptic, with sectional genus 2, and its complete linear series maps the fiber $2 : 1$ onto a general fiber of the scroll.*
5) *Y is a smooth rational normal scroll of dimension 3 in $\mathbf{P}^N$, $N \geq 6$, $n = 6$ and X is fibered over $\mathbf{P}^1$ with a smooth Abelian surface as a general fiber. The restriction of B to the general fiber of X is a $(1, 3)$ polarization, and its complete linear series maps the fiber $6 : 1$ onto a general fiber of the scroll.*
6) *Y is a cone over a conic in $\mathbf{P}^2$,*
7) *Y is a cone over a twisted cubic in $\mathbf{P}^3$,*
8) *Y is a cone over a Veronese surface.*

In addition if X has at worst canonical singularities and π is a quasi-cyclic cover, then $n = 2$ or 3.

Sketch of the proof. We want to first show that the inequality

$$n \leq \frac{6r + 6}{r - 2} \qquad (*)$$

holds if Y is a variety of minimal degree. This is proved using Riemann-Roch and [Mi], since $c_2(X) \cdot B \geq 0$, $r + 1 = h^0(B) \geq \frac{1}{6}B^3 = \frac{1}{6}n(r - 2)$, so we obtain $(*)$.

Now we describe all the possible types of varieties of minimal degree that may occur. The variety Y should be either $\mathbf{P}^3$, a smooth quadric hypersurface in

$\mathbf{P}^4$, a singular 3-dimensional rational normal scroll in $\mathbf{P}^4$, a (possibly singular) 3-dimensional rational normal scroll in $\mathbf{P}^r$, $r \geq 5$ or a cone in $\mathbf{P}^6$ over a Veronese surface in $\mathbf{P}^5$. We first rule out the possibility that Y is a 3-dimensional rational normal scroll singular along a single point. This would imply that X could be obtained by performing small contractions on some suitable space and X would not be smooth. For a detailed proof, we refer the reader to Proposition 1.6 in [GP4]. If Y is a cone over a rational normal curve then $r = 4$ or 5. This follows, after a little work, from (*) above. We deal now with Y a smooth rational normal scroll and $r \geq 6$. In such a case Y is fibered over $\mathbf{P}^1$ and so is X. Let φ the projection from Y to $\mathbf{P}^1$. Let us denote by F a general fiber of φ, and let G be a general fiber of $\varphi \circ \pi$. The idea is to show that $B(-G)$ is big, which follows after some work. So by Kawamata-Viehweg, $H^1(B(-G)) = 0$, which implies that $|B \otimes \mathcal{O}_G|$ maps G onto $\mathbf{P}^2$. Since B is ample, G is connected, and by Bertini, smooth. Then, by adjunction, G is either a smooth K3 surface or a smooth Abelian surface. If G is a smooth K3 surface, then $(G, B \otimes \mathcal{O}_G)$ is a genus 2, hyperelliptic polarized K3 surface, $\pi|G$ is $2:1$ and so is π. If G is a smooth Abelian surface, then $B \otimes \mathcal{O}_G$ is a $(1,3)$-polarization, hence $\pi|G$ has degree 6 and so has π.

Finally, the integers which can appear as degrees of π for items 1), 2) and 3) follows from (*), Corollary 5.4 and Corollary 5.5 mentioned in the preceding pages. □

References

[Be] A. Beauville, *L'application canonique pour les surfaces de type general*, Inventiones Math. **55** (1979), 121-140.

[Bo] E. Bombieri, *Canonical models of surfaces of general type*, Inst. Hautes. Et. Sci. Publ. Math. **42** (1973), 171–219.

[BS] M.C. Beltrametti and T. Szemberg, *On higher order embeddings of Calabi-Yau threefolds*, Arch. Math. (Basel). **74** (2000), 221-225.

[Ca1] F. Catanese, *Equations of pluriregular varieties of general type*, Geometry today (Rome, 1984), 47-67, Progr. Math., **60**, Birkhauser Boston, 1985.

[Ca2] F. Catanese, *Commutative algebra methods and equations of regular surfaces*, Algebraic Geometry, (Bucharest, 1982), 68-111, Lecture Notes in Math., **1056**, Springer Berlin, 1984.

[Ci] C. Ciliberto, *Sul grado dei generatori dell'anello di una superficie di tipo generale*, Rend. Sem. Mat. Univ. Politec. Torino **41** (1983)

[EH] D. Eisenbud and J. Harris, *On varieties of minimal degree (A centennial account)*, Algebraic Geometry, Bowdoin 1985, Amer. Math. Soc. Symp. in Pure and App. Math. **46** (1987), 1–14.

[G] M.L. Green, *The canonical ring of a variety of general type*, Duke Math. J. **49** (1982), 1087–1113.

[GP1] F.J. Gallego and B.P. Purnaprajna, *On the canonical ring of covers of surfaces of minimal degree*, To appear in Trans.Amer.Math.Soc.

[GP2] F.J. Gallego and B.P. Purnaprajna, *Triple canonical covers of varieties of minimal degree*, Preprint math.AG/0205010.

[GP3] F.J. Gallego and B.P. Purnaprajna, *Classification of quadruple canonical Galois covers*, in preparation.

[GP4] F.J. Gallego and B.P. Purnaprajna, *Very ampleness and higher syzygies for Calabi-Yau threefolds*, Math. Ann. **312** (1998), 133–149.

[GP5] F.J. Gallego and B.P. Purnaprajna, *Projective normality and syzygies of algebraic surfaces*, J. Reine Angew. Math. **506** (1999), 145–180.

[GP6] F.J. Gallego and B.P. Purnaprajna, *Some homogeneoues rings associated to finite morphisms*, To appear in Advances in Algebra and Geometry, Proceedings of the International conference in Hyderabad, India, Hindustan book Agency, New Delhi (2002).

[H1] E. Horikawa, *Algebraic surfaces of general type with small c_1^2. I*, Ann. of Math. (2) **104** (1976), 357–387.
[H2] E. Horikawa, *Algebraic surfaces of general type with small c_1^2, II*, Invent. Math. **37** (1976), 121–155.
[H3] E. Horikawa, *Algebraic surfaces of general type with small c_1^2, III*, Invent. Math. **47** (1978), 209–248.
[H4] E. Horikawa, *Algebraic surfaces of general type with small c_1^2, IV*, Invent. Math. **50** (1978/79), 103–128.
[HM] D. Hahn and R. Miranda, *Quadruple covers of algebraic varieties*, J. Algebraic Geom. **8** (1999), 1–30.
[K] K. Kodaira, *Pluricanonical systems on algebraic surfaces of general type*, J. Math. Soc. Japan **20** (1968), 170–192.
[Ko] K. Konno, *Algebraic surfaces of general type with $c_1^2 = 3p_g - 6$*, Math. Ann. **290** (1991), 77–107.
[Mi] Y. Miyaoka, *On the Chern numbers of surfaces of general type*, Invent. Math. **42** (1977), 255-237.
[MP] M. Mendes Lopes and R. Pardini, *Triple canonical surfaces of minimal degree*, International J. Math. **11** (2000), 553–578.
[OP] K. Oguiso and T. Peternell, *On polarized canonical Calabi-Yau threefolds*, Math. Ann. **301** (1995), 237–248.
[P] U. Persson, *Double coverings and surfaces of general type*, Algebraic Geometry. (Lect. Notes Math., vol 687. 168-195) **687** (1978).
[R] M. Reid, *Infinitesimal view of extending a hyperplane section- deformation theory and computer algebra*, Algebraic geometry, Proceedings L'Aquila 1988, 214–286, Lecture Notes in Math. **1417**, Springer-Verlag Berlin, 1990.
[T] S. Tan, *Surfaces whose canonical maps are of odd degrees*, Math. Annalen, **292** (1992), 13-29.
[Y] S. T. Yau, *On the Ricci-curvature of a complex Kahler manifold and the complex Monge Ampere equation*, Comment.Pure.Appl.Math. **31** (1978), 339-411.

B. P. Purnaprajna
Department of Mathematics
University of Kansas
Lawrence, KS 66045
U.S.A.
Email: purna@math.ukans.edu

Contemporary Mathematics
Volume **322**, 2003

Universal rings arising in geometry and group theory

Weiqiang Wang

ABSTRACT. Various algebraic structures in geometry and group theory have appeared to be governed by certain universal rings. Examples include: the cohomology rings of Hilbert schemes of points on projective surfaces and quasi-projective surfaces; the Chen-Ruan orbifold cohomology rings of the symmetric products; the class algebras of wreath products, as well as their associated graded algebras with respect to a suitable filtration. We review these examples, and further provide a new elementary construction and explanation in the case of symmetric products. We in addition show that the Jucys-Murphy elements can be used to clarify the Macdonald's isomorphism between the FH-ring for the symmetric groups and the ring of symmetric functions.

1. Introduction

1.1. The Hilbert scheme $X^{[n]}$ of n points on a complex (quasi-)projective surface X has provided a natural setup for an interplay among geometry, representation theory and combinatorics. The starting point is the construction of a Heisenberg algebra in terms of incidence varieties which acts irreducibly on the direct sum over all n of the cohomology groups $H^*(X^{[n]})$ with $\mathbb{C}$-coefficients [**38**] (cf. [**19**] for another construction and [**15, 45**] for motivations). In a nutshell, the Heisenberg algebra reveals the geometric structures of the Hilbert schemes in a way similar to the cycle structures for the symmetric groups. This has recently led to a new approach toward the cohomology rings of Hilbert schemes.

Being closely related yet complementary to each other, the results on these cohomology rings can be roughly divided as follows: the connections with vertex operators and $\mathcal{W}$ algebras [**27, 30, 33**]; the ring generators (and relations) for the rings $H^*(X^{[n]})$ [**30, 31, 34**] (also cf. [**10, 3, 36**] for a classical approach for some special surfaces); the constructions of the universal rings which govern the rings $H^*(X^{[n]})$ depending on whether or not X is compact [**32, 34**]; the intertwining relations with the study of Chen-Ruan orbifold cohomology rings [**6**] of the symmetric products [**28, 29, 40, 42, 44, 24, 17**] as well as the class algebras of the wreath products [**47, 14, 11, 26, 49, 50**]. We remark that there has been also remarkable connections of Hilbert schemes with combinatorics [**20**] and interesting work on the motives of Hilbert schemes [**7, 16**].

Much of the advances in these directions has been made after an earlier review [**48**] was written. The main purpose of the present paper is to review some of these recent progress complementary to *loc. cit.*, with an emphasis on the appearance of certain universal rings which govern various algebraic structures. Also it is interesting to compare with [**40**] where the axiomatic nature of the vertex operator approach developed for Hilbert schemes was formulated explicitly and similar framework was established for the Chen-Ruan orbifold cohomology rings of the symmetric products. The Sections 4 and 5.5 contain new constructions and results.

1.2. The Heisenberg algebra provides a distinguished linear basis for $H^*(X^{[n]})$ parameterized by multi-partitions (this basis will be referred to as the Heisenberg monomial basis). It is natural to ask how the structure constants of the cup product among these basis elements (up to some suitable scalings) behave. It turns out that the answer depends essentially on whether X is projective or quasi-projective[1].

When X is projective, a stability result formulated in [**32**] states that the modified structure constants with respect to a suitably chosen set of elements in $H^*(X^{[n]})$ (which include the Heisenberg monomial basis elements) can be uniquely taken to be independent of n. This has led to a universal ring (termed as the Hilbert ring associated to X) which governs the cohomology rings $H^*(X^{[n]})$ for all n. These are explained in Sect. 2. The Hilbert ring actually depends not on the projective surface X, but on the ring $H^*(X)$ and the canonical class of X. While all these statements are described in the classical language, the machinery used to establish them replies heavily on the connections with vertex operators.

The stability for the cohomology rings of $H^*X^{[n]})$ for X quasi-projective requires a quite different formulation. It is shown [**34**] that the structure constants with respect to the Heisenberg monomial basis of $H^*(X^{[n]})$ are independent of n, for a large class of quasi-projective surfaces (conjecturally [**50**] for all quasi-projective surfaces). This has led to a universal ring (called the FH-ring in [**34**]) which governs the cohomology rings $H^*(X^{[n]})$ for all n. This stability result can in turn be interpreted as giving rise to a surjective ring homomorphism from $H^*(X^{[n]})$ to $H^*(X^{[n-1]})$ for each n. This phenomenon has been rather unusual, since there is no natural algebraic embedding of $X^{[n-1]}$ into $X^{[n]}$ for $m < n$. In fact, an incidence variety relating $X^{[n]}$ and $X^{[n-1]}$ induces such a ring homomorphism. This will be explained in Sect. 3.

1.3. Recall that the Hilbert-Chow morphism $\pi_n : X^{[n]} \to X^n/S_n$ from the Hilbert schemes to the symmetric products is a crepant resolution of singularities, where S_n is the n-th symmetric group. A general principle [**9**] states that the geometry of an orbifold is "equivalent" to the geometry of its crepant resolutions. Originally, the word "geometry" here was expressed in terms of (orbifold) Betti numbers. Over the years, the "equivalence" extends to (orbifold) Hodge numbers, (orbifold) elliptic genera, motivic intergration, and (orbifold) cohomology rings etc, (see, for example, [**2, 4, 5, 6, 8, 23, 41, 45, 51, 52**] and the references therein). In particular, the Chen-Ruan cohomology ring $H^*_{\rm CR}(X^n/S_n)$ of the symmetric products has attracted much attention, as this is related to the cohomology rings of Hilbert schemes and vertex operators [**28, 29, 34, 40, 42, 44**]. This provides nice examples of testing and verifying Ruan's conjectures [**42, 43**]. We remark

[1]The use of the terminology "quasi-projective" excludes "projective" in the present paper.

that there has been some further reformulation and development of the orbifold cohomology rings [**12, 1**].

It developed that the structures of the rings $H^*_{\mathrm{CR}}(X^n/S_n)$ for a compact complex manifold X are very similar to the cohomology rings of Hilbert schemes of points on projective surfaces. A stable ring introduced in [**40**], which are analogous to the Hilbert ring mentioned above for Hilbert schemes, governs the Chen-Ruan cohomology rings $H^*_{\mathrm{CR}}(X^n/S_n)$. The proof therein relies on an axiomatic vertex operator approach which are parallel to the developments in Hilbert schemes. An additional bonus of such an approach is to provide a more conceptual proof of Ruan's conjecture on the ring isomorphism between $H^*(X^{[n]})$ and $H^*_{\mathrm{CR}}(X^n/S_n)$ when X is projective and has a numerically trivial canonical class (which was also established in [**12, 44**] using [**29**]).

However, the orbifold side is simpler than the crepant resolution side, and often group-theoretic techniques are applicable. A nice example is provided by the Jucys-Murphy elements of symmetric groups [**22, 37**] which feature significantly in the study of the rings $H^*_{\mathrm{CR}}(X^n/S_n)$ [**40**]. In Sect. 4, we shall provide a new and direct construction (without using vertex operators) which explains the stability for the rings $H^*_{\mathrm{CR}}(X^n/S_n)$. The idea here is to enlarge further the noncommutative orbifold cohomology rings [**12**] (also cf. [**29**]) of the symmetric products.

1.4. The Hilbert schemes are also resolutions of singularities of wreath product orbifolds (which are generalized symmetric products) [**47**], where the wreath product $\Gamma_n := \Gamma^n \rtimes S_n$ is the semidirect product of the product group Γ^n and S_n. The study of the class algebras $R(\Gamma_n)$ of the wreath products Γ_n (which can be regarded as the Chen-Ruan cohomology ring of the symmetric product of the orbifold pt/Γ) reveals a stability similar to the Hilbert schemes for *projective* surfaces [**49**]. We can further introduce a filtration on $R(\Gamma_n)$ and study its associated graded algebras $\mathcal{G}_\Gamma(n)$ [**50**] (also compare [**11, 13, 28, 46**]). It is shown [**50**] that the structure constants of these graded algebras $\mathcal{G}_\Gamma(n)$ with respect to the conjugacy classes (which are the analog of the Heisenberg monomial basis) are independent of n. This motivated and corresponds to the stability explained above of the cohomology rings of Hilbert schemes when the surface is *quasi*-projective. This allows one to introduce a universal ring $\mathcal{G}_\Gamma$ (called the FH-ring) which governs the algebras $\mathcal{G}_\Gamma(n)$ for all n. Both the Chen-Ruan cohomology ring of the symmetric product and the class algebra of wreath product are nontrivial generalizations in two directions of the class algebra of the symmetric group, yet they can be further unified under the common roof of the wreath product orbifolds.

Macdonald [**35**] has earlier constructed an isomorphism between the FH-ring $\mathcal{G}$ [**13**] of the symmetric groups (i.e. when $\Gamma = 1$) and the ring of symmetric functions. (Actually, the universal ring in each of the different setups above has been shown to be a polynomial ring which is isomorphic to a tensor product of several copies of the ring of symmetric functions.) In addition, he describes explicitly the symmetric functions corresponding to the *stable* conjugacy classes in $\mathcal{G}$. In Sect. 5 we will show how the Jucys-Murphy elements can be used to provide further explicit correspondences under Macdonald's isomorphism.

A better way of reading the present paper is to read first Sect. 4 by setting X

Acknowledgments. Special thanks go to Wei-Ping Li and Zhenbo Qin for very fruitful collaborations. I also benefit from stimulating discussions and remarks from many people, including M. de Cataldo, I. Dolgachev, L. Göttsche, M. Haiman, M. Lehn, E. Markman, H. Nakajima, Y. Ruan, and J.-Y. Thibon.

2. The Hilbert ring of Hilbert schemes for projective surfaces

2.1. Generalities on Hilbert schemes. Let X be a (quasi-)projective complex surface, and $X^{[n]}$ be the Hilbert scheme of n points on X. An element in the Hilbert scheme $X^{[n]}$ is represented by a length n 0-dimensional closed subscheme of X. According to Fogarty, $X^{[n]}$ is smooth. We denote by $H^*(X^{[n]})$ the cohomology group/ring of $X^{[n]}$ with complex coefficient. We denote

$$\mathbb{H}_X = \bigoplus_{n=0}^{\infty} H^*(X^{[n]}).$$

The element 1 in $H^0(X^{[0]}) = \mathbb{C}$ is called the *vacuum vector* and denoted by $|0\rangle$. A non-degenerate super-symmetric bilinear form $(,)$ on $\mathbb{H}_X$ is induced from the standard one on $H^*(X^{[n]})$ defined by $(\alpha, \beta) = \int_{X^{[n]}} \alpha\beta$ for $\alpha, \beta \in H^*(X^{[n]})$. For $\mathfrak{f} \in \mathrm{End}(\mathbb{H}_X)$ we denote by $\mathfrak{f}^\dagger \in \mathrm{End}(\mathbb{H}_X)$ its *adjoint* operator.

The Betti numbers of Hilbert schemes for an arbitrary (quasi-)projective surface was computed in [**15**]. As a corollary, one obtains the graded dimension of $\mathbb{H}_X$ to be

$$\dim_q \mathbb{H}_X := \sum_{n=0}^{\infty} \dim H^*(X^{[n]}) q^n = \prod_{r=1}^{\infty} \frac{(1+q^r)^{h^{odd}(X)}}{(1-q^r)^{h^{ev}(X)}}$$

where $h^{odd}(X)$ and $h^{ev}(X)$ denote the dimensions of the cohomology group of X of odd and respectively even degrees. Based on this formula, it is suggested [**45**] that the space $\mathbb{H}_X$ can be identified as the Fock space of a Heisenberg algebra associated to the lattice $H^*(X, \mathbb{Z})/tor$. This has been established firmly in terms of incidence varities [**38**]. Similar results were obtained in [**19**].

We recall the construction [**38, 39**] when X is projective. For $n > 0$ and $\ell \geq 0$, we define $Q^{[n+\ell,n]} \subset X^{[n+\ell]} \times X \times X^{[n]}$ to be the closed subset

$$\{(\xi, x, \eta) \in X^{[n+\ell]} \times X \times X^{[n]} \mid \xi \supset \eta \text{ and } \mathrm{Supp}(I_\eta/I_\xi) = \{x\}\}.$$

The linear operator $\mathfrak{a}_{-n}(\alpha) \in \mathrm{End}(\mathbb{H}_X)$ (called the creation operator) with $\alpha \in H^*(X)$ is defined by

$$\mathfrak{a}_{-n}(\alpha)(a) = pr_{1*}([Q^{[m+n,m]}] \cdot \rho^*\alpha \cdot pr_2^* a).$$

for all $a \in H^*(X^{[m]})$, where pr_1, ρ, pr_2 are the projections of $X^{[m+n]} \times X \times X^{[m]}$ to $X^{[m+n]}, X, X^{[m]}$ respectively. Here and below we use $\gamma \cdot \beta$ or $\gamma\beta$ to denote the cup product of two cohomology classes γ, β, and omit the Poincaré duality used to switch between a homology class and a cohomology class. The annihilation operator $\mathfrak{a}_n(\alpha)$ can be defined to be $\mathfrak{a}_n(\alpha) = (-1)^n \cdot \mathfrak{a}_{-n}(\alpha)^\dagger$. Also set $\mathfrak{a}_0(\alpha) = 0$. Nakajima's theorem [**39**] says that

$$[\mathfrak{a}_n(\alpha), \mathfrak{a}_m(\beta)] = -n \cdot \delta_{n,-m} \cdot \int_X (\alpha\beta) \cdot \mathrm{Id}.$$

Moreover, the space $\mathbb{H}_X$ is an irreducible representation of the Heisenberg algebra.

REMARK 2.1. The above construction remain to be valid when X is quasi-projective after some suitable modification, cf. [**38**]. That is, we have the creation operators $\mathfrak{a}_{-n}(\alpha)$ $(n > 0)$ associated to a cohomology class α of X as above. On the other hand, the annihilation operators $\mathfrak{a}_n(\gamma)$ $(n > 0)$ are associated to a cohomology class γ with compact support of X.

2.2. The Hilbert ring. Given a $\mathbb{Z}_2$-graded finite set $S = S_0 \sqcup S_1$, we denote by $\mathcal{P}(S)$ the set of partition-valued functions $\rho = (\rho(c))_{c\in S}$ on S such that for every $c \in S_1$, the partition $\rho(c)$ is required to be *strict* in the sense that $\rho(c) = (1^{m_1(c)}2^{m_2(c)}\ldots)$ with $m_r(c) = 0$ or 1 for all $r \geq 1$.

Now let us take a linear basis $S = S_0 \cup S_1$ of $H^*(X)$ such that $1_X \in S_0$, $S_0 \subset H^{\text{even}}(X)$ and $S_1 \subset H^{\text{odd}}(X)$. If we write a partition-valued function $\rho = (\rho(c))_{c\in S}$, where $\rho(c) = (r^{m_r(c)})_{r\geq 1} = (1^{m_1(c)}2^{m_2(c)}\ldots)$, then we define

$$\begin{aligned}
\ell(\rho) &= \sum_{c\in S} \ell(\rho(c)) = \sum_{c\in S, r\geq 1} m_r(c) \\
\|\rho\| &= \sum_{c\in S} |\rho(c)| = \sum_{c\in S, r\geq 1} r\cdot m_r(c) \\
\mathcal{P}_n(S) &= \{\rho \in \mathcal{P}(S) \mid \|\rho\| = n\}.
\end{aligned}$$

Given $\rho = (\rho(c))_{c\in S} = (r^{m_r(c)})_{c\in S, r\geq 1} \in \mathcal{P}(S)$ and $n \geq 0$, we define

$$\mathfrak{a}_{-\rho(c)}(c) = \prod_{r\geq 1} \mathfrak{a}_{-r}(c)^{m_r(c)}$$

$$\mathfrak{a}_\rho(n) = \mathbf{1}_{-(n-\|\rho\|)} \prod_{c\in S} \mathfrak{a}_{-\rho(c)}(c) \cdot |0\rangle \in H^*(X^{[n]}) \tag{2.1}$$

where we fix the order of the elements $c \in S_1$ appearing in $\prod_{c\in S}$ once and for all, and $\mathbf{1}_{-k}$ $(k \geq 1)$ denotes $\mathfrak{a}_{-1}(1_X)^k/k!$. By definition, $\mathfrak{a}_\rho(n) = 0$ for $n < \|\rho\|$.

THEOREM 2.2. [**32**] *Let X be a projective surface, and S be a linear basis of $H^*(X)$ which contains 1_X. Let $\rho, \sigma \in \mathcal{P}(S)$. Then the cup product $\mathfrak{a}_\rho(n)\cdot\mathfrak{a}_\sigma(n)$ in $H^*(X^{[n]})$ can be expressed uniquely as a linear combination*

$$\mathfrak{a}_\rho(n)\cdot\mathfrak{a}_\sigma(n) = \sum_{\nu\in\mathcal{P}(S)} A^\nu_{\rho\sigma}\mathfrak{a}_\nu(n) \tag{2.2}$$

where $\|\nu\| \leq \|\rho\| + \|\sigma\|$ and the structure coefficients $A^\nu_{\rho\sigma}$ are independent of n.

Note that the cohomology classes $\mathfrak{a}_\nu(n)$ with $\|\nu\| \leq n$ span $H^*(X^{[n]})$ but are not linearly independent for $n > 1$.

REMARK 2.3. Although it is not manifest in the formulation of the statement, the proof of Theorem 2.2 used the connections between the cup products in $H^*(X^{[n]})$ and vertex operators developed in [**27, 30**] in an essential way.

DEFINITION 2.4. The *Hilbert ring* associated to a projective surface X, denoted

Note that the Hilbert ring does not depend on the choice of a linear basis S of $H^*(X)$ containing 1_X since the operator $\mathfrak{a}_n(\alpha)$ depends on the cohomology class $\alpha \in H^*(X)$ linearly. The Hilbert ring $\mathfrak{H}_X$ captures all the information of the cohomology rings of $X^{[n]}$ for all n, as we easily recover the relations (2.2) from the ring $\mathfrak{H}_X$. These observations can be summarized into the following.

THEOREM 2.5. [**32**] *For a given projective surface X, the cohomology rings $H^*(X^{[n]})$, $n \geq 1$ give rise to a Hilbert ring $\mathfrak{H}_X$ which completely encodes the cohomology ring structure of $H^*(X^{[n]})$ for each n.*

We shall see in Sect. 4 that analogous results hold in the symmetric product setup which afford simple proofs and explanations.

3. The FH-ring of Hilbert schemes for quasi-projective surfaces

3.1. n-independence of the structure constants. Let X be a smooth quasi-projective surface. As before, let us take a linear basis $S = S_0 \cup S_1$ of $H^*(X)$ such that $1_X \in S_0$, $S_0 \subset H^{\text{even}}(X)$ and $S_1 \subset H^{\text{odd}}(X)$.

Fix $n \geq 1$. For a given $\rho \in \mathcal{P}(S)$, we set $\mathfrak{b}_\rho(n) = 0 \in H^*(X^{[n]})$ if $n < \|\rho\| + \ell(\rho(1_X))$. If $n \geq \|\rho\| + \ell(\rho(1_X))$, we define $\tilde{\rho} \in \mathcal{P}(S)$ by putting $m_r(\tilde{\rho}(c)) = m_r(\rho(c))$ for $c \in S - \{1_X\}$, $m_1(\tilde{\rho}(1_X)) = n - \|\rho\| - \ell(\rho(1_X))$, and $m_r(\tilde{\rho}(1_X)) = m_{r-1}(\rho(1_X))$ for $r \geq 2$. Note that $\|\tilde{\rho}\| = n$. We define $\mathfrak{b}_\rho(n) \in H^*(X^{[n]})$ by

$$\mathfrak{b}_\rho(n) = \frac{\mathbf{1}_{-(n-\|\rho\|-\ell(\rho(1_X)))}}{\prod_{r\geq 2}(r^{m_r(\tilde{\rho}(1_X))} m_r(\tilde{\rho}(1_X))!)} \left(\prod_{\substack{c\in S, r\geq 1 \\ c\neq 1_X \text{ or } r>1}} \mathfrak{a}_{-r}(c)^{m_r(\tilde{\rho}(c))} \right) |0\rangle.$$

where we fix the order of the elements $c \in S_1$ appearing in the product $\prod_{c\in S}$ once and for all. We remark that the only part in the definition of $\mathfrak{b}_\rho(n)$ involving n is the factor $\mathbf{1}_{-(n-\|\rho\|-\ell(\rho(1_X)))}$.

As a corollary to the construction of Heisenberg algebra [**39**], $H^*(X^{[n]})$ has a linear basis consisting of the classes $\mathfrak{b}_\rho(n)$ where $\rho \in \mathcal{P}(S)$ and $\|\rho\| + \ell(\rho(1_X)) \leq n$. Fix a positive integer n and $\rho, \sigma \in \mathcal{P}(S)$ satisfying $\|\rho\| + \ell(\rho(1_X)) \leq n$ and $\|\sigma\| + \ell(\sigma(1_X)) \leq n$. Then we can write the cup product $\mathfrak{b}_\rho(n) \cdot \mathfrak{b}_\sigma(n)$ as

$$\mathfrak{b}_\rho(n) \cdot \mathfrak{b}_\sigma(n) = \sum_{\nu\in\mathcal{P}(S)} B^\nu_{\rho\sigma}(n)\ \mathfrak{b}_\nu(n) \tag{3.1}$$

with structure constants $B^\nu_{\rho\sigma}(n)$, where $\|\nu\| + \ell(\nu(1_X)) \leq n$.

We shall need the following technical definition.

DEFINITION 3.1. A smooth quasi-projective surface X is said to satisfy the *S-property* if it can be embedded in a smooth projective surface $\overline{X}$ such that the induced ring homomorphism $H^*(\overline{X}) \to H^*(X)$ is surjective.

Let X be a smooth quasi-projective surface with the S-property. This class of quasi-projective surfaces is very large, including the minimal resolution of a simple singularity, the cotangent bundle of a smooth projective curve, and the surface obtained from a smooth projective surface by deleting a point, etc.

THEOREM 3.2. [**34**] *Let X be a smooth quasi-projective surface satisfying the S-property. Then all the structure constants $B^\nu_{\rho\sigma}(n)$ given in (3.1) are independent of n.*

REMARK 3.3. The statement in Theorem 3.2 was conjectured [**50**] to be valid for every quasi-projective surface. The proof of Theorem 3.2 given in [**34**] used in an essential way relations with Hilbert schemes of points on *projective* surfaces [**32, 33**]. Some new approach will be needed toward the general conjecture.

REMARK 3.4. The definitions of the elements $\mathfrak{a}_\rho(n)$ in Sect. 2 and $\mathfrak{b}_\rho(n)$ above (with different parametrization and scaling) are made in order to formulate two different types of stability depending on whether or not the surface X is projective. For example, Theorem 3.2 is invalid if X is projective.

3.2. The FH-ring of Hilbert schemes.

Recall from Remark 2.1 that the annihilation operators $\mathfrak{a}_n(\gamma)$ $(n > 0)$ are associated to a cohomology class γ with compact support of X, when X is quasi-projective. Set $[x]_c \in H_c^4(X)$ to be the Poincaré dual of the homology class in $H_0(X)$ represented by a point in X. Then $\mathfrak{A} = -\mathfrak{a}_1([x]_c)$ is induced from an incidence variety in $X^{[n-1]} \times X^{[n]}$. The following theorem is an equivalent formulation of Theorem 3.2.

THEOREM 3.5. [**34**] *Let X be a smooth quasi-projective surface which satisfies the S-property. Then the linear map $\mathfrak{A} : H^*(X^{[n]}) \to H^*(X^{[n-1]})$ is a surjective ring homomorphism. More explicitly, it sends $\mathfrak{b}_\rho(n)$ to $\mathfrak{b}_\rho(n-1)$ for all ρ.*

REMARK 3.6. In the case when X is the affine plane, the above theorem was also contained in [**28**]. The work of Farahat-Higman is very relevant here [**13, 50**] (also cf. Sect. 5 below).

DEFINITION 3.7. Let X be a smooth quasi-projective surface satisfying the S-property. We define the *FH-ring* $\mathcal{G}_X$ associated to X to be the ring with a linear basis given by the symbols $\mathfrak{b}_\rho$, $\rho \in \mathcal{P}(S)$, with the product given by

$$\mathfrak{b}_\rho \cdot \mathfrak{b}_\sigma = \sum_{\nu \in \mathcal{P}(S)} B_{\rho\sigma}^{\nu}\, \mathfrak{b}_\nu.$$

where the structure constants $B_{\rho\sigma}^{\nu}$ come from (3.1).

The above results have led to the following.

THEOREM 3.8. [**34**] *For a quasi-projective surface X which satisfies the S-property, the cohomology rings $H^*(X^{[n]})$, $n \geq 1$ give rise to the FH-ring $\mathcal{G}_X$ which completely encodes the cohomology ring structure of $H^*(X^{[n]})$ for each n.*

4. The stable ring of the symmetric products X^n/S_n for X compact

4.1. Generalities on Chen-Ruan cohomology rings. Let M be a complex manifold of complex dimension d with a finite group G action. Introduce

$$M \diamond G = \{(g, x) \in G \times M \mid gx = x\} = \bigsqcup_{g \in G} M^g,$$

with a G-action given by $h.(g, x) = (hgh^{-1}, hx)$. As a vector space, we define $H^*(M, G)$ to be the cohomology group of $M \diamond G$ with $\mathbb{C}$-coefficient, or equivalently,

The space $H^*(M,G)$ has a natural induced G action, which is denoted by $\mathrm{ad}\, h : H^*(M^g) \to H^*(M^{hgh^{-1}})$. As a vector space, the orbifold cohomology group $H^*_{\mathrm{CR}}(M/G)$ is the G-invariant part of $H^*(M,G)$, which is isomorphic to

$$\bigoplus_{[g]\in G_*} H^*(M^g/Z(g))$$

where G_* denotes the set of conjugacy classes of G and $Z(g)$ denotes the centralizer of g in G.

For $g \in G$ and $x \in M^g$, write the eigenvalues of the action of g on the complex tangent space TM_x to be $\mu_k = e^{2\pi i r_k}$, where $0 \le r_k < 1$. The *degree shift number* or *age* is the rational number $F^g_x = \sum_{k=1}^d r_k$, cf. [**52**]. It depends only on the connected component Z which contains x, so we can denote it by F^g_Z. Then associated to a cohomology class in $H^r(Z)$, we assign the corresponding element in $H^*(M,G)$ (and thus in $H^*_{\mathrm{CR}}(M/G)$) a degree of $r + 2F^g_Z$.

A graded ring structure on $H^*_{\mathrm{CR}}(M/G)$ was introduced by Chen and Ruan [**6**]. This was subsequently reformulated in [**12**] by introducing a ring structure on $H^*(M,G)$ first and then passing to $H^*_{\mathrm{CR}}(M/G)$ by restriction. We shall use $\circ$ to denote this product. The ring has the following property: $\alpha \circ \beta$ lies in $H^*(M^{gh})$ for $\alpha \in H^*(M^g)$ and $\beta \in H^*(M^h)$.

4.2. Enlarging the ring $H^*(X^n, S_n)$. Let X be a compact complex manifold of dimension d. Our main objects here are the Chen-Ruan cohomology ring $H^*_{\mathrm{CR}}(X^n/S_n)$. Let us fix a linear basis $S = S_0 \sqcup S_1$ of $H^*(X)$ such that $S_0 \subset H^{ev}(X)$, $1_X \in S_0$, and $S_1 \subset H^{odd}(X)$.

If U is a finite set, we denote by S_U the symmetric group of permutations on U. We denote by $\underline{n}$ the set $\{1, 2, \ldots, n\}$, so $S_{\underline{n}}$ is just our usual symmetric group S_n. For $U \subset \underline{n}$, we regard S_U as a subgroup of S_n. We define $X^U = \{f : U \to X\}$ with a natural action by S_U. In particular, $X^{\underline{n}} = X^n$.

Following [**21**], we introduce the semigroup of "partial permutations" PS_n as follows. A *partial permutation* of the set $\underline{n}$ is a pair (σ, U) which consists of a finite subset $U \subset \underline{n}$ and an element $\sigma \in S_U$. Denote by PS_n the set of all partial permutations of $\underline{n}$. The finite set PS_n is endowed with a natural semigroup structure by letting the product of two elements (σ, U) and (τ, V) in PS_n to be $(\sigma\tau, U \cup V)$. We denote by $\mathbb{Z}[PS_n]$ the semigroup algebra over $\mathbb{Z}$. The S_n acts on PS_n and thus on $\mathbb{Z}[PS_n]$ by $\mathrm{ad}\, g : (\sigma, U) \mapsto (g\sigma g^{-1}, gU)$.

We introduce

$$PH^*(X^n, S_n) = \bigoplus_{(\sigma,U)\in PS_n} H^{*-2F_\sigma}((X^U)^\sigma) \tag{4.1}$$

and define a product $\bullet$ on $PH^*(X^n, S_n)$ as follows. Let (σ, U) and (τ, V) be in PS_n. Given $\alpha \in H^{*-2F_\sigma}((X^U)^\sigma)$ and $\beta \in H^{*-2F_\tau}((X^V)^\tau)$, we may identify α as an element in $H^{*-2F_\sigma}((X^{U\cup V})^\sigma)$ via the natural embedding

$$H^{*-2F_\sigma}((X^U)^\sigma) \subset H^{*-2F_\sigma}((X^{U\cup V})^\sigma)$$

where $\sigma \in S_U$ is regarded as a permutation in $S_{U\cup V}$ by fixing $V \backslash U \cap V$ pointwise. Similarly, we may regard β as an element in $H^{*-2F_\tau}((X^{U\cup V})^\tau)$. We define the product $\alpha \bullet \beta$ to be $\alpha \circ \beta \in H^{*-2F_{\sigma\tau}}((X^{U\cup V})^{\sigma\tau})$. Then the associativity of $H^*(X^n, S_n)$ [**6, 12**] (together with the trivial associativity of the union operation of sets) implies that $PH^*(X^n, S_n)$ thus defined is an associative algebra.

Note that ad $g : H^*((X^U)^\sigma) \to H^*((X^{gU})^{g\sigma g^{-1}})$ for $g \in S_n$ and $(\sigma, U) \in PS_n$ defines an action of S_n on $PH^*(X^n, S_n)$. The product structure on $PH^*(X^n, S_n)$ is clearly compatible with the S_n-action. We denote by $(H_{X,n}, \bullet)$ the S_n-invariant subalgebra of $PH^*(X^n, S_n)$.

4.3. The stability of the rings $H^*_{\mathbf{CR}}(X^n/S_n)$. By associating a partition of $|U|$ (the cardinality of U) to an element (σ, U) in PS_n, we have a natural parametrization of the S_n-orbits of PS_n by partitions λ with $\|\lambda\| \le n$. Just as $H^*_{\mathrm{CR}}(X^n/S_n)$ has a linear basis given by $\mathcal{P}_n(S)$ (cf. e.g. [**40**]), we see that $H_{X,n}$ affords a linear basis parameterized by $\rho \in \mathcal{P}(S)$ with $\|\rho\| \le n$ as follows. Given $\rho = (\rho(s))_{s \in S} \in \mathcal{P}(S)$ with $\|\rho\| \le n$, define a partition $\bar{\rho}$ of $\|\rho\|$ to be $\cup_{s \in S} \rho(s)$; that is, $\bar{\rho}$ is obtained from $\rho \in \mathcal{P}(S)$ by forgetting the indices $s \in S$ and then rearranging the parts in descending order. Take an element $(\sigma, U) \in PS_n$ with $|U| = \|\rho\|$ and σ of cycle type $\bar{\rho}$, and take an $x \in H^*((X^U)^\sigma) \cong H^*(X)^{\otimes \ell(\bar{\rho})}$ with $\ell(\rho(s))$ factors equal to s for each $s \in S$ such that the cycles of σ associated to s correspond to the partition $\rho(s)$. Denote by $c_\rho(n) \in H_{X,n}$ the sum of the S_n-orbit of x. It is easy to see that $c_\rho(n)$ does not depend on the choice of (σ, U) and x, but only on ρ and n. The $c_\rho(n)$, where $\rho \in \mathcal{P}(S)$ and $\|\rho\| \le n$, form a linear basis for $H_{X,n}$.

Let us write

$$c_\rho(n) \bullet c_\sigma(n) = \sum_{\nu \in \mathcal{P}(S)} C^\nu_{\rho\sigma}(n)\, c_\nu(n)$$

summed over ν with $\|\nu\| \le n$, where $C^\nu_{\rho\sigma}(n)$ denotes the structure constants of the ring $H_{X,n}$.

Keeping (4.1) in mind, we define a linear map (for $m \le n$)

$$\theta_{n,m} : PH^*(X^n, S_n) \to PH^*(X^m, S_m)$$

to be the identity map on the component $H^{*-2F_\sigma}((X^U)^\sigma)$ associated to $(\sigma, U) \in PS_n$ if $U \subset \underline{m}$, and 0 if $U \not\subset \underline{m}$. It is clear that $\theta_{n,m}$ is a ring homomorphism and it is surjective. Since the homomorphisms $\theta_{n,m}$ for all $m \le n$ are compatible, we can define an inverse limit

$$PH^*(X^\infty, S_\infty) := \varprojlim_n PH^*(X^n, S_n)$$

which has an induced algebra structure and an induced action by the group $S_\infty = \cup_n S_n$. Denote by $H_{X,\infty}$ the algebra of S_∞-invariants in $PH^*(X^\infty, S_\infty)$. One can construct a linear basis c_ρ with $\rho \in \mathcal{P}(S)$ of $H_{X,\infty}$, in the same way as constructing the basis $c_\rho(n)$'s for $H_{X,n}$. Write

$$c_\rho \bullet c_\sigma = \sum_\nu C^\nu_{\rho\sigma} c_\nu. \tag{4.2}$$

The homomorphisms $\theta_{n,m}$ induces surjective ring homomorphisms, which will be also denoted by $\theta_{n,m}$, from $H_{X,n}$ to $H^*_{X,m}$. This in turn is compatible with a surjective ring homomorphism $\theta_n : H_{X,\infty} \to H_{X,n}$ by letting $\theta_n(c_\rho) = c_\rho(n)$ if $\|\rho\| \le n$ and $\theta_n(c_\rho) = 0$ otherwise. It follows that

The 'forgetful' map $f_n : PH^*(X^n, S_n) \to H^*(X^n, S_n)$ which, when restricted to the component $H^{*-2F_\sigma}((X^U)^\sigma)$ for $(\sigma, U) \in PS_n$, is given by the inclusion map $H^{*-2F_\sigma}((X^U)^\sigma) \subset H^{*-2F_\sigma}((X^n)^\sigma)$ where σ in the latter is regarded as an element of S_n by fixing pointwise $\underline{n}\backslash U$. Clearly, f_n is indeed a surjective S_n-equivariant ring homomorphism, and thus induces a surjective ring homomorphism from $H_{X,n} \to H^*_{\mathrm{CR}}(X^n/S_n)$ which will be again denoted by f_n. We define $\mathfrak{p}_\rho(n) = f_n(c_\rho(n))$. It follows from (4.2) that

$$\mathfrak{p}_\rho(n) \circ \mathfrak{p}_\sigma(n) = \sum_\nu C^\nu_{\rho\sigma} \mathfrak{p}_\nu(n). \tag{4.3}$$

REMARK 4.2. It is quite straightforward to define $\mathfrak{p}_\rho(n)$ directly, just as we defined $c_\rho(n)$. We denote $\mathcal{F}_X = \oplus_{n=0}^\infty H^*_{\mathrm{CR}}(X^n/S_n)$. It is well known that one can construct a Heisenberg algebra associated to the lattice $H^*(X)/tor$ acting irreducibly on $\mathcal{F}_X$ (cf. e.g. [**40**]). One can show that the $\mathfrak{p}_\rho(n)$ in the present paper coincides with the one defined in [**40**], Sect. 4.4, in terms of the Heisenberg monomials (up to a scale multiple independent of n). The $\mathfrak{p}_\rho(n)$'s span $H^*_{\mathrm{CR}}(X^n/S_n)$ but are not linearly independent when $n > 1$.

We summarize the above into a diagram of surjective ring homomorphisms:

$$\begin{array}{cccc}
H_{X,\infty} = \varprojlim_n H_{X,n} & \cdots\cdots \longrightarrow H_{X,n} & \stackrel{\theta_{n,n-1}}{\longrightarrow} & H_{X,n-1} \longrightarrow \cdots \\
 & \downarrow f_n & & f_{n-1} \downarrow \\
 & H^*_{\mathrm{CR}}(X^n/S_n) & & H^*_{\mathrm{CR}}(X^{n-1}/S_{n-1})
\end{array}$$

In this way, the algebra $H_{X,\infty}$ with the multiplication (4.2) governs the structures of the rings $H^*_{\mathrm{CR}}(X^n/S_n)$. The identity (4.3) with structure constants $C^\nu_{\rho\sigma}$ shows that the algebra $H_{X,\infty}$ can be identified with the *stable ring* $\mathfrak{R}_X$ associated to a compact complex manifold X introduced in [**40**]. Thus, we have established the following theorem which first appeared in [**40**] with the assumption that the complex dimension of X is even. The proof in *loc. cit.* was highly nontrivial and used a vertex operator method developed therein for the rings $H^*_{\mathrm{CR}}(X^n/S_n)$.

THEOREM 4.3. *For a compact complex manifold X, the Chen-Ruan cohomology rings $H^*_{CR}(X^n/S_n)$, $n \geq 1$ give rise to a stable ring $H_{X,\infty}$ which completely encodes the cohomology ring structure of $H^*_{CR}(X^n/S_n)$ for each n.*

REMARK 4.4. In the case when X is a point, our constructions of $H_{X,n}$ etc simplify greatly and specialize to that of [**21**] which were used to explain the stability in the class algebras of symmetric groups discovered in [**25**] (also see [**49**] for the extension to the wreath product setup.)

REMARK 4.5. Assume that $H^*(X, \mathbb{Z})$ has no torsion. We can refine the definition of $H^*_{\mathrm{CR}}(X^n/S_n)$ to define the orbifold cohomology ring of the symmetric product with integer coefficient, which has no torsion. If we further choose S to be an integral linear basis for $H^*(X, \mathbb{Z})$, then the structure constants $C^\nu_{\rho\sigma}$'s above can be shown to be integers. Also compare with [**1**] where the orbifold cohomology rings of a general orbifold can be defined over integers.

5. The FH-ring of wreath products

5.1. Preliminaries on the wreath products. Let Γ be a finite group, and Γ_* be the set of conjugacy classes of Γ. We will denote the identity of Γ by 1 and the identity conjugacy class in Γ by c^0. The symmetric group S_n acts on the product group $\Gamma^n = \Gamma \times \cdots \times \Gamma$ by permutations: $\sigma(g_1, \cdots, g_n) = (g_{\sigma^{-1}(1)}, \cdots, g_{\sigma^{-1}(n)})$. The wreath product of Γ with S_n is defined to be the semidirect product

$$\Gamma_n = \{(g, \sigma) | g = (g_1, \cdots, g_n) \in \Gamma^n, \sigma \in S_n\}$$

with the multiplication $(g, \sigma) \cdot (h, \tau) = (g\,\sigma(h), \sigma\tau)$. The i-th factor subgroup of the product group Γ^n will be denoted by $\Gamma^{(i)}$. The wreath product Γ_n embeds in Γ_{n+1} as the subgroup $\Gamma_n \times 1$, and the union $\Gamma_\infty = \cup_{n\geq 1}\Gamma_n$ carries a natural group structure. When Γ is trivial, Γ_∞ reduces to $S_\infty = \cup_{n\geq 1} S_n$.

The space $R_{\mathbb{Z}}(\Gamma)$ of the class functions of a finite group Γ (which is often called the *class algebra* of Γ) is closed under the convolution. In this way, $R_{\mathbb{Z}}(\Gamma)$ is also identified with the center of the group algebra $\mathbb{Z}[\Gamma]$. We will denote by $R(\Gamma_n)$ (resp. $R_{\mathbb{Z}}(\Gamma_n)$) the *class algebra* of complex (resp. integer) class functions on Γ_n endowed with the convolution product.

The conjugacy classes of Γ_n can be described in the following way (cf. [**35, 53**]). Let $x = (g, \sigma) \in \Gamma_n$, where $g = (g_1, \ldots, g_n) \in \Gamma^n$, $\sigma \in S_n$. The permutation σ is written as a product of disjoint cycles. For each such cycle $y = (i_1 i_2 \cdots i_k)$, the element $p_y = g_{i_k} g_{i_{k-1}} \cdots g_{i_1} \in \Gamma$ is determined up to conjugacy in Γ by g and y, and will be called the *cycle-product* of x corresponding to the cycle y. For any conjugacy class $c \in \Gamma_*$ and each integer $i \geq 1$, the number of i-cycles in σ whose cycle-product lies in c will be denoted by $m_i(\rho(c))$, and $\rho(c)$ denotes the partition $(1^{m_1(c)} 2^{m_2(c)} \ldots)$. Then each element $x = (g, \sigma) \in \Gamma_n$ gives rise to a partition-valued function $\rho = (\rho(c))_{c\in\Gamma_*} \in \mathcal{P}(\Gamma_*)$ such that $\|\rho\| := \sum_{r,c} r m_r(\rho(c)) = n$. The ρ is called the *type* of x. It is known that any two elements of Γ_n are conjugate in Γ_n if and only if they have the same type.

5.2. The modified types. Let x be an element of Γ_n of type $\rho = (\rho(c))_{c\in\Gamma_*} \in \mathcal{P}_n(\Gamma_*)$. If we regard it as an element in Γ_{n+k} by the natural inclusion $\Gamma_n \leq \Gamma_{n+k}$, then x has the type $\rho \cup (1^k) \in \mathcal{P}_{n+k}(\Gamma_*)$, where $(\rho \cup (1^k))(c) = \rho(c)$ for $c \neq c^0$ and $(\rho \cup (1^k))(c^0) = (\rho(c^0), 1, \ldots, 1) = \rho(c^0) \cup (1^k)$. It is convenient to introduce the *modified type* of x to be $\widetilde{\rho} \in \mathcal{P}_{n-r}(\Gamma_*)$, where $r = \ell(\rho(c^0))$, as follows: $\widetilde{\rho}(c) = \rho(c)$ for $c \neq c^0$ and $\widetilde{\rho}(c^0) = (\rho_1 - 1, \ldots, \rho_r - 1)$ if we write the partition $\rho(c^0) = (\rho_1, \ldots, \rho_r)$. Two elements in Γ_∞ are conjugate if and only if their modified types coincide.

Given $\mu \in \mathcal{P}(\Gamma_*)$, we denote by $\mathcal{K}_\mu$ the conjugacy class in Γ_∞ of elements whose modified type is μ. For each $n \geq 0$ and each $\mu \in \mathcal{P}(\Gamma_*)$, let $K_\mu(n)$ be the characteristic function of the conjugacy class in Γ_n whose modified type is μ, i.e. the sum of all $\sigma \in \Gamma_n \cap \mathcal{K}_\mu$. The nonzero $K_\mu(n)$'s form a $\mathbb{Z}$-basis for $R_{\mathbb{Z}}(\Gamma_n)$. If $x \in \mathcal{K}_\mu$, the *degree* $\|x\|$ of x is defined to be the degree $\|\mu\|$ of its *modified* type.

Given $g \in \Gamma$ and $1 \leq i \neq j$, we denote by $(i \xrightarrow{g} j)$ the element $((g_1, g_2, \ldots), (i, j))$ in Γ_∞, where $(i, j) \in S_\infty$ is a transposition, $g_j = g, g_i = g^{-1}$, and $g_k = 1$ for $k \neq i, j$. Note that $(i \xrightarrow{g} j) = (j \xrightarrow{g^{-1}} i)$ and it is of order 2. The elements in Γ_∞ of the form $(i \xrightarrow{g} j)$, where g runs over Γ and (i, j) (where $i < j$) runs over all transpositions of product is called a *reduced expression* for x [**50**] if x cannot be written as a product of fewer such elements. For a general element in Γ_∞, a reduced expression can be constructed cycle-by-cycle.

The number of elements appearing in a reduced expression of x is shown in *loc. cit.* to be $\|\lambda\|$ for a given element $x \in \Gamma_\infty$ of modified type λ. It follows that $\|xy\| \leq \|x\| + \|y\|$, and $\|\cdot\|$ defines an algebra filtration on $R_{\mathbb{Z}}(\Gamma_n)$. The associated graded ring will be denoted by $\mathcal{G}_\Gamma(n)$.

5.3. The FH-ring of wreath products. Given $\lambda, \mu \in \mathcal{P}(\Gamma_*)$, we write the convolution product $K_\lambda(n)\, K_\mu(n)$ in $R_{\mathbb{Z}}(\Gamma_n)$ as a linear combination of $K_\nu(n)$:

$$K_\lambda(n) K_\mu(n) = \sum_\nu D^\nu_{\lambda\mu}(n)\; K_\nu(n)$$

where the structure constants $D^\nu_{\lambda\mu}(n) \in \mathbb{Z}_+$ are zero unless $\|\nu\| \leq \|\lambda\| + \|\mu\|$.

THEOREM 5.1. [**50**] *Let $\lambda, \mu, \nu \in \mathcal{P}(\Gamma_*)$. Then,*

(1) *there is a unique polynomial $f^\nu_{\lambda\mu}(x)$ such that $f^\nu_{\lambda\mu}(n) = D^\nu_{\lambda\mu}(n)$ for all positive integers n.*
(2) *the polynomial $f^\nu_{\lambda\mu}(x)$ is a constant if $\|\nu\| = \|\lambda\| + \|\mu\|$.*

The next theorem is a reformulation of Theorem 5.1 (2).

THEOREM 5.2. *The graded $\mathbb{Z}$-ring $\mathcal{G}_\Gamma(n)$ has a multiplication given by*

$$K_\lambda(n)\, K_\mu(n) = \sum_{\|\nu\| = \|\lambda\| + \|\mu\|} a^\nu_{\lambda\mu} K_\nu(n)$$

where the structure constants $a^\nu_{\lambda\mu}$ are nonnegative integers independent of n.

The results above specialize to the results of Farahat-Higman [**13**] (also cf. [**35**]) in the symmetric group case. Thus, we make the following definition.

DEFINITION 5.3. The Farahat-Higman ring (or FH-ring) of the wreath products associated to Γ, denoted by $\mathcal{G}_\Gamma$, is a $\mathbb{Z}$-ring with a $\mathbb{Z}$-basis K_λ, where $\lambda \in \mathcal{P}(\Gamma_*)$, and a multiplication (called the FH-product)

$$K_\lambda\, K_\mu = \sum_{\|\nu\| = \|\lambda\| + \|\mu\|} a^\nu_{\lambda\mu} K_\nu. \tag{5.1}$$

There is a natural surjective ring homomorphism $\mathcal{G}_\Gamma \to \mathcal{G}_\Gamma(n)$ for each n which is compatible with the surjective ring homomorphism $res_n : \mathcal{G}_\Gamma(n) \to \mathcal{G}_\Gamma(n-1)$ by restriction. The restriction map res_n clearly sends $K_\lambda(n)$ to $K_\lambda(n-1)$ for all λ. The ring $\mathcal{G}_\Gamma$ can be regarded as the inverse limit of the family of rings $\{\mathcal{G}_\Gamma(n)\}_{n\geq 1}$ under the restriction maps. Theorem 5.2 is equivalent to the statement that the restriction maps are ring homomorphisms. We summarize the sequence of surjective ring homomorphisms above into the following diagram:

$$\mathcal{G}_\Gamma = \lim_{\overleftarrow{n}} \mathcal{G}_\Gamma(n) \qquad \ldots\ldots \longrightarrow \mathcal{G}_\Gamma(n) \stackrel{res_n}{\longrightarrow} \mathcal{G}_\Gamma(n-1) \longrightarrow \ldots$$

5.4. Macdonald's interpretation of the FH-ring. We restrict ourselves to the symmetric group case (i.e. when $\Gamma = 1$) in the remainder of this section. In this case, we shall omit the subscript Γ from the notations for the rings $\mathcal{G}_\Gamma$, $\mathcal{G}_\Gamma(n)$.

Denote by $\Lambda = \oplus_k \Lambda^k$ the $\mathbb{Z}$-ring of symmetric functions (in infinitely many variables), and Λ^k denotes the subspace of symmetric functions of degree k. Denote by $e_k, h_k, p_k \in \Lambda^k$ respectively the k-th elementary, complete, and power-sum symmetric functions. There is a standard bilinear form $\langle -, - \rangle$ on Λ and the Schur functions form an orthnormal $\mathbb{Z}$-basis of Λ. Denote by λ' the transpose of a partition λ. An involution $\omega : \Lambda \to \Lambda$ is defined by $\omega(s_\lambda) = s_{\lambda'}$ for all λ, and it can also be characterized as the ring isomorphism of Λ by switching e_n and h_n for all n.

A (less standard) involution on the ring Λ is defined as follows [**35**]. Let

$$u = t + h_1 t^2 + h_2 t^3 + \ldots .$$

Then t can be expressed as a power series in u, say

$$t = u + h_1^* u^2 + h_2^* u^3 + \ldots ,$$

with coefficients $h_k^* \in \Lambda^k$ $(k \geq 1)$. The ring homomorphism $\psi : \Lambda \to \Lambda$ defined by $\psi(h_k) = h_k^*$, $k \geq 1$, is an involution. We denote $f^* = \psi(f)$. For example, $h_\lambda^* = h_{\lambda_1}^* h_{\lambda_2}^* \ldots$ for each partition $\lambda = (\lambda_1, \lambda_2, \ldots)$, and the h_λ^*'s form a $\mathbb{Z}$-basis of Λ. We denote by (g_λ) the dual basis, i.e. $\langle g_\lambda, h_\mu^* \rangle = \delta_{\lambda,\mu}$.

The following theorem is due to Macdonald [**35**], Ex. 25, pp. 132 (also cf. Goulden-Jackson [**18**] for another proof).

THEOREM 5.4. (Macdonald) *The linear map $\varphi : \Lambda \to \mathcal{G}$ defined by $\varphi(g_\lambda) = K_\lambda$ for all partitions λ is a ring isomorphism.*

5.5. Jucys-Murphy elements and Macdonald's isomorphism. Recall [**22, 37**] that the Jucys-Murphy elements $\xi_{j;n}$ of the symmetric group S_n are defined to be the sums of transpositions:

$$\xi_{j;n} = \sum_{i<j} (i, j), \quad j = 1, \ldots, n.$$

In particular, we have $\xi_{1;n} = 0$. When it is clear from the text, we may simply write $\xi_{j;n}$ as ξ_j. An important property of the JM elements is that any symmetrization of $\{\xi_1, \ldots, \xi_n\}$ lies in $R_{\mathbb{Z}}(S_n)$. It is further known that the k-th elementary symmetric function $e_k(\Xi_n)$ of $\Xi_n = \{\xi_1, \ldots, \xi_n\}$ is equal to the sum of all permutations in S_n having exactly $(n - k)$ cycles. For example, $e_1(\Xi_n)$ is exactly the characteristic function $K_{(1)}(n)$ of the conjugacy classes of transpositions in S_n. We will use $\star$ to denote the product in the FH-ring $\mathcal{G}(n)$. We define

$$(-\Xi)^{\star k}(n) = \sum_{j=1}^{n} (-1)^k \underbrace{\xi_{j;n} \star \ldots \star \xi_{j;n}}_{k} .$$

PROPOSITION 5.5. In the ring $\mathcal{G}(n)$, we have the equality:

PROOF. Since we have $\|x\| \leq n-1$ for every element x in $R_{\mathbb{Z}}(S_n)$, the statement is clear for $k \geq n$.

Now assume $k < n$. We first note that $(i\ j) \star (i\ j) = 0$ for any transposition $(i\ j)$, since $(i\ j)^2 = 1$, $\|(i\ j)\| = 1$, and $\|1\| = 0$. It follows from the definition of the Jucys-Murphy elements $\xi_{j;n}$ that the only permutations which survive in $\underbrace{\xi_{j;n} \star \ldots \star \xi_{j;n}}_{k}$ is the product (under the ordinary group multiplication) of k distinct transpositions and any such a product is necessarily a $(k+1)$-cycle. Since $(-\Xi)^{\star k}(n)$ lies in $R_{\mathbb{Z}}(S_n)$, it is a multiple of $K_{(k)}(n)$. It remains to show that the multiple is indeed 1.

The total number of such $(k+1)$-cycles in $\underbrace{\xi_{j;n} \star \ldots \star \xi_{j;n}}_{k}$ is $k! \cdot \binom{j-1}{k}$, and thus the total number of such products in $(-\Xi)^{\star k}(n)$ is $\sum_{j=2}^{n} k! \cdot \binom{j-1}{k}$. The number of $(k+1)$-cycles appearing in $K_{(k)}(n)$ is $\frac{n!}{(n-k-1)! \cdot (k+1)}$. We conclude that $(-\Xi)^{\star k}(n) = K_{(k)}(n)$ from the identity (which can be proved easily by induction on n)

$$\sum_{j=2}^{n} k! \cdot \binom{j-1}{k} = \frac{n!}{(n-k-1)! \cdot (k+1)}.$$

$\square$

Now, the elements $(-\Xi)^{\star k}(n)$, $n \geq 1$, give rise to an element $(-\Xi)^{\star k}$ in $\mathcal{G}$. Let us denote by $\phi : \mathcal{G} \to \Lambda$ the isomorphism inverse to φ.

PROPOSITION 5.6. We have $\phi(K_{(k)}) = -p_k$, and $\omega\phi((-\Xi)^{\star k}) = p_k$.

PROOF. The first part was established in **[35]**, Ex. 25 (b), pp. 133. The second part now follows from the first part, Proposition 5.5, and the fact that $\omega(p_k) = (-1)^{k-1} p_k$. $\square$

Denote by $e_k(-\Xi_n)$, $h_k(-\Xi_n)$ and $s_\lambda(-\Xi_n)$ respectively the k-th elementary, complete, and Schur symmetric function in the variables $-\xi_1, \ldots, -\xi_n$ (using the FH-product). We denote by $e_k^\star(-\Xi)$, $h_k^\star(-\Xi)$ and $s_\lambda^\star(-\Xi)$ the corresponding elements in $\mathcal{G}$.

THEOREM 5.7. *We have the following correspondence under the isomorphisms ϕ and $\omega\phi$ from $\mathcal{G}$ to Λ:*

(1) $\phi(K_\lambda) = g_\lambda$.
(2) $\omega\phi((-\Xi)^{\star k}) = p_k$.
(3) $\omega\phi(e_k^\star(-\Xi)) = e_k$.
(4) $\omega\phi(h_k^\star(-\Xi)) = h_k$.
(5) $\omega\phi(s_\lambda^\star(-\Xi)) = s_\lambda$.

PROOF. Part (1) is Theorem 5.4 and part (2) is Proposition 5.6. The rest follows from part (2) and the definitions of $e_k^\star(-\Xi)$, $h_k^\star(-\Xi)$ and $s_\lambda^\star(-\Xi)$. $\square$

REMARK 5.8. The Jucys-Murphy elements have been vital in understanding the rings $H^*_{\mathrm{CR}}(X^n/S_n)$ for compact X **[40]**, and they are intimately related to vertex operators, cf. **[26, 49, 40]**. The natural role of the Jucys-Murphy elements in $\mathcal{G}$ indicates that they should play also an important role in the study of the rings $H^*_{\mathrm{CR}}(X^n/S_n)$ for noncompact X even without vertex operators.

REMARK 5.9. For an arbitrary finite group Γ, it has been shown [**50**] that the FH-ring $\mathcal{G}_\Gamma$ is isomorphic to a tensor product of several copies (parametrized by Γ_*) of the ring Λ of symmetric functions. It will be very interesting to generalize Macdonald's symmetric function interpretation of $\mathcal{G}$ to the general FH-ring $\mathcal{G}_\Gamma$.

References

[1] D. Abramovich, T. Graber and A. Vistoli, *Algebraic orbifold quantum products*, math.AG/0112004.

[2] V. Batyrev, *Non-Archimedean integrals and stringy Euler numbers of log-terminal pairs*, J. Eur. Math. Soc. **1** (1999), 5–33.

[3] A. Beauville, *Sur la cohomologie de certains espaces de modules de fibrés vectoriels*, Geometry and Analysis (Bombay, 1992), 37–40, Tata Inst. Fund. Res., Bambay, 1995.

[4] L. Borisov and A. Libgober, *McKay correspondence for elliptic genera*, Preprint, math.AG/0206241.

[5] T. Bridgeland, A. King and M. Reid, *The McKay correspondence as an equivalence of derived categories*, J. Amer. Math. Soc. **14** (2001), 535–554.

[6] W. Chen and Y. Ruan, *A new cohomology theory for orbifold*, math.AG/0004129.

[7] M. de Cataldo and L. Migliorini, *The Chow groups and the motive of the Hilbert scheme of points on a surface*, math.AG/0005249.

[8] J. Denef and F. Loeser, *Motivic integration, quotient singularities and the McKay correspondence*, Compositio Math. **131** (2002), 267–290.

[9] L. Dixon, J.A. Harvey, C. Vafa, and E. Witten, *Strings on orbifolds*, Nuclear Phys. **B 261** (1985), 678–686.

[10] G. Ellingsrud and S. Strømme, *Towards the Chow ring of the Hilbert scheme of* $\mathbb{P}^2$, J. Reine Angew. Math. **441** (1993), 33-44.

[11] P. Etingof and V. Ginzburg, *Symplectic reflection algebras, Calogero-Moser space, and deformed Harish-Chandra homomorphism*, Invent. Math. **147** (2002), 243–348.

[12] B. Fantechi and L. Göttsche, *Orbifold cohomology for global quotients*, Preprint, math.AG/0104207.

[13] H. Farahat and G. Higman, *The centres of symmetric group rings*, Proc. Roy. Soc. (A) **250** (1959), 212–221.

[14] I. Frenkel and W. Wang, *Virasoro algebra and wreath product convolution*, J. Alg. **242** (2001), 656–671.

[15] L. Göttsche, *The Betti numbers of the Hilbert scheme of points on a smooth projective surface*, Math. Ann. **286** (1990), 193–207.

[16] ———, *On the motive of the Hilbert scheme of points on a surface*, Math. Res. Lett. **8** (2001), 613–627.

[17] ———, in preparation.

[18] I. Goulden and D. Jackson, *Symmetric functions and Macdonald's result for top connexion coefficients in the symmetric group*, J. Alg. **166** (1994), 364–378.

[19] I. Grojnowski, *Instantons and affine algebras I: the Hilbert scheme and vertex operators*, Math. Res. Lett. **3** (1996), 275–291.

[20] M. Haiman, *Hilbert schemes, polygraphs, and the Macdonald positivity conjecture*, J. Amer. Math. Soc. **14** (2001), 941–1006.

[21] V. Ivanov and S. Kerov, *The algebra of conjugacy classes in symmetric groups and partial permutations*, J. Math. Sciences **107** (2001), 4212–4230.

[22] A. Jucys, *Symmetric polynomials and the center of the symmetric group rings*, Rep. Math. Phys. **5** (1974), 107–112.

[23] D. Kaledin, *McKay correspondence for symplectic quotient singularities*, Invent. Math. **148** (2002), 151–175.

[24] R. Kaufmann, *Second quantized Frobenius algebras*, math.AG/0206137.

[25] S. Kerov and G. Olshanski, *Polynomial functions on the set of Young diagrams*, C.R. Acad. Sci. Paris Sér. I **319** (1994), 121–126.

[26] A. Lascoux and J.-Y. Thibon, *Vertex operators and the class algebras of symmetric*

[28] M. Lehn, C. Sorger, *Symmetric groups and the cup product on the cohomology of Hilbert schemes*, Duke Math. J. **110** (2001), 345–357.
[29] ———, *The cup product of the Hilbert scheme for $K3$ surfaces*, Preprint 2000, math.AG/0012166.
[30] W.-P. Li, Z. Qin and W. Wang, *Vertex algebras and the cohomology ring structure of Hilbert schemes of points on surfaces*, Math. Ann (to appear), math.AG/0009132.
[31] ———, *Generators for the cohomology ring of Hilbert schemes of points on surfaces*, Internat. Math. Res. Notices **20** (2001), 1057–1074.
[32] ———, *Stability of the cohomology rings of Hilbert schemes of points on surfaces*, J. Reine Angew. Math. (to appear), arXiv:math.AG/0107139.
[33] ———, *Hilbert schemes and $\mathcal{W}$ algebras*, Intern. Math. Res. Notices **27** (2002), 1427–1456.
[34] ———, *Ideals of the cohomology rings of Hilbert schemes and their applications*, math.AG/0208070.
[35] I. G. Macdonald, *Symmetric functions and Hall polynomials*, 2nd Ed., Clarendon Press, Oxford, 1995.
[36] E. Markman, *Generators of the cohomology ring of moduli spaces of sheaves on symplectic surfaces*, J. Reine Angew. Math. **544** (2002), 61–82.
[37] G. Murphy, *A new construction of Young's seminormal representation of the symmetric group*, J. Alg. **69** (1981), 287–291.
[38] H. Nakajima, *Heisenberg algebra and Hilbert schemes of points on projective surfaces*, Ann. Math. **145** (1997), 379–388.
[39] ———, *Lectures on Hilbert schemes of points on surfaces*, Univ. Lect. Ser. **18**, Amer. Math. Soc. (1999).
[40] Z. Qin and W. Wang, *Hilbert schemes and symmetric products: a dictionary*, Mathematical Aspects of Orbifold String Theory, Madison, Wisconsin, 2001, Contemp. Math. (to appear), math.AG/0112070.
[41] M. Reid, *La correspondance de McKay*, Séminaire Bourbaki, Vol. 1999/2000. Astrisque **276** (2002), 53–72.
[42] Y. Ruan, *Stringy geometry and topology of orbifolds*, math.AG/0011149.
[43] ———, *Cohomology ring of crepant resolutions of orbifolds*, math.AG/0108195.
[44] B. Uribe, *Orbifold cohomology of the symmetric product*, math.AT/0109125.
[45] C. Vafa and E. Witten, *A strong coupling test of S-duality*, Nucl. Phys. **B 431** (1994), 3–77.
[46] E. Vasserot, *Sur l'anneau de cohomologie du schéma de Hilbert de $\mathbb{C}^2$*, C. R. Acad. Sci. Paris, Sér. I Math. **332** (2001), 7–12.
[47] W. Wang, *Equivariant K-theory, wreath products, and Heisenberg algebra*, Duke Math. J. **103** (2000), 1–23.
[48] ———, *Algebraic structures behind Hilbert schemes and wreath products*, Contemp. Math. **297** (2002), 271–295.
[49] ———, *Vertex algebras and the class algebras of wreath products*, Preprint, arXiv:math.QA/0203004.
[50] ———, *The Farahat-Higman ring of wreath products and Hilbert schemes*, Preprint, arXiv:math.QA/0205071.
[51] W. Wang and J. Zhou, *Orbifold Hodge numbers of the wreath product orbifolds*, J. Geom. Phys. **38** (2001), 153-170.
[52] E. Zaslow, *Topological orbifold models and quantum cohomology rings*, Commun. Math. Phys. **156** (1993), 301–331.
[53] A. Zelevinsky, *Representations of finite classical groups. A Hopf algebra approach*, Lect. Notes in Math. **869**, Springer-Verlag, Berlin-New York, 1981.

Department of Mathematics, University of Virginia, Charlottesville, VA 22904.
E-mail address: `ww9c@virginia.edu`

Contemporary Mathematics
Volume **322**, 2003

HyperKähler Manifolds and Birational Transformations in dimension 4

Dan Burns, Yi Hu, Tie Luo

ABSTRACT. We show that any birational map between hyperKähler manifolds of dimension 4 is composed of a sequence of simple flops or elementary Mukai transformations under the assumption that each irreducible component of the indeterminacy of the birational map is normal.

§1 Introduction and statement of main theorem

A holomorphic symplectic manifold is a complex manifold equipped with a holomorphic symplectic form ω. A hyperKähler manifold is a projective holomorphic symplectic manifold. The canonical bundles of such manifolds are trivial. It is expected, following Mori's minimal model program, that any birational map between two smooth varieties with trivial canonical bundles can be decomposed as a finite sequences of elementary ones-flops. A particularly interesting class of flops consists of the so-called Mukai's elementary transformations [Mu2].

More precisely, let (X, ω) be a holomorphic symplectic manifold (may not be projective) and P an embedded smooth subvariety. Assume further that P is a $\mathbb{P}^r$-bundle over a smooth variety Σ such that the codimension of P coincides with r. Then it is known [Mu2] that one can blow up X along P to get a smooth variety $\widetilde{X}$ and the exceptional divisor can be blow down along a different ruling to get a smooth variety X'. Moreover, X' comes equipped with a holomorphic symplectic form ω' which coincides with ω away from the exceptional locus. Such a simple birational process $X - - \to X'$ is called a Mukai elementary transformation (MET).

The main result of this paper is a solution to an open problem posed by Mukai more than a decade ago (Problem 4.5, [Mu2]) under a normality assumption.

THEOREM 1.1. *Let $\Phi : X - - \to X'$ be a birational transformation between two hyperKähler fourfolds and $B \subset X$ the indeterminacy of Φ. Assume each irreducible component of B is normal. Then, up to an isomorphism, Φ can be decomposed as a sequence of birational maps and each map consists of the Mukai elementary transformations along some disjoint $\mathbb{P}^2$s which are birational transformations of some components in B.*

Remark. Note that the birational maps between Calabi-Yau threefolds have been classified in [Ko1]. After this paper was written, J. Wisniewski informed us recently that jointly with J. Wierzba they could verify the normal condition on components of B.

The main ideas of the proof of Theorem 1.1 are as following. It is clear that one needs to isolate a copy of $\mathbb{P}^2$ (or copies disjoint $\mathbb{P}^2$s) to be able to perform a Mukai elementary transformation. For this purpose we need to classify the exceptional locus of a logterminal contraction, which contracts part of the indeterminacy. Here existence of ample divisors (hence projectivity) is required. It turns out to be a formidable task to fully achieve this goal. Even if assuming the normality of each irreducible component, the task is still quite involved. The relative relation of various components can be very complex. One key point is to check that the normality of each irreducible component in B is kept after METs. To this end, a useful result (2.6) of Wierzba [W] is applied. It is also important to notice that one may need to perform several Mukai elementary transformations simultaneously to keep the projectivity of the new variety. The proof is completed by using an induction process and termination of the process based on the number of irreducible components of indeterminacy. Once again projectivity plays a crucial role on termination.

Naturally one may ask whether it is possible to do a bit better than Theorem 1.1. For example, is it possible to make each map involved in the sequence actually a MET on a single $\mathbb{P}^2$? Unfortunately the answer is no. Y. Namikawa [N] shows an example X which is the Hilbert scheme of length 2 zero cycles of a projective K3 surface admitting an elliptic fibration over $\mathbb{P}^1$ with two singular fibers of cycles of three $\mathbb{P}^1$. X is projective and has six disjoint $\mathbb{P}^2$s coming from those $\mathbb{P}^1$s. One can show that flopping any one or two $\mathbb{P}^2$s leads to a new hyperKähler fourfold. However flopping three $\mathbb{P}^2$s coming from the $\mathbb{P}^1$s in the same fiber would lead to a nonhyperKähler fourfold. Using similar arguments one can construct a birational map between two hyperKähler fourfolds (which are $S^{[2]}$where S is a particular quartic in $\mathbb{P}^3$) with indeterminacy as some disjoint $\mathbb{P}^2$s (corresponding to lines on the quartic). Any sequence METs which flop just one of the $\mathbb{P}^2$ at a time in the indeterminacy will result a nonhyperKähler fourfold at some point. It implies that at some step in the decomposition according to Theorem 1.1 one needs to do METs on at least two $\mathbb{P}^2$s simultaneously to keep projectivity.

There are also examples of birational maps between two hyperKähler fourfolds with indeterminacy consisting a chain of rational surfaces as shown below.

Example. Let S be the K3 surface defined by the quartic homogeneous polynomial

$$xy(x^2 + y^2 + z^2 + w^2) + w(x^3 + y^3 + z^3 + w^3)$$

in $\mathbb{P}^3$ with coordinates $[x, y, z, w]$. Note that there are three $\mathbb{P}^1$s on S in the $\mathbb{P}^2$ defined by $w = 0$, i.e; two lines and a conic. Let $S^{[2]}$ be the Hilbert scheme of points of length 2 on S. There is a chain of three $\mathbb{P}^2$, say B_1, B_2, B_3 on $S^{[2]}$ corresponding to the three $\mathbb{P}^1$ on S. Let B_1, B_3 be the $\mathbb{P}^2$s corresponding to the two lines. B_1, B_3 are disjoint and they both have one point in common with B_2. Now we perform a MET to B_2 to get Y and METs to B_1, B_3 to get X. Let f be the birational map from X to Y. The indeterminacy on X is a chain of three surfaces with the two ends isomorphic to $\mathbb{P}^2$s and the middle one isomorphic to blowup of $\mathbb{P}^2$ at two points. The indeterminacy on Y is a chain of three surfaces with two ends isomorphic to blowup of $\mathbb{P}^2$ at a point and the middle one isomorphic to $\mathbb{P}^2$. The projectivity of X (or Y) can be checked based on the fact that there is a projective morphism from S contracting the two lines (or the conic).

This example indicates that our situation is more complicated than the picture demonstrated in [Ka1] where the indeterminacy is a disjoint union of $\mathbb{P}^2$s.

ACKNOWLEDGMENTS. . We thank János Kollár for his help, Nick Shepherd-Barron for pointing out an error and Jan Wierzba for sharing his results in [W].

§2 Some General Lemmas

In this section we put together some useful results which are essential to the proof of the main theorem. Some of them are known to experts and their proofs are included for lack of references. Projectivity is not assumed unless it is necessary.

DEFINITION 2.1. *Let X be a holomorphic symplectic fourfold. A MET from (X,B) to (X',B') is the Mukai elementary transformation*

$$\begin{array}{ccccc} & & E \subset \bar{X} & & \\ & p \swarrow & & \searrow q & \\ B \simeq \mathbb{P}^2 \subset X & & - - \to & & X' \supset B' \simeq (\mathbb{P}^2)^* \end{array}$$

where E is the incidence correspondence between B and B'.

Regarding a MET, one has the following basic result due to Shokurov. The proof may be found in Proposition 5-1-11 in [KMM].

LEMMA 2.2. *Let $f: X - - \to X'$ be a MET with exceptional loci $B \subset X$, $B' \subset X'$. Let $E \subset \bar{X}$ be the exceptional divisor. Let H be an divisor such that $H \cdot C < 0$ for any curve $C \subset B$ and H' the proper transform of H in X'. Then $p^*H - q^*H' = aE$ for $a > 0$. Moreover, H' is numerically positive on B', i.e., $H' \cdot C' > 0$ for any curve $C' \subset B'$.*

Several other technical results are also needed.

First, one expects that the exceptional locus of $\Phi : X - - \to X'$ contains a rational curve. The following lemma asserts that in the case of symplectic variety, any rational curve moves in one more family than the Riemann-Roch theorem predicts.

LEMMA 2.3 (RAN). *Assume a symplectic manifold X^n contains a rational curve C, then C deforms at least in $(n-2)$ families.*

PROOF. See the proof in [R] when C is smooth.

When C is singular, we consider the graph of $f : \mathbb{P}^1 \to C \subset X$:

$$\bar{f} : \mathbb{P}^1 \to \bar{C} \subset \mathbb{P}^1 \times X = \bar{X}.$$

$\bar{C}$ is smooth. Let $\bar{\mathcal{H}}$ $(\mathcal{H})$ be the Hilbert scheme containing $\bar{C}$ in $\bar{X}$ (C in X), one has the following estimate by [R] (see also [Ka2])

$$\dim \bar{\mathcal{H}} \geq \chi(N_{\bar{C}/\bar{X}}) + \dim \operatorname{im}(\pi)$$

$$= n + \dim \operatorname{im}(\pi)$$

where π is the semi-regularity map, whose dual is

$$\pi^t : H^0(\bar{X}, \Omega^2_{\bar{X}}) \to H^0(\bar{C}, N^*_{\bar{C}/\bar{X}} \otimes \Omega_{\bar{C}}).$$

To see that π^t is nontrivial, we consider the image of $\beta^*\omega$ where ω is the holomorphic symplectic form on X and α, β are projections from $\bar{X}$ to $\mathbb{P}^1$ and X. We shall show that $\beta^*\omega$ is not zero at any point $y = (x, f(x)) \in \bar{C}$ as long as $f(x)$ is a smooth point on C. Around an analytic neighborhood of y in $\bar{X}$ which is viewed as the product of analytic neighborhoods of x and $f(x)$ in $\mathbb{P}^1$ and X (respectively), one has a (non-canonical) isomorphism

$$\Omega^2_{\bar{X}} \simeq \beta^*\Omega^2_X + \beta^*\Omega_X \wedge \alpha^*\Omega_{\mathbb{P}^1}$$

$$\simeq \beta^* \wedge^2 N^*_{C/X} + \beta^*(\Omega_C \otimes N^*_{C/X}) + \beta^* N^*_{C/X} \otimes \alpha^*\Omega_{\mathbb{P}^1} + \beta^*\Omega_C \wedge \alpha^*\Omega_{\mathbb{P}^1}.$$

Under this identification, we have

$$\wedge^2 N^*_{\bar{C}/\bar{X}} \simeq \beta^* \wedge^2 N^*_{C/X} + \beta^* N^*_{C/X} \otimes \alpha^*\Omega_{\mathbb{P}^1}.$$

Then the non-degenerate property of ω implies that the component of $\beta^*\omega$ in $\beta^*(N^*_{C/X} \otimes \Omega_C)$ is not trivial. Thus $\pi^t(\beta^*(\omega)) \neq 0$.

So the semi-regularity map π is nontrivial. Hence

$$\dim \bar{\mathcal{H}} \geq n + 1.$$

Notice that a nontrivial automorphism of $\mathbb{P}^1$ from the first factor of $\bar{X}$ gives a nontrivial deformation of $\bar{C}$, which however does not move C in X. Therefore

$$\dim \mathcal{H} \geq n + 1 - 3 = n - 2.$$

□

When $n = 4$, we obtain that every rational curve in a symplectic fourfold moves in a at least 2-dimensional family.

The next lemma was pointed out to us by J. Kollár.

LEMMA 2.4. *Let S be a normal surface, proper over $\mathbb{C}$. Then S satisfies exactly one of the following:*

1. *Every morphism $f : \mathbb{P}^1 \to S$ is constant;*
2. *There is a morphism $f : \mathbb{P}^1 \to S$ such that f is rigid.*
3. *$S \cong \mathbb{P}^2$;*
4. *$S \cong \mathbb{P}^1 \times \mathbb{P}^1$, or S is isomorphic to a minimal ruled surface over a curve of positive genus or a minimal ruled surface with a negative section contracted.*

PROOF. If we have either (1) or (2), we are done. Otherwise, there is a morphism $f : \mathbb{P}^1 \to S$ deforming in a 1-parameter family. Thus S is uniruled.

Let $p : \bar{S} \to S$ be the minimal desingularization with the exceptional curve E. $\bar{S}$ is also uniruled, hence there is an extremal ray R. There are 3 possibilities for R.

1. $\bar{S} \cong \mathbb{P}^2$, thus also $S \cong \mathbb{P}^2$ (which implies (3)).
2. $\bar{S}$ is a minimal ruled surface (which implies (4)).
3. R is spanned by a (-1)-curve C_0 in $\bar{S}$.

But (3) is impossible, because the image of C_0 in S would have been rigid. The proof goes as follows. Assume the contrary that $f_0 : \mathbb{P}^1 \cong C_0 \subset \bar{S} \to S$ is not rigid and let $f_t : \mathbb{P}^1 \to S$ be a 1-parameter deformation. For general t, f_t lifts to a family of morphisms $\bar{f}_t : \mathbb{P}^1 \to \bar{S}$. As $t \to 0$, the curves $\bar{f}_t(\mathbb{P}^1)$ degenerate and we obtain a cycle

$$\lim_{t \to 0} \bar{f}_t(\mathbb{P}^1) = C_0 + F$$

where Supp$F \subset$ Supp$E$. $\bar{S}$ is the minimal resolution, thus $K_{\bar{S}} \cdot F \geq 0$. Therefore,

$$K_{\bar{S}} \cdot \bar{f}_t(\mathbb{P}^1) \geq K_{\bar{S}} \cdot C_0 = -1.$$

On the other hand, for a general t the morphism $\bar{f}_t$ is free, thus

$$K_{\bar{S}} \cdot \bar{f}_t(\mathbb{P}^1) \leq -2$$

by II.3.13.1, [Ko2]. This contradication shows that f_0 is rigid. □

We will also use the following lemma which is essentially from 2.19 of [Ketal].

LEMMA 2.5. *Let $\Phi : X - - \to X'$ be a birational map between two hyperKähler fourfolds. Assume H' is ample on X' and H its proper transform on X. Φ is a morphism if H is nef. Φ is an isomorphism if H is ample (or numerically positive).*

PROOF. See the proof of Proposition 2.7 of [C] for the details when H is ample. The same proof goes through when H is numerically positive. □

Finally, the following proposition due to J. Wierzba [W] will be very useful for checking the invariance of the normality of the exceptional locus under METs.

PROPOSITION 2.6. *(Theorem 1.3 and 1.4 in[W]) Let $\pi : \hat{X} \to X$ be an isolated symplectic singularity of dimension 4 and E the exceptional locus. E is a union of irreducible projective surfaces B_i whose normalizations are $\mathbb{P}^2$. Then*

1) if E_i meets E_j along a curve C, C is in the singular locus of either E_i or E_j;

2) if E_i is nonsingular in codimension 1 for all i, E_i is normal for all i.

§3 Proof of the main theorem

PROOF. We now start to prove Theorem 1.1. To begin with, let

$$\Phi : X - - \to X'$$

be a birational map between two hyperKähler fourfolds. This map is necessarily isomorphic in codimension 1. Our goal is to show that the indeterminancy of Φ is a union of disjoint $\mathbb{P}^2$s with other rational surfaces and a moeromorphic map consisting of METs can be performed on some of the disjoint $\mathbb{P}^2$s.

Let H' be a very ample divisor on X'. Let H be its proper transform in X.

We divide the proof into a few steps.

1. First, we consider the pair $(X, \epsilon H)$. It is log-terminal for $\epsilon << 1$. Since H is not nef on X (actually not nef on $B = \cup B_i$, the union of all irreducible subvarieties where f is not defined), by Lemma 2.5, there is a curve $C \subset B$ such that $C \cdot H < 0$.

of g is covered by rational curves. We focus on one connected component of the exaceptional locus in the following. By abusing notation a little, we still use the letter C to denote a rational curve contracted by g. By Lemma 2.2 C moves in at least two famlies and it can not move out of B because $C \cdot H < 0$. Let B_1 be an irreducible component of B which is generically swept out by C. B_1 is contracted by the map g that contracts C. We argue in the following lemma that B_1 has to be contracted to a point.

LEMMA 3.1. *B_1 is contracted to a point by the map g. (In particular, the resulting variety Y has only an isolated singular point locally)*

PROOF. Here the proof uses holomorphic Hamiltonian flows. First note that by Kawamata (Theorem 2, [Ka2]), the exceptional locus of $g : X \to Y$ is covered by a families of rational curves. B_1 is actually covered by at least a two dimensional family of rational curves. Hence B_1 is unirational and thus rational. Clearly, B_1 is a surface. This implies that B_1 is (generically) Lagrangean.

Assume the contrary that the map g mapped B_1 onto a curve D in Y, rather than onto a point as we wish. Let f be a holomorphic function defined in a neighborhood of a general point of D, and which has $df \neq 0$ when restricted to the curve D (locally around the point). We pull this function back up to a neighborhood of a fiber F in the original variety X. Let H_f be the Hamiltonian holomorphic vector field determined by f and the symplectic structure in a neighborhood of the fiber curve F. Since the differential of f on B_1 near F is nonzero, by what we assumed about f and D, and since B_1 is Lagrangean, it follows that H_f is transverse to B_1 along F. Now flowing F along the integral curves of H_f, we will get a holomorphic deformation of F outside of B_1, which contradicts the fact that the contraction of the extremal curve F contracted only B_1 locally around F. □ of the lemma.

It could happen that a rational curve in B_1 moves into another component which is also contracted by g. To show the normalization of B_1 is $\mathbb{P}^2$, we need to know additionally that every rational curve moves within B_1. Assume otherwise. Let C' be a rigid rational curve in B_1 and C'' a general rational curve in B_1 meeting C'. Then C'' moves in only one family which is against Lemma 2.3. Hence every rational curves moves in B_1. By Lemma 2.4, there is a morphism

$$\nu : \mathbb{P}^2 \to B_1$$

such that ν is the normalization of B_1. So $B_1 \simeq \mathbb{P}^2$.

Since there may be several connected components, the exceptional locus of g is a disjoint union of $\mathbb{P}^2$s by using the same argument to each connected component. Let us denote them by $\cup B_{1s}$. We now continue from the step 1 of the proof of the main theorem.

2. Next, we perform METs to $(X, \cup B_{1s})$ to get $(X^1, \cup B^1_{1s})$. Denote the map by ϕ_1. Let H^1 be the proper transform of H. It is easy to see that X_1 remains projective because of the diagram

$$\begin{array}{ccc} & \cup E_{1s} \subset \bar{X} & \\ & p \swarrow \quad \searrow q & \\ \cup B_{1s} \subset X & - - \to & X^1 \supset \cup B^1_{1s} \\ & g \searrow \quad \swarrow g_1 & \\ & Y & \end{array}$$

where $\cup E_{1s}$ is the disjoint union of incidence correspondence between B_{1s} and B^1_{1s}. Note $\bar{X}$ is projective.

By Lemma 2.2, H^1 is numerically positive on all B^1_{1s}. Let $\cup_{i\geq k_1} B^1_i$ be the image of $\cup_{i\geq k_1} B_i$ under ϕ_1, those B_i other than B_{1s}. We are done if H^1 is numerically positive. Otherwise there is a curve $C^1 \subset \cup_{i\geq k_1} B^1_i$ such that $C^1 \cdot H^1 < 0$. Obviously C^1 is not contained in $\cup B^1_{1s}$ because of the positivity of H^1 on $\cup B^1_{1s}$. Again, we apply the contraction-rationality theorem of [Ka2] to the log-terminal pair $(X^1, \epsilon_1 H^1)$ for $\epsilon_1 << 1$ to get a rational curve C_1 which deforms in an at least two-dimensional family. Again let us consider one connected component of the exceptional locus. Let B^1_2 ($\neq B^1_1$) be the irreducible component which contains the family.

B^1_2 normalizes to a $\mathbb{P}^2$ and is a $\mathbb{P}^2$ if it is normal by the same arguement used before. There are two situations: 1) B^1_2 does not intersect B^1_1. 2) B^1_2 does intersect B^1_1. In case 1) $B_2 \simeq B^1_2$, so B^1_2 is normal.

In case 2), the intersection must be a set of finitely many points since H^1 is positive on B^1_1 and negative on B^1_2. This implies that B_1 intersects B_2 along some $\mathbb{P}^1$s before the MET. Moreover the nonnormal locus of B^1_2 is contained in $B^1_1 \cap B^1_2$, hence isolated. If B^1_i is another irreducible component which is also contracted, again B^1_i intersects B^1_1 at isolated points. This implies that B^1_i could be nonnormal only at finitely many points since the original B_i is assumed to be normal. Proposition 2.6 says that nonnonormal locus must be empty. After all $B^1_2 \simeq \mathbb{P}^2$.

Applying the argument to all connected components, we see that the exceptional locus is a disjoint union of $\mathbb{P}^2$s. Let us denote them by $\cup B^1_{2s}$.

3. Perform METs to $(X^1, \cup B^1_{2s})$ to get $(X^2, \cup B^2_{2s})$. After k steps of doing MET, we arrive at $(X^k, \cup B^k_{ks})$. If the proper transform H^k of H is positive on B^k, we are done. Otherwise there is a disjoint union $\cup B^k_{(k+1)s}$, each of them normalizing to $\mathbb{P}^2$ and H^k is negative on them. We may assume that $\cup B^k_{(k+1)s}$ is contracted by a logterminal contraction assoiated with $(X^k, \epsilon_k H^k)$ for $\epsilon_k << 1$. A comparison between $\cup B_{(k+1)s} \subset X$ (each component is assumed to be normal) and $\cup B^k_{(k+1)s}$ shows that the nonnormal locus of each $B^k_{(k+1)s}$ is a set of finitely many points. More precisely, assume some $B^k_{(k+1)s}$ is singular along a curve C^k, we backtrack to a previous i-th step ($i > 1$) after which the cprresponding $B^i_{(k+1)s}$ become singular along the curve C^i, the proper transform of C^k. Note that $H^i.C < 0$ and $B^i_{(k+1)s}$ has to intersect $\cup B^i_{is}$. Since H^i is positive on $\cup B^i_{is}$, we conclude that $B^i_{(k+1)s}$ intersects $\cup B^i_{is}$ at finitely many points. But this implies that $B^i_{(k+1)s}$ can not be singular along C^i. Again Proposition 2.6 says that $B^k_{(k+1)s}$ is normal and hence is a $\mathbb{P}^2$. So $\cup B^k_{(k+1)s}$ is a disjoint union of $\mathbb{P}^2$s. METs can be performed on $(X^k, \cup B^k_{(k+1)s})$.

Since each step reduces the number of irreducible components in the indeterminancy by at least one, the above process has to terminate. After finitely many steps we obtain

$$\Phi^m : X^m - - \rightarrow X'$$

The above proof has the following consequence on the uniqueness of METs.

COROLLARY 3.2. *Let $\Phi : X - - \to X'$ be a birational transformation of two hyperKähler fourfolds which is obtained by blowing-up a smooth center in each of X, X'. Then Φ is a Mukai elementary transformation up to isomorphism.*

§4 Rational Hodge structure

As an Application we show that Theorem 1.1 yields an isomorphism between the rational Hodge structures of two hyperKähler fourfolds. Similar and related results have been proved by other authors (for example [D] Cor. 5.3) using different methods.

COROLLARY 4.1. *Let $\Phi : X - - \to X'$ be a birational morphism between two hyperKähler fourfolds. Then Φ induces an isomorphism between the rational Hodge structures of X and X'.*

PROOF. Theorem 1.1 reduces the proof to the case when $\Phi : X - - \to X'$ consists of several METs done to a disjoint union of $\mathbb{P}^2$s. Let Z be the common blowdown of X and X' by collapsing the exceptional loci of Φ and Φ^{-1} to several isolated points $\{z's\}$. Then the contraction $g : X \to Z$ ($g' : X' \to Z$) is strictly semi-small in the sense that $\dim g^{-1}(\{z's\}) = \frac{1}{2}\operatorname{codim}\{z's\}$. By the Beilinson-Bernstein-Deligne-Gabber decomposition theorem (applied to the contraction g and g'), we have the following quasi-isomorphisms between perverse sheaves

$$Rg_*\mathbb{C}^\bullet_X \cong IC^\bullet(Z) \oplus IC^\bullet(\{z's\})[-4]$$

and

$$Rg'_*\mathbb{C}^\bullet_{X'} \cong IC^\bullet(Z) \oplus IC^\bullet(\{z's\})[-4].$$

Both isomorphisms are compatible with rational Hodge decompositions by M. Saito's results. This proves the corollary. □

References

[C] A. Corti, *Factoring birational maps of threefolds after Sarkisov*, J. Alg. Geom. **4** (1995), 223-254.

[H] D. Huybrecht, *Birational symplectic manifolds and their deformations*, J. Diff. Geom. **45** (1997), 488-513.

[Ka1] Y. Kawamata, *Small contractions of four dimensional algebraic manifolds*, Math. Ann. **284** (1989), 595-600.

[Ka2] Y. Kawamata, *On the length of an extremal rational curves*, Invent. Math. **105** (1991), 609-611.

[Ka3] Y. Kawamata, *Unobstructed deformations, II*, J. Alg. Geom. **4** (1995), 277-279.

[KMM] Y. Kawamata, K, Matsuda and K. Matsuki, *Introduction to the minimal model program*, Adv. Stud. Pure Math., vol. 10, 1987, pp. 283-360.

[Ko1] J. Kollár, *Flops,*, Nagoya Math. J. **113** (1989), 15-36.

[Ko2] J. Kollár, *Rational curves on algebraic varieties*, Ergeb. der Math. und ihrer Grenzg., Springer-Verlag, Berlin, New York, 1995.

[Ketal] J. Kollár et al, Astérisque **217** (1992).

[KMM] Y. Kawamata, K. Matsuda, K. Matsuki, *Introduction to the minimal model problem*, Algebraic geometry, Sendai, Adv. Stud. Pure Math., vol. 10, 1985, pp. 283-360.

[Mu1] S. Mukai, *Symplectic structure of the moduli space of sheaves on an abelian or K3 surface*, Invent. math. **77** (1984), 101-116.

[Mu2] S. Mukai, *Moduli of vector bundles on K3 surfaces and symplectic manifolds*, Sugaku Expositions **1** (1988), 139-174.

[N] Y. Namikawa, *Deformation theory of singular symplectic n-folds*, math.AG/0010113 (2000).
[R] Z. Ran, *Hodge theory and Hilbert schemes*, J. Diff. Geom. **37** (1993), 191-198.
[W] J. Wierzba, *On 4-dimensional isolated symplectic singularities*, preprint (1999).

Department of Mathematics, University of Michigan, Ann Arbor, MI 48109

Department of Mathematics, University of Arizona, Tucson, AZ 85721

Department of Mathematics, University of Texas, Arlington, TX 76019

Contemporary Mathematics
Volume **322**, 2003

Standard vector bundle deformations on $\mathbb{P}^n$

N. Mohan Kumar, Chris Peterson, and A. Prabhakar Rao

ABSTRACT. This paper illustrates how certain construction methods for low rank vector bundles on projective spaces can be modified to produce interesting deformations of bundles.

1. Introduction

The problem of constructing enough vector bundles on projective spaces is an old one. The case of rank one vector bundles (line bundles) on $\mathbb{P}^n$ is well understood. The only line bundles on $\mathbb{P}^n$ are the twists (or powers), $\mathcal{O}_{\mathbb{P}^n}(\nu)$, of the hyperplane bundle $\mathcal{O}_{\mathbb{P}^n}(1)$. The case of rank r vector bundles on $\mathbb{P}^n$ with $r > 1$ is not very well understood. Of course, one can obtain some rank r bundles by taking the direct sum of bundles of rank less than r. This, in general, yields only a small subset of the set of all bundles of rank r. A direct sum of line bundles will be called a free bundle, and a direct sum of bundles, all equal to $\mathcal{O}_{\mathbb{P}^n}$ will be called a trivial bundle. We will be concerned with the problem of producing non-free bundles. For any free bundle $\mathcal{G}$ on $\mathbb{P}^n$, it is true that the cohomology modules $H^i_*(\mathbb{P}^n, \mathcal{G})$ (defined as $\oplus_{\mu \in \mathbb{Z}} H^i(\mathbb{P}^n, \mathcal{G}(\mu))$, where $\mathcal{G}(\mu)$ is the twist $\mathcal{G} \otimes \mathcal{O}_{\mathbb{P}^n}(\mu)$ of $\mathcal{G}$), are zero for $1 \leq i \leq n-1$. A theorem of Horrocks asserts the converse. This combination of facts will provide a useful tool for establishing that a bundle is not free.

It is easy to obtain non-free bundles on $\mathbb{P}^n$ if the rank of the bundle is allowed to be large enough. By a theorem of Serre, given any bundle $\mathcal{G}$ on $\mathbb{P}^n$, of rank m, where $m > n$, then for μ large enough, the twist $\mathcal{G}(\mu)$ contains a trivial sub-bundle of rank $m - n$. The quotient will be non-free because of the following argument that we insert here (since it will subsequently appear often). The dual map $\mathcal{G}^\vee(-\mu) \to (m-n)\mathcal{O}_{\mathbb{P}^n} \to 0$ is not surjective on global sections since when μ is large, $\mathcal{G}^\vee(-\mu)$ will have no global sections while clearly, $(m-n)\mathcal{O}_{\mathbb{P}^n}$ has $m - n$ global sections. Hence the bundle which is the kernel, call it $\mathcal{K}$, has $H^1_*(\mathbb{P}^n, \mathcal{K}) \neq 0$ by the long exact sequence of cohomology. The bundle, $\mathcal{K}$, is therefore not free, and neither is the quotient which is its dual, $\mathcal{K}^*$.

Since in this argument, $\mathcal{G}$ could have been a free bundle, there is no obstruction to the construction of non-free bundles of rank n (or more) on $\mathbb{P}^n$. The problem

---

2000 *Mathematics Subject Classification.* Primary 14F05, 13D02, 14J60, 32L05, 13A35.

hence is really one of constructing non-free bundles of rank less than n on $\mathbb{P}^n$. We will indicate by *corank* the difference between the dimension of the space and the rank of the bundle. Hence one looks for non-free bundles on $\mathbb{P}^n$ of positive corank. The case of $\mathbb{P}^2$ is trivial, and corank one bundles on $\mathbb{P}^3$ have been well studied, with many non-free bundles of rank two, both stable and unstable, available in large families.

The outstanding question in this area is whether there exists on $\mathbb{P}^6$ a rank two bundle which is not a sum of two line bundles. This question was posed by Mumford and Horrocks when they constructed the rank two Horrocks-Mumford bundle on $\mathbb{P}^4$. In characteristic zero, the Horrocks-Mumford bundle is the only known rank two bundle on $\mathbb{P}^4$ which is not free, except for minor variations on it like the pull-back via morphisms from $\mathbb{P}^4$ to $\mathbb{P}^4$. In rank three, there are bundles on $\mathbb{P}^4$ of at least two types known: a construction due to Trautmann and Vetter and independently by Tango, and a construction of Sasakura. On $\mathbb{P}^5$, there is a rank three bundle that was constructed by Horrocks called the parent bundle, and there are variations of this bundle. Apart from the Horrocks-Mumford bundle and Horrocks' parent bundle (together with their modifications), in characteristic zero, all other known positive corank, non-free bundles are of corank one. There are two main methods for the construction of corank one bundles on $\mathbb{P}^n$. One construction gives null-correlation bundles and more generally mathematical instanton bundles on $\mathbb{P}^n$ where n is odd. The other construction, of Trautmann-Vetter and also Tango, gives corank one bundles on $\mathbb{P}^n$ for any n. No corank two bundles are known on $\mathbb{P}^n$ (with $n \geq 6$) in any characteristic.

Mohan Kumar in [**7**], [**8**] discussed the construction of bundles on $\mathbb{P}^n$ as well as the construction of bundles on $\mathbb{P}^{n-1} \times \mathbb{A}^1$. In his discussion, he addressed a problem (due to Peskine) about the existence of a family of rank two bundles on $\mathbb{P}^3$, where the general bundle is free and the special one is non-free. Both this question and the question of the existence of non-free rank two bundles on $\mathbb{P}^4$ (other than the Horrocks-Mumford bundle) were translated into one of constructing certain bundles on $\mathbb{P}^3$ with nilpotent endomorphisms. In positive characteristic, Mohan Kumar, through this translation, succeeded in obtaining examples for each of the two questions, in much the same way. In addition, in any characteristic, his approach gives rank three bundles on $\mathbb{P}^4$ as well as families of rank three bundles on $\mathbb{P}^3$, where the general bundle is free and the special one is non-free. These were also studied in [**10**] and [**11**].

In view of this, in this paper, we look at Peskine's question in higher rank, on $\mathbb{P}^n$. The question is whether there are families of bundles on $\mathbb{P}^n$, where the general one is free, and the special one is not. Once again, as in the construction of bundles on $\mathbb{P}^n$, this question is not of interest if the rank is too large. For example, the bundle of 1-forms on $\mathbb{P}^n$ fits into the Euler exact sequence on $\mathbb{P}^n$

$$0 \to \Omega^1_{\mathbb{P}^n} \to (n+1)\mathcal{O}_{\mathbb{P}^n}(-1) \to \mathcal{O}_{\mathbb{P}^n} \to 0.$$

Since this is a non-split sequence, we immediately that the family parametrized by $\mathbb{A}^1 = \mathrm{Ext}^1(\mathcal{O}_{\mathbb{P}^n}, \Omega^1_{\mathbb{P}^n})$ has generic member equal to $(n+1)\mathcal{O}_{\mathbb{P}^n}(-1)$ and special member (when the extension class is zero) equal to $\Omega^1_{\mathbb{P}^n} \oplus \mathcal{O}_{\mathbb{P}^n}$.

Hence the Peskine question is interesting only when we consider families of bundles on $\mathbb{P}^n$ of corank greater than or equal to zero. Our purpose in this paper is to restrict ourselves to arbitrary characteristic, and review the constructions of some of the low rank and low corank bundles on $\mathbb{P}^n$ listed above. The ones that we review allow also a modified construction that creates a bundle of the same rank, not on $\mathbb{P}^n$, but on $\mathbb{P}^{n-1} \times \mathbb{A}^1$. Our modification is quite an easy one. The bundles we review are all constructed out of the standard Koszul complex on $\mathbb{P}^n$, which is the complex built out of a regular sequence of $n+1$ linear forms on $\mathbb{P}^n$. We modify the constructions to include also regular sequences on $\mathbb{P}^{n-1} \times \mathbb{A}^1$, where the degrees of the terms in the regular sequence can vary, with one term being the zero degree affine parameter. This modification, which is straightforward, has the pleasant consequence that since, when the affine parameter is non-zero, the restriction of the Koszul complex to the $\mathbb{P}^{n-1}$ fibre splits, we get a family of bundles for which the general bundle is free but the special bundle is non-free.

Our results are as follows:

(1) a family of rank three bundles on $\mathbb{P}^3$, where the general bundle is free (and non-trivial) while the special bundle is non-free.
(2) a family of rank three bundles on $\mathbb{P}^3$, where the general bundle is trivial while the special bundle is non-free.
(3) a family of rank three bundles on $\mathbb{P}^4$, where the general bundle is free (and either trivial or non-trivial) while the special bundle is non-free.
(4) a family of corank zero bundles on $\mathbb{P}^n$ (n even), where the general bundle is free (and either trivial or non-trivial) while the special bundle is non-free.
(5) a family of corank zero bundles on $\mathbb{P}^n$ (n odd, $n \geq 5$), where the general bundle is free (and non-trivial) while the special bundle is non-free.

Missing from our results are:

- a family of corank zero bundles on $\mathbb{P}^n$ (n odd, $n \geq 5$), where the general bundle is trivial while the special bundle is non-free. From our point of view, the case $n = 3$ is special because we can exploit many different constructions on $\mathbb{P}^4$, that of Trautmann-Vetter and that of Sasakura, as well as the Horrocks construction on $\mathbb{P}^5$ (which can be restricted to $\mathbb{P}^4$) and the method of Mohan Kumar. For higher even dimensional spaces, we have only the construction of Trautmann-Vetter to exploit.
- a family of rank two bundles on $\mathbb{P}^3$, as in Peskine's original question, in zero characteristic. From our point of view, in characteristic zero, there is only one known rank two bundle on $\mathbb{P}^4$, and it resists the kind of exploitation that worked for our other cases. G. Ottaviani has pointed out an explanation for this resistance. The modified constructions we make below are similar to Horrocks' notion of a weighted pull-back which can be created for bundles with a $\mathbf{C}^*$ group of symmetries. The Horrocks-Mumford bundle has only a discrete group of symmetries.
- on $\mathbb{P}^3$, and given three degrees, a family of bundles on $\mathbb{P}^3$ where the general bundle is a sum of line bundles of those degrees while the special one is non-free. Both Trautmann-Vetter and Horrocks constructions give arithmetic sequences for the degrees, while Sasakura's construction gives even degrees.

The authors would like to thank G. Ottaviani for correspondence on these topics. The third author would like to thank Jun-Muk Hwang for an invitation to KIAS, and for his enquiry about deforming rank three bundles which started this

2. The Koszul complex

Let X represent either $\mathbb{P}^n$ or $\mathbb{P}^{n-1} \times \mathbb{A}^1$ over a field k. We will denote by S_X the graded polynomial ring which in the first case is $k[X_0, X_1, \ldots, X_n]$ (where the X_i's of degree one give a basis of linear forms on $\mathbb{P}^n$) and in the second case is $k[X_0, X_1, \ldots, X_{n-1}, t]$ (where the X_i's of degree one give a basis of linear forms on $\mathbb{P}^{n-1}$ and t of degree zero is the parameter on $\mathbb{A}^1 = \operatorname{Spec} k[t]$).

We will let $\mathcal{O}_X(\nu)$ denote the power of the hyperplane line bundle on $\mathbb{P}^n$ or the pull-back of $\mathcal{O}_X(\nu)$ on $\mathbb{P}^{n-1}$. A global section of $\mathcal{O}_X(\nu)$ will be a form in S_X of degree ν. For a bundle $\mathcal{E}$ on X, we denote by $H^i_*(\mathcal{E})$ the graded S_X module $\bigoplus_{\nu \in \mathbb{Z}} H^i(X, \mathcal{E}(\nu))$.

Let us establish notation for Koszul complexes.

Let $a_0, a_1, \ldots, a_r$ be a sequence of homogeneous forms on X. By this we mean a sequence of homogeneous elements chosen from S_X. Let d_i equal the degree of a_i. Let $K_1 = \oplus_{i=0}^r e_i \mathcal{O}_X(-d_i)$, where $e_0, e_1, \ldots, e_r$ is a basis for the direct sum. Let $K_q = \wedge^q K_1$. We denote $e_{i_1} \wedge e_{i_2} \wedge \cdots \wedge e_{i_q}$ by $e_{i_1 i_2 \ldots i_q}$ for short and when $i_1 < i_2 < \ldots i_q$, these will be our basis vectors of choice for K_q, appearing with $\mathcal{O}_X(-d_{i_1} - d_{i_2} - \cdots - d_{i_q})$. The Koszul map from K_{q+1} to K_q is given by $\sum a_i e_i^*$, where $\{e_i^*\}$ is the dual basis to $\{e_i\}$. We will use a matrix representation of the Koszul map, where the bases of K_{q+1} and K_q are ordered lexicographically by multi-index, calling the matrix M_q. So for example, e_{13} is mapped to $a_3 e_1 - a_1 e_3$, and if $r = 4$, then the sixth column of M_1 is $\begin{bmatrix} 0 & a_3 & 0 & -a_1 & 0 \end{bmatrix}^\vee$.

The resulting complex $(K_\bullet, M_\bullet)$ is called the Koszul complex on X corresponding to the sequence $a_0, a_1, \ldots, a_r$. We denote by $\mathcal{K}^q$ the image of the Koszul map given by M_q.

There are four cases we will be interested in.

- If $X = \mathbb{P}^n$ and the sequence is $a_0, a_1, \ldots, a_n$, a regular sequence of $n+1$ homogeneous forms (necessarily of positive degree) on X, then the Koszul complex is exact, and the image $\mathcal{K}^q$ of the Koszul map M_q will be a vector sub-bundle of K_q of rank $\binom{n}{q}$. $\mathcal{K}^q$ is generated by its global sections (in various twists) $e_{i_1 i_2 \ldots i_{q+1}}$. Each $\mathcal{K}^q$ is non-free, for $0 < q < n$.
- If $X = \mathbb{P}^n$ and the sequence is $a_0, a_1, \ldots, a_{n+1}$ where $a_0, a_1, \ldots, a_n$ is a regular sequence of $n+1$ homogeneous forms on X and a_{n+1} is the unit element 1, then the Koszul complex is split exact, and each $\mathcal{K}^q$ is a direct summand of K_q, hence free.
- If $X = \mathbb{P}^n$ and the sequence is $a_0, a_1, \ldots, a_{n+1}$ where $a_0, a_1, \ldots, a_n$ is a regular sequence of $n+1$ homogeneous forms on X and a_{n+1} is the zero element, then the Koszul complex is the direct sum of two copies of the Koszul complex on $a_0, a_1, \ldots, a_n$, where the second is shifted in grading by one. In particular, the Koszul complex is exact and $\mathcal{K}^q$ is a vector bundle which is non-free in the range $0 < q < n+1$.
- The *standard Koszul complex* is when $X = \mathbb{P}^n$ and the regular sequence is $\{l_0, l_1, \ldots, l_n\}$, where these are independent linear forms on $\mathbb{P}^n$. The bundles $\mathcal{K}^q$ are then the usual bundles of q-forms $\Omega^q_{\mathbb{P}^n}$ on $\mathbb{P}^n$.

When we are looking for deformations, we will be considering the case $X = \mathbb{P}^n \times \mathbb{A}^1$, and deal with regular sequences of the form $a_0, a_1, \ldots, a_{n+1}$ where $a_{n+1} = t$ is the affine parameter of degree 0 and $a_0, a_1, \ldots, a_n$ is a regular sequence on $\mathbb{P}^n$, not depending on t. This will give rise to bundles $\mathcal{K}^q$ on $\mathbb{P}^n \times \mathbb{A}^1$, which will be viewed as a family of bundles on the fibres $\mathbb{P}^n$ of the product. When we look at a fibre $\mathbb{P}^n$ over a nonzero value, say 1, of t, the restriction of $\mathcal{K}^q$ will be a sum of line bundles, since the Koszul complex will split. But the restriction to the special fibre $t = 0$ will be non-split. (When $q = 1$, this re-creates the banal example obtained from the Euler sequence that we described in the introduction.)

3. Null-correlation bundles.

Let $X = \mathbb{P}^n$ be odd dimensional, with $n \geq 3$. Let $S_X = k[X_0, X_1, \ldots, X_n]$ be the underlying polynomial ring. Choose a regular sequence of linear forms $l_0, l_1, \ldots, l_n$. Then the following is the *monad* for a rank $n-1$ bundle, $\mathcal{N}$, on $\mathbb{P}^n$:

$$0 \to \mathcal{O}_{\mathbb{P}^n}(-1) \xrightarrow{[-l_n, +l_{n-1}, -l_{n-2}, \ldots, -l_1, +l_0]^\vee} (n+1)\mathcal{O}_{\mathbb{P}^n} \xrightarrow{[l_0, l_1, \ldots l_{n-1}, l_n]} \mathcal{O}_{\mathbb{P}^n}(1) \to 0.$$

In other words,

- this sequence is a complex (obvious from the matrices of the maps),
- the left hand map is an injection of vector bundles (since the map is never zero at any point of $\mathbb{P}^n$),
- the right hand map is a surjection of vector bundles (since again the map is never zero at any point of $\mathbb{P}^n$).

Hence the homology at the middle is a vector bundle, $\mathcal{N}$, of rank $n-1$. (The oddness of n is required because this monad is pairing the linear forms.) The bundle is non-free because the right hand map is not surjective on the level of global sections in degree -1, and hence the S-module $H^1_*(\mathbb{P}^n, \mathcal{N})$ is non-zero and Horrocks' theorem applies.

This construction is easily mimicked over $X = \mathbb{P}^n \times \mathbb{A}^1$, when $\mathbb{P}^n$ is *even* dimensional. Let $S_X = k[X_0, X_1, \ldots, X_n, t]$ be the underlying polynomial ring of X. Let $a_0, a_1, \ldots, a_n, t$ be a regular sequence in S_X of homogeneous forms of degree $n+1, n, n-1, \ldots, 1, 0$, where $a_0, a_1, \ldots, a_n$ are forms on $\mathbb{P}^n$ itself, independent of t. We construct a monad on X as follows:

$$(3.1) \qquad 0 \to \mathcal{O}_X\left(-\frac{n+1}{2}\right) \xrightarrow{[-t, +a_n, -a_{n-1}, \ldots, +a_0]^\vee}$$
$$\oplus_{i=0}^{n+1} \mathcal{O}_X\left(-\frac{n+1}{2} + i\right) \xrightarrow{[a_0, a_1, \ldots, a_n, t]} \mathcal{O}_X\left(\frac{n+1}{2}\right) \to 0.$$

Our choice of degrees makes this a valid complex. By inspection of the ranks of the two matrices at each point of X, we see that this is the monad of a vector bundle $\mathcal{N}$ on $\mathbb{P}^n \times \mathbb{A}^1$, of rank n. Consider the general member of this family of rank n bundles on $\mathbb{P}^n$, when $\mathbb{P}^n$ is the fibre over a point $t \neq 0$. Both the left hand map and the right hand map of the monad are easily seen to be split in this case. Hence for $t \neq 0$, say $t = 1$, $\mathcal{N}_{t=1}$ is the sum of line bundles $\oplus_{i=1}^n \mathcal{O}_{\mathbb{P}^n}(-\frac{n+1}{2} + i)$. On the other hand, when $t = 0$, the monad is a minimal monad on $\mathbb{P}^n$. We conclude that the bundle, $\mathcal{N}_{t=0}$, is non-free because the right hand map is not surjective on the level of global sections (and like before, the S-module $H^1_*(\mathbb{P}^n, \mathcal{N}_{t=0})$ is non-zero equal to $\frac{n+1}{2}$, and a_0 of degree $n+1$ as before. Now the monad looks like

$$(3.2) \qquad 0 \to \mathcal{O}_X(-\frac{n+1}{2}) \xrightarrow{[-t,+a_n,-a_{n-1},\ldots,+a_0]^\vee}$$

$$\mathcal{O}_X(-\frac{n+1}{2}) \oplus (n-1)\mathcal{O}_X \oplus \mathcal{O}_X(\frac{n+1}{2}) \xrightarrow{[a_0,a_1,\ldots,a_n,t]} \mathcal{O}_X(\frac{n+1}{2}) \to 0.$$

It follows in this case that the general fibre $\mathcal{N}_{t=1} = (n-1)\mathcal{O}_{\mathbb{P}^n}$, a trivial bundle. So, on even dimensional projective spaces, we can find families of corank zero bundles where the special bundle is non-free and the general bundle is free, or even trivial. In fact, after normalizing, we can make the general fibre to be anything we want as long as the degrees of the summands are symmetric around zero. However, this construction does not yield free general fibres where, for example, the first Chern class is odd.

4. Trautmann-Vetter bundles.

The construction of null-correlation bundles, described above, works only on odd dimensional projective spaces. A construction due to Trautmann-Vetter and, independently by Tango, creates bundles of rank $n-1$ on any dimensional projective space. Tango's approach is to perform Chern class calculations on bundles appearing in the standard Koszul complex, which we cannot easily mimick in our setting for deformations. However, Trautmann-Vetter give a more explicit construction in terms of the bases and matrices of the Koszul complex and this readily carries over to the setting for deformations, provided some care is taken to ensure homogeneity of maps. (Cascini [**3**] has also made variations of these bundles to construct 'weighted Tango bundles'.)

The construction of the Trautmann-Vetter bundle is achieved by considering global sections of $\Omega^1_{\mathbb{P}^n}$ on $\mathbb{P}^n$. Using the regular sequence $X_0, X_1, \ldots, X_n$ on $\mathbb{P}^n$, and the standard Koszul complex $K_\bullet$ built out of it, $\Omega^1_{\mathbb{P}^n}(2)$ has a basis of $\binom{n+1}{2}$ global sections given by e_{ij} (where $i<j$) which generate the bundle at each point of $\mathbb{P}^n$. However, Trautmann-Vetter observe that just $2n-1$ sections suffice to generate the bundle $\Omega^1_{\mathbb{P}^n}(2)$ at each point, given by the following set:

$$\{\sum_{\substack{i<j \\ i+j=k}} e_{ij} | k = 1, 2, \ldots, 2n-1\}.$$

So for example, on $\mathbb{P}^3$, five sections $e_{01},\ e_{02},\ e_{03}+e_{12},\ e_{13},\ e_{23}$ generate $\Omega^1_{\mathbb{P}^3}(2)$. The proof is obvious when we look at the representation of these sections under the inclusion $\Omega^1_{\mathbb{P}^3} \subseteq K_1$. For example, in the case of $\mathbb{P}^3$, we get the 5 columns of the matrix

$$\begin{bmatrix} X_1 & X_2 & X_3 & 0 & 0 \\ -X_0 & 0 & X_2 & X_3 & 0 \\ 0 & -X_0 & -X_1 & 0 & X_3 \\ 0 & 0 & -X_0 & -X_1 & -X_2 \end{bmatrix}.$$

Now, to verify that these 5 sections generate the rank 3 bundle $\Omega^1_{\mathbb{P}^3}(2)$ on $\mathbb{P}^3$ at each point, it is enough to show that some 3×3 minor of the matrix is non-vanishing at each point of $\mathbb{P}^3$. But it is evident that modulo $X_0, X_1, \ldots, X_{i-1}$, a 3×3 upper triangular matrix can be extracted with X_i along the diagonal. This argument clearly also works in the case of general n, giving Trautmann-Vetter's observation.

Let $\mathcal{T}$ be the kernel of the induced surjection:

$$0 \to \mathcal{T} \to \oplus_{k=1}^{2n-1}\{\sum_{\substack{i<j\\ i+j=k}} e_{ij}\}\mathcal{O}_{\mathbb{P}^n}(-2) \to \Omega^1_{\mathbb{P}^n} \to 0.$$

In this case, $\mathcal{T}$ is a rank $n-1$ bundle on $\mathbb{P}^n$, and is not a sum of line-bundles since the map to $\Omega^1_{\mathbb{P}^n}(2)$ does not surject onto all the global sections, hence $H^1(\mathbb{P}^n, \mathcal{T}(-2)) \neq 0$.

With this example to follow, we have a Trautmann-Vetter type construction of a family of rank n bundles on $\mathbb{P}^n$, where the generic bundle is a sum of line bundles and the special one is not, obtained as follows.

First choose a regular sequence $a_0, a_1, \ldots, a_n, a_{n+1} = t$ on $\mathbb{P}^n \times \mathbb{A}^1$, where $a_0, a_1, \ldots, a_n$ is a regular sequence on $\mathbb{P}^n$, independent of t, with degree a_i equal to $d_i = n+1-i$. Consider the Koszul complex $K_\bullet$ on $\mathbb{P}^n \times \mathbb{A}^1$ given by this sequence, and consider $\mathcal{K}^1$. Each sum, $\sum_{\substack{i<j\\ i+j=k}} e_{ij}$, is a *homogeneous* section of $\mathcal{K}^1(k)$, for $1 \leq k \leq 2n+1$. These $2n+1$ sections generate the rank $n+1$ bundle $\mathcal{K}^1$, by an argument based upon matrices that is identical to the argument above. The kernel is therefore a bundle $\mathcal{T}$ on $X = \mathbb{P}^n \times \mathbb{A}^1$, of rank n, where

$$0 \to \mathcal{T} \to \oplus_{k=1}^{2n+1}\{\sum_{\substack{i<j\\ i+j=k}} e_{ij}\}\mathcal{O}_X(-k) \to \mathcal{K}^1 \to 0.$$

When $t \neq 0$, say $t = 1$, the restriction to the fibre $\mathbb{P}^n$ is trivial because $\mathcal{K}^1_{t=1}$ is trivial on this fibre and the matrix of the representation of the $2n+1$ sections shows explicitly a splitting because of the uppermost diagonal with $n+1$ copies of t. When $t = 0$, the restriction $\mathcal{K}^1_{t=0}$ has a summand of $\mathcal{O}_{\mathbb{P}^n}$, yet the $2n+1$ sections we have chosen do not map onto the generating section of this $\mathcal{O}_{\mathbb{P}^n}$. Hence $\mathcal{T}_{t=0}$ picks up a non-zero element in H^1_* from the exact sequence connecting $\mathcal{T}_{t=0}$ and $\mathcal{K}^1_{t=0}$. Hence, by Horrocks' theorem, $\mathcal{T}_{t=0}$ is non-free.

The general bundle in this case is seen to be $\oplus_{k=n+2}^{2n+1}\mathcal{O}_{\mathbb{P}^n}(-k)$. So we have obtained a family of rank n bundles on $\mathbb{P}^n$ where the general bundle is free (but not trivial), and the special bundle is not free. The combinatorics of Trautmann-Vetter's construction forces the degrees of the general bundles to be always in a non-trivial arithmetic sequence (provided $X = \mathbb{P}^n \times \mathbb{A}^1$ has dimension ≥ 4). Hence the types of degrees allowed for the general bundle in the family are quite restricted. In particular, it does not provide an example where the general bundle is trivial.

5. Sasakura's bundle

The Trautmann-Vetter construction above gives a family of rank 3 bundles on $\mathbb{P}^3$, where the general bundle is free, and the special bundle is not. After normalizing, the general fibre is of the form $\mathcal{O}_{\mathbb{P}^3} \oplus \mathcal{O}_{\mathbb{P}^3}(1) \oplus \mathcal{O}_{\mathbb{P}^3}(2)$. Clearly, by pull-backs via maps from $\mathbb{P}^3$ to $\mathbb{P}^3$, one can arrange the general fibre to be of the form $\mathcal{O}_{\mathbb{P}^3} \oplus \mathcal{O}_{\mathbb{P}^3}(a) \oplus \mathcal{O}_{\mathbb{P}^3}(2a)$, *ie.* any arithmetic progression of degrees is achievable. To get examples where the general fibre is not of this form, a different rank three bundle contruction on $\mathbb{P}^4$ due to Sasakura can be used. A treatment of this construction can be found in [**1**]. The treatment described below is analogous difference is that while the Trautmann-Vetter rank three bundle appears as the kernel of a surjection

$$7\mathcal{O}_{\mathbb{P}^4}(-2) \to \Omega^1_{\mathbb{P}^4},$$

Sasakura's bundle appears as the homology of a monad of the form

$$0 \to 2\mathcal{O}_{\mathbb{P}^4} \to 9\mathcal{O}_{\mathbb{P}^4} \to \mathcal{K}^1 \to 0.$$

where we do not specify here the regular sequence, nor the degrees in the monad above.

In what follows, X will be either $\mathbb{P}^4$ or $\mathbb{P}^3 \times \mathbb{A}^1$, and we will take a regular sequence a_0, a_1, a_2, a_3, a_4 on X. (In a change from previous examples, when $X = \mathbb{P}^3 \times \mathbb{A}^1$, it will be a_2 which will be t, and the others a regular sequence on $\mathbb{P}^3$.)

In the Koszul complex on the above regular sequence, subject to the conditions of homogeneity, replace the ten basis elements of K_2 by a new basis where e_{04} and e_{13} are replaced by $e_{04}+e_{13}, e_{04}-e_{13}$. $\mathcal{K}^1$ is generated by the nine basis elements after leaving out $e_{04}-e_{13}$ (this follows for instance, from the stronger observation of Trautmann-Vetter). Since $\mathcal{K}^1$ has rank four, the kernel of this surjection by nine copies of $\mathcal{O}_X$ has rank five.

Consider the basis vectors of K_3 consisting of $e_{024}+e_{123}, e_{234}+a_1e_{013}-a_0e_{014}$, once again subject to conditions of homogeneity. It is an easy calculation that these two vectors map to the subspace given by the nine generating sections, hence are maps to the rank five kernel. Their 9×2 matrix in this basis is of rank two at each point of the space. These follow from

$$\begin{aligned}
M_2(e_{024}+e_{123}) &= a_0e_{24} - a_2e_{04} + a_4e_{02} + a_1e_{23} - a_2e_{13} + a_3e_{12} \\
&= -a_2(e_{04}+e_{13}) + a_0e_{24} + a_1e_{23} + a_4e_{02} + a_3e_{12} \\
M_2(e_{234}+a_1e_{013}-a_0e_{014}) &= a_2e_{34} - a_3e_{24} + a_4e_{23} + a_1a_0e_{13} - a_1^2e_{03} \\
&\quad + a_1a_3e_{01} - a_0^2e_{14} + a_0a_1e_{04} - a_0a_4e_{01} \\
&= a_0a_1(e_{04}+e_{13}) - a_3e_{24} + a_4e_{23} + a_2e_{34} - a_1^2e_{03} \\
&\quad + a_1a_3e_{01} - a_0^2e_{14} - a_0a_4e_{01}.
\end{aligned} \tag{5.1}$$

So indeed we have a rank two sub-bundle of the rank 5 kernel of the surjection $9\mathcal{O}_X \to \mathcal{K}^1$, delivering our rank three bundle $\mathcal{S}$ as the quotient.

The homogeneity conditions are as follows:

$$\begin{aligned}
d_0+d_4 &= d_1+d_3 \\
d_0+d_2+d_4 &= d_1+d_2+d_3 \\
d_2+d_3+d_4 &= (d_1)+d_0+d_1+d_3 = (d_0)+d_0+d_1+d_4.
\end{aligned}$$

These work out to $d_0+d_4 = d_1+d_3, d_2+d_4 = d_0+2d_1, d_2+d_3 = 2d_0+d_1$.

Sasakura's bundle on $X=\mathbb{P}^4$ is obtained if we choose $d_2 = 2$, $d_0 = d_1 = d_3 = d_4 = 1$.

COROLLARY 5.1. *Given even positive integers $2k, 2l$, there exists a family of rank three bundles on $\mathbb{P}^3$ (obtained through Sasakura's construction) such that the special bundle is non-free and the general bundle is $\mathcal{O}_{\mathbb{P}^3}\oplus\mathcal{O}_{\mathbb{P}^3}(2k)\oplus\mathcal{O}_{\mathbb{P}^3}(2l)$.*

PROOF. We choose on $X=\mathbb{P}^3\times\mathbb{A}^1$, the regular sequence a_0, a_1, t, a_3, a_4 where the a_i's are forms on $\mathbb{P}^3$. We get the monad of a rank three bundle $\mathcal{S}$ on X, and when $t=0$, the restriction $\mathcal{S}_{t=0}$ is non-free as usual.

When t is say equal to 1, the restriction of $\mathcal{K}^1$ splits as $\mathcal{O}_{\mathbb{P}^3}(-d_0) \oplus \mathcal{O}_{\mathbb{P}^3}(-d_1) \oplus \mathcal{O}_{\mathbb{P}^3}(-d_3) \oplus \mathcal{O}_{\mathbb{P}^3}(-d_4)$, and the monad described above can be inspected to see the explicit splitting

$$\mathcal{S}_{t=1} = \mathcal{O}_{\mathbb{P}^3}(-d_0 - d_1) \oplus \mathcal{O}_{\mathbb{P}^3}(-d_0 - d_3) \oplus \mathcal{O}_{\mathbb{P}^3}(-d_1 - d_4).$$

By the homogeneity conditions, $d_4 = d_0 + 2d_1, d_3 = 2d_0 + d_1$, hence the dual degrees are $d_0 + d_1, 3d_0 + d_1, d_0 + 3d_1$. Twisting down by $d_0 + d_1$ gives the degrees $0, 2d_0, 2d_1$, where d_0, d_1 can be freely chosen as any k, l. $\square$

6. Horrocks Bundles

We will present the rank three bundle on $\mathbb{P}^5$ discovered by Horrocks (called the parent bundle). Variants of this bundle, obtained by altering the degrees of the forms used in the construction, have been studied by Horrocks, and in [**2**] and [**4**]. Our purpose is to display the construction in such a way that the existence of a similar construction on $\mathbb{P}^4 \times \mathbb{A}^1$ becomes quite apparent. The upshot will be a family of rank 3 bundles on $\mathbb{P}^4$, generically a sum of line bundles, with the specialization at $t = 0$ non-split.

Accordingly, let X stand for either $\mathbb{P}^5$ or $\mathbb{P}^4 \times \mathbb{A}^1$. Let $a_0, a_1, a_2, a_3, a_4, a_5$ be a regular sequence on X, with d_i equal to the degree of a_i. Assume that

$$d_0 + d_5 = d_1 + d_4 = d_2 + d_3 = 2d, \text{ an even number}$$
$$d_0 + d_1 + d_2 = d_3 + d_4 + d_5 = D.$$

Consider the Koszul complex $K_\bullet$ built on this regular sequence. Let $V = K_1(d)$, with basis vectors $e_0, \ldots, e_5$. Its dual $V^\vee$ has a dual basis $e_0^*, \ldots, e_5^*$. The element $\eta = e_0^* \wedge e_5^* + e_1^* \wedge e_4^* + e_2^* \wedge_3^*$, a homogeneous element of degree 0, gives rise to three maps:

$$\mathcal{O}_X \xrightarrow{\eta} \wedge^2 V^\vee$$
$$V \xrightarrow{\eta} V^\vee, \text{skew-symmetric},$$
$$\wedge^2 V \xrightarrow{\wedge^2 \eta} \wedge^2 V^\vee, \text{symmetric}.$$

Next, let $\mathcal{E} = \mathcal{K}^1(d)$, with $\iota : \mathcal{E} \hookrightarrow V$ the inclusion. The induced map

$$\mathcal{E} \xrightarrow{\iota^\vee \eta \iota} \mathcal{E}^\vee$$

is skew-symmetric, and corresponds to the map

$$\rho : \mathcal{O}_X \xrightarrow{\eta} \wedge^2 V^\vee \xrightarrow{\wedge^2 \iota^\vee} \wedge^2 \mathcal{E}^\vee.$$

Claim 1: The induced map $\wedge^2 \rho : \wedge^2 \mathcal{E} \to \wedge^2 \mathcal{E}^\vee$ is symmetric, with image $\mathcal{A}$, a vector bundle of rank 6.

The symmetry of the map is obvious. In order to check the rank statement, consider the rank 4 bundle $\mathcal{N}$ obtained from the homology of the monad

$$0 \to \mathcal{O}_X(-d) \xrightarrow{[-a_5, -a_4, -a_3, a_2, a_1, a_0]^\vee} V \xrightarrow{[a_0, a_1, a_2, a_3, a_4, a_5]} \mathcal{O}_X(d) \to 0,$$

hence given as

$$0 \to \mathcal{O}_X(-d) \to \mathcal{E} \to \mathcal{N} \to 0.$$

is zero. Hence $\mathcal{E} \xrightarrow{\iota^\vee \eta\iota} \mathcal{E}^\vee$ factors through a map of rank 4 bundles $\mathcal{N} \xrightarrow{\psi} \mathcal{N}^\vee$. ψ is forced to be skew-symmetric as well. On the other hand, since $\eta\iota : \mathcal{E} \to V^\vee$ has image a bundle of rank 5, and since $V^\vee \to \mathcal{E}^\vee \to 0$ has kernel of rank 1, the image of $\mathcal{E} \xrightarrow{\iota^\vee \eta\iota} \mathcal{E}^\vee$ has rank at least 4 at each point.

Hence $\mathcal{E} \xrightarrow{\iota^\vee \eta\iota} \mathcal{E}^\vee$ factors through a skew-symmetric isomorphism

$$\psi : \mathcal{N} \to \mathcal{N}^\vee.$$

In particular, $\wedge^2\rho : \wedge^2\mathcal{E} \to \wedge^2\mathcal{E}^\vee$ has image isomorphic to $\wedge^2\mathcal{N}$, hence $\mathcal{A} = \wedge^2\mathcal{N}$, of rank 6.

Claim 2: $\mathcal{A} = \wedge^2\mathcal{N}$ has $\mathcal{O}_X$ as a summand.

In fact, the skew-symmetric isomorphism $\psi : \mathcal{N} \to \mathcal{N}^\vee$ induces a symmetric isomorphism $\wedge^2\psi : \wedge^2\mathcal{N} \to \wedge^2\mathcal{N}^\vee$. Now compose the inverse of this with the map $\psi : \mathcal{O}_X \to \wedge^2\mathcal{N}^\vee$ and its dual. The resulting composite

$$\mathcal{O}_X \to \wedge^2\mathcal{N}^\vee \xrightarrow{(\wedge^2\psi)^{-1}} \wedge^2\mathcal{N} \to \mathcal{O}_X$$

is a constant map since locally, at a point, we may calculate it using the local representation $\begin{bmatrix} 0 & 0 & 0 & -1 \\ 0 & 0 & -1 & 0 \\ 0 & 1 & 0 & 0 \\ 1 & 0 & 0 & 0 \end{bmatrix}$ for ψ, and the calculation of the composite is seen to be the multiplication by the constant 2. This shows the splitting of the copy of $\mathcal{O}_X$.

(Note: Certainly, on $X = \mathbb{P}^5$, this shows a splitting. When we deal with the case of $X = \mathbb{P}^4 \times \mathbb{A}^1$, this will induce a map from $\mathcal{O}_X$ to $\mathcal{O}_X$ which is non-zero at each point. This map must also be a constant (independent of t).)

Let $\mathcal{S}$ be the kernel of the map $\wedge^2\mathcal{N} \to \mathcal{O}_X$. $\mathcal{S}$ is a vector bundle of rank 5 and

$$\wedge^2\mathcal{N} = \mathcal{S} \oplus \mathcal{O}_X.$$

Consider the basis element e_{012} of $K_3 = (\wedge^3 V)(-3d)$ giving the inclusion $\mathcal{O}_X(-D) \to (\wedge^3 V)(-3d)$. Under the Koszul map $\sum a_i e_i^* : (\wedge^3 V)(-3d) \to (\wedge^2 V)(-2d)$, e_{012} maps to $a_0e_{12} - a_1e_{02} + a_2e_{01}$. Since

$$0 \to \mathcal{E} \xrightarrow{\iota} V \to \mathcal{O}_X(d) \to 0,$$

we have also the exact sequence

$$0 \to \wedge^2\mathcal{E} \xrightarrow{\wedge^2\iota} \wedge^2 V \to V \otimes \mathcal{O}_X(d).$$

In turn, under the map $\wedge^2 V \to V \otimes \mathcal{O}_X(d)$, the image of e_{012} is sent to zero since $e_{12} \mapsto a_1e_2 - a_2e_1$ etc.

Hence, under the Koszul map, e_{012} maps to $(\wedge^2\mathcal{E})(-2d) \subseteq (\wedge^2 V)(-2d)$. On the other hand, the composite map $\wedge^2\mathcal{E} \to \wedge^2\mathcal{N} \to \mathcal{O}_X$ factors through $\wedge^2 V \xrightarrow{\eta^\vee} \mathcal{O}_X$. A calculation shows that e_{012} is sent to $a_0\eta^\vee e_{12} - a_1\eta^\vee e_{02} + a_2\eta^\vee e_{01} = 0$ in $\mathcal{O}_X$.

Hence $e_{012} : \mathcal{O}_X(-D + 2d) \to \mathcal{S} \subseteq \wedge^2\mathcal{N}$.

Likewise e_{345}.

There is a similar dual picture: the basis element e_{012} also induces $K_3^\vee \to \mathcal{O}_X(D)$. The composite under the dual Koszul map gives

$$(\wedge^2 V^\vee)(2d) \to (\wedge^3 V^\vee)(3d) \to \mathcal{O}_X(D)$$

seen to be given by $a_2e_{01} - a_1e_{02} + a_0e_{12}$. It is an easy check that this composite vanishes on the image of $V^\vee \otimes \mathcal{O}_X(-d)$ in $(\wedge^2 V^\vee)$, giving an induced map

$$\wedge^2\mathcal{E}^\vee \to \mathcal{O}_X(D-2d).$$

Hence there is an induced a map

$$\wedge^2\mathcal{N}^\vee \to \mathcal{O}_X(D-2d).$$

The composite of this map with the inclusion $\mathcal{O}_X \xrightarrow{\psi} \wedge^2\mathcal{N}^\vee$ is calculated via the composite $\mathcal{O}_X \xrightarrow{\psi} \wedge^2 V^\vee \to \mathcal{O}_X(D-2d)$, and is easily seen to be zero. Hence this map given by e_{012} from $\wedge^2\mathcal{N}^\vee \to \mathcal{O}_X(D-2d)$ factors through the subset $\mathcal{S} \subseteq \wedge^2\mathcal{N}$.

Likewise e_{345}.

Claim 3: The complex

$$0 \to \mathcal{O}_X(-D+2d) \xrightarrow{e_{012}+e_{345}} \mathcal{S} \xrightarrow{e_{012}-e_{345}} \mathcal{O}_X(D-2d) \to 0$$

is a monad of vector bundles.

Viewing the image of $\mathcal{O}_X(-D+2d) \xrightarrow{e_{012}+e_{345}} \mathcal{S}$ as an element $a_0e_{12} - a_1e_{02} + a_2e_{01} + a_3e_{45} - a_4e_{35} + a_5e_{34}$ of $\wedge^2 V$, it is evident that the left hand map is an inclusion of bundles. Likewise the right hand map can be viewed as induced from $\wedge^2 V^\vee \to \mathcal{O}_X(D-2d)$, and is given by $a_0e_{12} - a_1e_{02} + a_2e_{01} + a_3e_{45} - a_4e_{35} + a_5e_{34}$, hence is a surjection of bundles. The composite is calculated via the isomorphism $\wedge^2\eta : \wedge^2 V \to \wedge^2 V^\vee$. We get $\wedge^2\eta(a_0e_{12} - a_1e_{02} + a_2e_{01} + a_3e_{45} - a_4e_{35} + a_5e_{34}) = a_0e^*_{43} - a_1e^*_{53} + a_2e^*_{54} + a_3e^*_{10} - a_4e^*_{20} + a_5e^*_{21}$, in $\wedge^2 V^\vee$, and hence the final image in $\mathcal{O}_X(D-2d)$ is $-a_0a_5 - a_1a_4 - a_2a_3 + a_3a_2 + a_4a_1 + a_5a_0 = 0$. Thus we have a monad.

The homology of this bundle is called a Horrocks rank 3 bundle. It is quite plainly not a sum of line bundles because the map

$$\mathcal{S} \xrightarrow{e_{012}-e_{345}} \mathcal{O}_X(D-2d) \to 0$$

as calculated above, is not surjective on global sections.

The *parent bundle* on $\mathbb{P}^5$ is obtained if we take the standard Koszul complex on $\mathbb{P}^5$. Variants on $\mathbb{P}^5$ can be obtained with other choices of degrees for the regular sequence.

COROLLARY 6.1. There exist on $\mathbb{P}^4$ families of rank 3 bundles, where the general bundle is a sum of line bundles, and the special one is not so. The general bundle can be chosen as trivial or with degrees in any arithmetic sequence.

PROOF. Perform the construction of a Horrocks rank 3 bundle on $\mathbb{P}^4 \times \mathbb{A}^1$, choosing a sequence $a_0, a_1, a_2, a_3, a_4, t$, where a_i is a form on $\mathbb{P}^4$ of degree d_i, and where $d_0 = d_1 + d_4 = d_2 + d_3$ and $d_0 + d_1 + d_2 = d_3 + d_4$. On a fibre $\mathbb{P}^4$ for which $t \neq 0$, the restrictions of $\mathcal{E}$ and $\mathcal{N}$ are split as sums of line bundles. Hence, so are the restrictions of $\mathcal{A} = \wedge^2\mathcal{N}$ and its summand $\mathcal{S}$. The map

$$0 \to \mathcal{O}_X(-D+2d) \xrightarrow{e_{012}+e_{345}} \mathcal{S}$$

as calculated above is split on this fibre because it splits as a map to $\wedge^2 V$. Likewise, the map

$$\mathcal{S} \xrightarrow{e_{012}-e_{345}} \mathcal{O}_X(D-2d) \to 0$$

monad is a rank three bundle which is not the sum of line bundles for the reason mentioned at the end of the main construction.

Now choose degrees $4, 1, 1, 3, 3$ for d_0, d_1, d_2, d_3, d_4 (with $d_5 = 0$). The family obtained has general fibre trivial.

Choose degrees $6, 1, 2, 4, 5$ for d_0, d_1, d_2, d_3, d_4 (with $d_5 = 0$). The family obtained has general fibre free but non-trivial. In fact, it is $\mathcal{O}_{\mathbb{P}^4}(-1) \oplus \mathcal{O}_{\mathbb{P}^4} \oplus \mathcal{O}_{\mathbb{P}^4}(1)$, so any non-trivial arithmetic sequence of degrees can be obtained by pull-backs of this example. □

7. An example where the general bundle on $\mathbb{P}^3$ is trivial

This last example is different from all the earlier ones in that it does not rely on properties of the Koszul complex. It produces a family of rank three bundles on $\mathbb{P}^3$ where the general bundle is trivial and the special bundle is non-free. Modifications of this method might be the way to show that any three integers are achievable as the degrees of the general bundle in a family on $\mathbb{P}^3$. We have not yet been able to establish if this is true.

On $\mathbb{P}^3$, consider the matrix

$$\varphi = \begin{bmatrix} 0 & -X_0^3 & -X_1^3 & -\alpha \\ X_0^3 & 0 & -\beta & -X_2^6 \\ X_1^3 & \beta & 0 & -X_3^6 \\ \alpha & X_2^6 & X_3^6 & 0 \end{bmatrix},$$

where $\alpha = X_0X_3^2 - X_1X_2^2$ and $\beta = X_0^2X_3^4 + X_0X_1X_2^2X_3^2 + X_1^2X_2^4$ are chosen to make the pfaffian of the matrix vanish. φ gives a map from $\mathcal{O}_{\mathbb{P}^3} \oplus 3\mathcal{O}_{\mathbb{P}^3}(-3) \to \mathcal{O}_{\mathbb{P}^n} \oplus 3\mathcal{O}_{\mathbb{P}^3}(3)$, which is checked to be of rank two at each point on $\mathbb{P}^3$, hence we get its image to be a rank two sub-bundle A. A very similar matrix

$$\psi = \begin{bmatrix} 0 & X_3^6 & -X_2^6 & \beta \\ -X_3^6 & 0 & \alpha & -X_1^3 \\ X_2^6 & -\alpha & 0 & X_0^3 \\ -\beta & X_1^3 & -X_0^3 & 0 \end{bmatrix}$$

gives a second rank two bundle B and since $\psi\varphi = 0$, we get the short exact sequence

$$0 \to A \to \mathcal{O}_{\mathbb{P}^3} \oplus 3\mathcal{O}_{\mathbb{P}^3}(3) \to B \to 0.$$

A, B are non-free bundles and hence this is a nontrivial extension $\eta \in \operatorname{Ext}^1(B, A)$. By choosing as $\mathbb{A}^1$, the line $\{t\eta | t \in k\}$ spanned by η in this vector space, we get a family of rank four bundles on $\mathbb{P}^3$ where the general one is $\mathcal{O}_{\mathbb{P}^3} \oplus 3\mathcal{O}_{\mathbb{P}^3}(3)$ and the special one (with $t = 0$) is $A \oplus B$. It is clear that both A and B have a section in degree zero, say s and s', where s maps to the column vector $\begin{bmatrix} 0 & X_0^3 & X_1^3 & \alpha \end{bmatrix}^\vee$, and s' is the image of $\begin{bmatrix} 1 & 0 & 0 & 0 \end{bmatrix}^\vee$. Hence the section $\begin{bmatrix} 1 & X_0^3 & X_1^3 & \alpha \end{bmatrix}^\vee$ of the general fibre $\mathcal{O}_{\mathbb{P}^3} \oplus 3\mathcal{O}_{\mathbb{P}^3}(3)$ becomes, in the limit, the section (s, s') of $A \oplus B$. It is evident from inspection that s (the first column of φ) and s' (the first column of ψ) do not both vanish at the same point of $\mathbb{P}^3$. Hence we have a nowhere vanishing section of the rank four bundle on $\mathbb{P}^3 \times \mathbb{A}^1$, and the quotient is a rank three bundle on $\mathbb{P}^3 \times \mathbb{A}^1$. When $t \neq 0$, the rank three bundle obtained is $3\mathcal{O}_{\mathbb{P}^3}(3)$, and when $t = 0$, we get the rank three bundle $(A \oplus B)/(s, s')$, which is clearly non-free.

References

[1] H. Abo, W. Decker and N. Sasakura, *An elliptic conic bundle in* $\mathbf{P}^4$ *arising from a stable rank-3 vector bundle*, Mathematische Zeitschrift **229** no.4 (1998), 725–741.

[2] V. Ancona and G. Ottaviani, *The Horrocks bundles of rank three on* P^5, J. Reine Angew. Math. **460** (1995), 69–92.

[3] P. Cascini, *Weighted Tango bundles on* P^n *and their moduli spaces*, Forum Math. **13** (2001), 251–260.

[4] W.Decker, N.Manolache and F.O.Schreyer, *Geometry of the Horrocks bundle on* $\mathbf{P}^5$, Complex Projective Geometry, London Math. Soc. Lecture Note Series 179, 128–148, Cambridge 1992.

[5] G. Horrocks, *Examples of rank three vector bundles on five-dimensional projective space*, Journal of the London Mathematical Society **18** (1978), 15–27.

[6] G. Horrocks and D. Mumford, *A rank 2 vector bundle on* $\mathbf{P}^4$ *with 15,000 symmetries*, Topology **12** (1973), 63–81.

[7] N. Mohan Kumar, *Smooth Degeneration of Complete Intersection Curves in Positive Characteristic*, Inventiones Mathematicae **104** (1991), 313–319.

[8] N. Mohan Kumar, *Construction of rank two vector bundles on* $\mathbf{P}^4$ *in positive characteristic*, Inventiones Mathematicae **130** (1997), 277–286.

[9] N. Mohan Kumar, C. Peterson and A.P. Rao, *Hilbert scheme components in characteristic 2*, Communications in Algebra **28** (2000), 5735–5744.

[10] N. Mohan Kumar, C. Peterson and A.P. Rao, *Constructing low rank vector bundles on* $\mathbb{P}^4$ *and* $\mathbb{P}^5$, J. Algebraic Geometry **11** (2002), 203–217.

[11] N. Mohan Kumar, C. Peterson and A.P. Rao, *Degenerating families of rank two bundles*, To Appear in: *Proceedings of the AMS.*

[12] H. Tango, *An example of indecomposable vector bundle of rank* $n-1$ *on* $\mathbb{P}^n$, Journal of Mathematics of Kyoto University **16** no. 1 (1976), 137–141.

[13] G. Trautmann, *Darstellung von Vektorraumbündeln über* $\mathbf{C^n} - \{\mathbf{0}\}$, Arch. Math. **24** (1973), 303–313.

[14] U. Vetter, *Zu einem Satz von G. Trautmann über den Rang gewisser kohärenter analytischer Moduln*, Archiv der Mathematik (Basel) **24** (1973), 158- -161.

DEPT. OF MATHEMATICS, WASHINGTON UNIVERSITY, SAINT LOUIS, MO 63130
E-mail address: `kumar@math.wustl.edu`

DEPT. OF MATHEMATICS, COLORADO STATE UNIVERSITY, FORT COLLINS, CO 80523
E-mail address: `peterson@math.colostate.edu`

DEPT. OF MATHEMATICS, UNIVERSITY OF MISSOURI - ST. LOUIS, SAINT LOUIS, MO 63121
E-mail address: `rao@arch.umsl.edu`

Contemporary Mathematics
Volume **322**, 2003

The S^1 fixed points in Quot-schemes and mirror principle computations

Bong H. Lian, Chien-Hao Liu, Kefeng Liu, Shing-Tung Yau

ABSTRACT. We describe the S^1-action on the Quot-scheme $Quot(\mathcal{E}^n)$ associated to the trivial bundle $\mathcal{E}^n = \mathbb{C}\mathrm{P}^1 \times \mathbb{C}^n$. In particlular, the topology of the S^1-fixed-point components in $Quot(\mathcal{E}^n)$ and the S^1-weights of the normal bundle of these components are worked out. Mirror Principle, as developed by three of the current authors in the series of work [L-L-Y1, I, II, III, IV], is a method for studying certain intersection numbers on a stable map moduli space. As an application, in Mirror Principle III, Sec 5.4, an outline was given in the case of genus zero with target a flag manifold. The results on S^1 fixed points in this paper are used here to do explicit Mirror Principle computations in the case of Grassmannian manifolds. In fact, Mirror Principle computations involve only a certain distinguished subcollection of the S^1-fixed-point components. These components are identified and are labelled by Young tableaus. The S^1-equivariant Euler class e_{S^1} of the normal bundle to these components is computed. A diagrammatic rule that allows one to write down e_{S^1} directly from the Young tableau is given. From this, the aforementioned intersection numbers on the moduli space of stable maps can be worked out. Two examples are given to illustrate our method. Using the method, the A-model for Calabi-Yau complete intersections in a Grassmannian manifold can now also be computed explicitly. This work is motivated by the intention to provide further details of mirror principle and to understand the relation of mirror principle to physical theory. Some related questions are listed for further study.

2000 *Mathematics Subject Classification.* Primary 14N35, 14D20; Secondary 55N91, 55R91, 81T30.

Key words and phrases. Mirror principle, Grassmannian manifold, Quot-scheme, Young tableau, equivariant Euler class, homogeneous bundle.

The work is supported by DOE grant DE-FG02-88ER25065, NSF grants DMS-9803347, and DMS-0072158.

[1] String theorists may refer to the eprint version of this work: arXiv:math.AG/0111256, where more backgrounds are provided for their convenience.

Outline.

1. Backgrounds and notations for physicists.

Essential backgrounds or their main references used in this article and notations for objects involved are collected in this section for the convenience of readers.

• **Schemes, coherent sheaves, Hilbert polynomials, and Quot-schemes.** (See [E-H], [Fri], [Gr], [Ke], [Kol], [Ha], [H-L], [H-M], [Mu1], and [Mu2].)

• **Quot-scheme compactification of** $\mathit{Hom}(\mathrm{CP}^1, Gr_r(\mathbb{C}^n))$**.** (Cf. [Str].) Let $C = \mathrm{CP}^1$ with the very ample line bundle $\mathcal{O}_{\mathrm{CP}^1}(1)$, $\mathcal{E}^n$ be a trivialized trivial bundle of rank n over C, $Gr_r(\mathbb{C}^n)$ be the Grassmannian manifold that parameterizes r-planes in $\mathbb{C}^n$, and $\mathit{Hom}(\mathrm{CP}^1, Gr_r(\mathbb{C}))$ be the space of morphisms from CP^1 to $Gr_r(\mathbb{C}^n)$. Then an element $(f : \mathrm{CP}^1 \to Gr_r(\mathbb{C}^n))$ in $\mathit{Hom}(\mathrm{CP}^1, Gr_r(\mathbb{C}))$ determines a unique rank-r subbundle $\mathcal{V}$ in $\mathcal{E}^n$, which corresponds in turn to the element $\mathcal{E}^n \to \mathcal{E}^n/\mathcal{V}$ in $\mathit{Quot}(\mathcal{E}^n)$. This gives a natural embedding of $\mathit{Hom}(\mathrm{CP}^1, Gr_r(\mathbb{C}^n))$ in $\mathit{Quot}(\mathcal{E}^n)$. The component of $\mathit{Hom}(\mathrm{CP}^1, Gr_r(\mathbb{C}^n))$ that contains degree-d image curves in $Gr_r(\mathbb{C}^n)$ is embedded in $\mathit{Quot}_P(\mathcal{E}^n)$ with the Hilbert polynomial $P = P(t) = (n-r)t+d+(n-r)$. This gives a compatification of $\mathit{Hom}(\mathrm{CP}^1, Gr_r(\mathbb{C}^n))$ via Quot-schemes, other than the moduli space of stable maps. Recall also that $\mathit{Quot}_P(\mathcal{E}^n)$ is a smooth, irreducible, rational projective variety of dimension $dn + (n-r)r$, cf. [Ch] and [Kim]. The S^1-action on CP^1 induces S^1-actions on $\mathit{Hom}(\mathrm{CP}^1, Gr_r(\mathbb{C}^n))$ and $\mathit{Quot}(\mathcal{E}^n)$ respectively. The two actions coincide under the natural embedding of $\mathit{Hom}(\mathrm{CP}^1, Gr_r(\mathbb{C}^n))$ in $\mathit{Quot}(\mathcal{E}^n)$.

• **Mirror principle for Grassmannian manifolds.** For the details of Mirror Principle, readers are referred to [L-L-Y1: I, II, III, IV]. Some survey is given in [L-L-Y2]. To avoid digressing too far away, here we shall take [L-L-Y1, III: Sec. 5.4] as our starting point and restrict to the case that the target manifold of stable maps is $X = Gr_r(\mathbb{C}^n)$. Recall first the Plücker embedding $\tau : X = Gr_r(\mathbb{C}^n) \to Y = \mathrm{CP}^{\binom{n}{r}}$, which induces an isomorphism between the divisor class groups $\tau^* : A^1(Y) \xrightarrow{\sim} A^1(X)$.

Recall next the setup of Mirror Principle for $X = Gr_r(\mathbb{C}^n)$. The geometric objects involved are contained in the following diagram:

$$\begin{array}{ccccccc}
V & & U_d & & V_d & & \\
\downarrow & & \downarrow & & \downarrow & & \\
X & \xleftarrow{ev} & M_{0,1}(d,X) & \xrightarrow{\rho} & M_{0,0}(d,X) & \xleftarrow{\pi} &
\end{array}$$

$$\begin{array}{ccccccc}
 & \mathcal{U}_d & & & & & \\
 & \downarrow & & & & & \\
\xleftarrow{\pi} & M_d & \xrightarrow{\varphi} & W_d & \xleftarrow{\psi} & Quot_{(d)} & \\
 & \cup & & \cup & & \cup & \\
 & F_0 & \xrightarrow{ev^Y} & Y_0\,(\supset X_0 = X) & \xleftarrow{g} & E_0 & = \cup_s E_{0s}\,, \\
 & \downarrow ev^X & & \wr\wr & & & \\
 & X & \xrightarrow{\tau} & Y & & &
\end{array}$$

where

(1) *Moduli spaces*: $M_{0,0}(d,X)$ is the moduli space of genus-0 stable maps of degree d into X, $M_{0,1}(d,X)$ is the moduli space of genus-0, 1-pointed stable maps of degree d into X, $M_d = M_{0,0}(\mathbb{CP}^1 \times X, (1,d))$, W_d is the linear sigma model at degree d, which can be chosen to be the projective space $\mathbb{P}(H^0(\mathbb{CP}^1, \mathcal{O}_{\mathbb{CP}^1}(d)) \otimes \Lambda^r\mathbb{C}^n)$ for $X = Gr_r(\mathbb{C}^n)$, and $Quot_{(d)} = Quot_P(\mathcal{E}^n)$ with $P = P(t) = (n-r)t + d + (n-r)$;

(2) *Group actions*: there are $\mathbb{C}^\times$-actions on M_d, W_d, and $Quot_{(d)}$ respectively that are compatible with the morphisms among these moduli spaces; these $\mathbb{C}^\times$-actions induce S^1-actions on these moduli spaces by taking the subgroup $U(1) \subset \mathbb{C}^\times$;

(3) *Morphisms*: ev is the evaluation map, ρ is the forgetful map, π is the contracting morphism, φ is the collapsing morphism, and ψ is an S^1-equivariant resolution of singularities of $\varphi(M_d)$, which will be discuessed in detail in Sec. 3.1;

(4) *Bundles*: V is a vector bundle over X, $V_d = \rho_! ev^* V$, $U_d = \rho^* V_d$, and $\mathcal{U}_d = \pi^* V_d$;

(5) *Special S^1-fixed-point locus*: $F_0 \simeq M_{0,1}(d,X)$ is the special S^1-fixed-point component in M_d that corresponds to gluing stable maps (C', f', x') to $\mathbb{CP}^1$ at $x' \in C'$ and $\infty \in \mathbb{CP}^1$, Y_0 is the special S^1-fixed-point component in W_d such that $\varphi^{-1}(Y_0) = F_0$, and E_0 is the S^1-fixed-point locus in $\psi^{-1}(Y_0)$ and is called the distinguished S^1-fixed-point locus or components in $Quot_P(\mathcal{E}^n)$.

Associated to each (V,b), where b is a multiplicative characteristic class, is the Euler series $A(t) \in A^*(X)(\alpha)[t]$:

$$\begin{array}{lclcl}
A(t) & = & A^{V,b} & = & e^{-H\cdot t/\alpha} \sum_d A_d\, e^{d\cdot t}\,, \\
A_d & = & i_0^*\, b(\mathcal{U}_d) & := & ev_*^X \left(\frac{\rho^* b(V_d) \cap [M_{0,1}(d,X)]}{e_{\mathbb{C}^\times}(F_0/M_d)} \right) = \frac{(i_{X_0}^* \varphi_* b(\mathcal{U}_d)) \cap [X_0]}{e_{\mathbb{C}^\times}(X_0/W_d)}\,, \\
 & & & & \text{denoted } \frac{\Theta_d}{e_{\mathbb{C}^\times}(X_0/W_d)}\,,
\end{array}$$

$$\left(= \quad \frac{(i_{E_{0s}}^* g^* i_{Y_0}^* \varphi_* b(\mathcal{U}_d)) \cap [E_{0s}]}{\cdots}\right)$$

where $\alpha = c_1(\mathcal{O}_{\mathbb{CP}^\infty})(1)$ is the generator for $H^*_{\mathbb{C}^\times}(pt)$. On the other hand, one has the intersection numbers and their generating function

$$\begin{array}{lllll} K_d & = & K_d^{V,b} & = & \int_{M_{0,0}(d,X)} b(V_d)\,, \\ \Phi & = & \Phi^{V,b} & = & \sum_d K_d\, e^{d\cdot t}\,. \end{array}$$

In the good cases, K_d and Φ can be obtained from A_d and $A(t)$ by appropriate integrals of the form $\int_X e^{-H\cdot t/\alpha} A_d$, where H is the hyperplane class on Y restricted to X, e.g. [L-L-Y1, III: Theorem 3.12]. This integral can be turned into an integral on E_0:

$$\begin{aligned} \int_X \tau^* e^{H\cdot t} \cap A_d &= \int_{Y_0} e^{H\cdot t} \cap g_* \left(\sum_s \frac{\Xi_{d,s}}{e_{\mathbb{C}^\times}(E_{0s}/Quot_{(d)})} \right) \\ &= \sum_s \int_{E_{0s}} \frac{g^* e^{H\cdot t} \cap \widehat{\Xi}_{d,s}}{e_{\mathbb{C}^\times}(E_{0s}/Quot_{(d)})}\,, \end{aligned}$$

where $\widehat{\Xi}_{d,s}$ is the Poincaré dual of $\Xi_{d,s}$ with respect to $[E_{0,s}]$. As will be discussed in Sec. 3.1, E_{0s} is a flag manifold fibred over X and, hence, $g^*e^{H\cdot t}$ can be read off from the natural fibration of flag manifolds $E_{0s} \to X$.

Following [L-L-Y1, III: Sec. 5.4], in the case that $b = 1$ the above integral is reduced to the integral

$$\sum_s \int_{E_{0s}} \frac{g^*\psi^* e^{\kappa\cdot\zeta}}{e_{\mathbb{C}^\times}(E_{0s}/Quot_{(d)})}\,,$$

where κ is the hyperplane class in W_d. In this article, we work out all the equivariant Euler classes $e_{\mathbb{C}^\times}(E_{0s}/Quot_{(d)})$ and hence this integral.

- **Conventions and notation.**
  (1) All the dimensions are *complex* dimensions unless otherwise noted.
  (2) The S^1-actions involved in this article are induced from $\mathbb{C}^\times$-actions and both have the same fixed-point locus. In many places, it is more convenient to phrase things in term of $\mathbb{C}^\times$-action and we will not distinguish the two actions when this ambiguity causes no harm.
  (3) A locally free sheaf and its associated vector bundle are denoted the same.
  (4) An $I \times J$ matrix whose (i,j)-entry is a_{ij} is denoted by $(a_{ij})_{i,j}$ when the position of an entry is emphasized and by $[a_{ij}]_{I\times J}$ when the size of the matrix is emphasized.
  (5) From Section 2 on, the smooth curve C will be $\mathbb{CP}^1$ unless other noted.

2. The S^1-action on Quot-schemes.

Let $\mathcal{E}^n$ be a trivialized trivial bundle of rank n over C. The S^1-action on the Quot-schemes $Quot(\mathcal{E}^n)$, the topology of the S^1-fixed-point components, and the S^1-weights of the normal bundle to these components are studied in this section.

2.1. The S^1-fixed-point components. We recall first two basic facts that will be needed in the discussion.

Fact 2.1.1 [modules over P.I.D.]. (Cf. [Ja].)

(1) *Let D be a principal ideal domain and $D^{\oplus k}$ be a free module of rank k over D. Then any submodule of $D^{\oplus k}$ is free with basis of $\leq k$ elements.*

(2) *If $A \in M_{k\times k}(D)$ be a $k \times k$ matrix with entries in D, then A is equivalent to a diagonal matrix $Diag\{\,d_1, \ldots, d_s, 0, \ldots, 0\,\}$ for some s, where $d_i \neq 0$ and $d_i | d_j$ if $i \leq j$. (Recall that $A_1, A_2 \in M_{k\times k}(D)$ are called equivalent if $A_2 = PA_1Q$ for some invertible $P, Q \in M_{k\times k}(D)$.)*

Recall the embedding $S^1 = U(1) \hookrightarrow \mathbb{C}^\times$, which acts on $C = \mathbb{C} \cup \{\infty\}$ via $z \to tz$. This lifts to an S^1-action (i.e. a linearization) on the trivialized trivial bundle $\mathcal{E}^n \simeq \mathcal{O}_C \otimes \mathbb{C}^n$ given by $(z, v) \mapsto (t \cdot z, v)$. This induces then an S^1-action $\mathcal{S} \mapsto t \cdot \mathcal{S}$ on the set of coherent subsheaves $\mathcal{S}$ in $\mathcal{E}$ by pulling back local sections: $(t \cdot s)(z) = s(t^{-1}z)$, where $s \in \mathcal{S}(U)$ and $t \cdot s \in (t \cdot \mathcal{S})(t \cdot U)$ with U an open subset in C. Since each subsheaf in $\mathcal{E}^n$ corresponds to a point in the Quot-scheme $Quot(\mathcal{E}^n)$, this gives an S^1-action $Quot(\mathcal{E}^n)$. (Cf. [Ak], [Ch], and [Str].)

When restricted to the set of rank-r subbundles in $\mathcal{E}^n$, each holomorphic subbundle in $\mathcal{E}^n$ corresponds to a holomorphic map f from C to a Grassmannian manifold $Gr_r(\mathbb{C}^n)$ and the above S^1-action is the S^1-action on $Hom(C, Gr_r(\mathbb{C}^n))$ given by $f \mapsto t \cdot f := f \circ t^{-1}$, cf. [Ak].

In the following, we first characterize the S^1-fixed-point in $Quot(\mathcal{E}^n)$ and then give a description of the topology of the S^1-fixed-point components in $Quot(\mathcal{E}^n)$.

The followinmg two lemmas are immediate.

Lemma 2.1.2 [coherent subsheaf of $\mathcal{E}^n$]. *Any coherent subsheaf $\mathcal{V}$ of $\mathcal{E}^n$ is locally free.*

Lemma 2.1.3 [S^1-fixed-point $=$ $\mathbb{C}^\times$-fixed-point]. *A coherent subsheaf $\mathcal{V}$ of $\mathcal{E}^n$ on C is S^1-invariant if and only if it is $\mathbb{C}^\times$-invariant.*

The following lemma strengthens Statement (2) of Fact 2.1.1 in the case of $\mathbb{C}^\times$-invariant submodules in $\mathbb{C}[z]^{\oplus k}$.

Lemma 2.1.4 [$\mathbb{C}^\times$-invariant submodule]. *Let $D = \mathbb{C}[z]$, $A = A(z) \in GL(l, \mathbb{C}[z])$ be an invertible $l \times l$-matrix with entries in $\mathbb{C}[z]$, in Fact 2.1.1. If, furthermore, the column vectors of $A(tz)$ generate the same $\mathbb{C}[z]$-module for all $t \in \mathbb{C}^\times$, then there exist invertible $P \in GL(l, \mathbb{C})$ and $Q(z) \in GL(l, \mathbb{C}[z])$ such that $d_i = z^{\alpha_i}$ in Fact 2.1.1 and*

$$A(z) \;=\; P\, Diag\{\, z^{\alpha_1}, \ldots, z^{\alpha_l}\,\}\, Q(z)\,,$$

where $0 \leq \alpha_1 \leq \cdots \leq \alpha_l$.

Proof. By a sequence of elementary column transformations (e.g. [Ja]), which correspond to multiplications from the right by a sequence of elementary matrices in $GL(l, \mathbb{C}[z])$, together with permutations of rows, which corresponds to a multipli-

(2) $\mathit{deg}\, b_{ii}(z) \leq \mathit{deg}\, b_{i+1,i+1}(z)$ for all i, and
(3) $\mathit{deg}\, b_{ij}(z) < \mathit{deg}\, b_{ii}(z)$ for all $i > j$,

where $\mathit{deg}\,(\,\cdot\,)$ is the degree of the polynomial $(\,\cdot\,)$ with respect to the variable z and $\mathit{deg}\,(0) = -\infty$ by convention.

The assumption that the column vectors of $A(tz)$ generate the same $\mathbb{C}[z]$-module for all $t \in \mathbb{C}^\times$ implies that the column vectors of $B(tz)$ generate the same $\mathbb{C}[z]$-module as the module generated by the column vectors of $B(z)$ for all $t \in \mathbb{C}^\times$. In terms of matrices, this is equivalent to the existence of $\widehat{Q}(z,t) \in GL\,(k, \mathbb{C}[z])$ such that $B(tz) = B(z)\widehat{Q}(z,t)$, $t \in \mathbb{C}^\times$. The fact that both $B(tz)$ and $B(z)$ are lower triangular implies that $\widehat{Q}(z,t)$ is also lower triangular.

On the other hand, $\mathit{deg}\ b_{ij}(tz) = \mathit{deg}\ b_{ij}(z)$ for all i,j. This puts a strong constraint in the form of $B(z)$ in order that $B(tz) = B(z)\widehat{Q}(z,t)$ always admits a solution for $\widehat{Q}(z,t)$ in $GL\,(l, \mathbb{C}[z])$. Together with the Inequality (3) above: $\mathit{deg}\ b_{ij}(tz) < \mathit{deg}\ b_{ii}(z)$ for all $i > j$, and a tedious yet straightforward induction argument, one can show that $B(z)$ must be of the form

$$B(z) \;=\; B(1)\,\mathit{Diag}\{\, z^{\alpha_1},\, \ldots,\, z^{\alpha_l}\,\}$$

with $0 \leq \alpha_1 \leq \ldots \leq \alpha_l$ and $B(1)_{ij} = 0$ if $i < j$ or $\alpha_i = \alpha_j$.

This proves the lemma. □

Proposition 2.1.5 [**S^1-fixed coherent subsheaf**]. *Let $\mathcal{V}$ be a rank r coherent subsheaf of $\mathcal{E}^n$ on C. Then $\mathcal{V}$ is a locally free $\mathcal{O}_C$-module. When $\mathcal{V}$ is in addition S^1-invariant, then $\mathcal{V}$ determines a unique enlarged sheaf $\widehat{\mathcal{V}}$ such that*

(1) *$\widehat{\mathcal{V}}$ is a constant subsheaf in the globally trivialized $\mathcal{E}^n$ of the same rank r as $\mathcal{V}$, (thus $\widehat{\mathcal{V}} \simeq \mathcal{O}_C^{\oplus r}$).*
(2) *$\mathcal{V}$ is a subsheaf of $\widehat{\mathcal{V}}$.*
(3) *Let $\{\, U_0 = C - \{\infty\} = \mathit{Spec}\,\mathbb{C}[z]\,,\, U_\infty = C - \{0\} = \mathit{Spec}\,\mathbb{C}[w]\,\}$ be an atlas of affine charts on C. Then there exists a constant re-trivialization*

$$\widehat{\mathcal{V}}|_{U_0} \;=\; \mathcal{O}|_{U_0}^{\oplus r} \;=\; (\mathbb{C}[z]^{\oplus r})^\sim$$

such that

$$\mathcal{V}|_{U_0} \;=\; (\mathbb{C}[z]\, z^{\alpha_1} \oplus \cdots \oplus \mathbb{C}[z]\, z^{\alpha_r})^\sim \quad \textit{with} \quad 0 \leq \alpha_1 \leq \cdots \leq \alpha_r$$

with respect to this new trivialization, where $(\,\cdot\,)^\sim$ is the sheaf of modules over the affine scheme $U = \mathit{Spec}\, R$ in question associated to the R-module $(\,\cdot\,)$, cf. [Ha]. *Similarly for $\widehat{\mathcal{V}}|_{U_\infty}$ and $\mathcal{V}|_{U_\infty}$. (Corresponding to $0 \leq \beta_1 \leq \ldots, \leq \beta_r$.)*

Proof. For Claim (1) and Claim (2). Since $\mathcal{V}$ is an S^1-fixed subsheaf in $\mathcal{E}^n$, $\mathcal{V}|_{C-\{0,\infty\}}$ admits a unique trivial extension to a subsheaf of $\mathcal{E}^n$ on the whole C. By construction, it has the same rank as $\mathcal{V}$. We shall choose $\widehat{\mathcal{V}}$ to be this extension sheaf of $\mathcal{V}_{C-\{0,\infty\}}$. If $\mathcal{V}$ is not contained in $\widehat{\mathcal{V}}$ as a subsheaf, then there exists an affine chart U of C such that $\mathcal{V}|_U$ has a section s not contained in $\widehat{\mathcal{V}}$. Since $\widehat{\mathcal{V}}|_U = \mathcal{V}|_U$, this implies that s must restrict to the zero-section when localized to $U - \{0,\infty\}$. In other words, it is a torsion section. This contradicts with Lemma 2.1.2, which says that $\mathcal{V}$ must be torsion-free. Consequently, $\mathcal{V}$ must be a subsheaf of $\widehat{\mathcal{V}}$ as well.

For Claim (3). Recall Lemma 2.1.4, with l replaced by r. Since the right multiplication of $A(z)$ by matrices in $GL(r, \mathbb{C}[z])$ does not change the $\mathbb{C}[z]$-module generated by the column vectors of $A(z)$, while the left multiplication by a constant matrix in $GL(r, \mathbb{C})$ can by interpreted as a change of coordinates without changing the notion of being a constant section in the associated sheaf, this concludes Claim (3) and hence the proposition.

□

The remaining problem is to decide when two diagonalized forms of $\mathcal{O}_{U_0}$-modules (resp. $\mathcal{O}_{U_\infty}$-modules) of rank r determine the same submodule in $\mathcal{V}|_{U_0}$ (resp. $\mathcal{V}|_{U_\infty}$). To determine this, let two diagonal forms of $\mathbb{C}[z]$-modules be given by

$$B_1(z) = B_1(1)Diag\{z^{\alpha_1}, \ldots, z^{\alpha_r}\} \quad \text{and} \quad B_2(z) = B_2(1)Diag\{z^{\alpha_1}, \ldots, z^{\alpha_r}\}$$

respectively. Then $B_1(z)$ and $B_2(z)$ determine the same $\mathbb{C}[z]$-module if and only if there exists a $Q(z) \in GL(r, \mathbb{C}[z])$ such that $B_1(z)Q(z) = B_2(z)$. From this, one obtains that

$$\begin{aligned} Q(z) &= Diag\{z^{-\alpha_1}, \ldots, z^{-\alpha_r}\} B_1(1)^{-1} B_2(1)\, Diag\{z^{\alpha_1}, \ldots, z^{\alpha_r}\} \\ &= Diag\{z^{-\alpha_1}, \ldots, z^{-\alpha_r}\} B\, Diag\{z^{\alpha_1}, \ldots, z^{\alpha_r}\} \\ &= (z^{-\alpha_i+\alpha_j} b_{ij})_{i,j} \in GL(r, \mathbb{C}[z]), \end{aligned}$$

where $B = B_1(1)^{-1}B_2(1) = (b_{ij})_{i,j}$. This implies that $b_{ij} = 0$ if $\alpha_i > \alpha_j$. Consequently, $Q(z)$ is a *block upper triangular matrix, whose block form is determined by the multiplicity of elements in* $(\alpha_1, \ldots, \alpha_r)$. (For example, if this sequence is $(1, 1, 4, 4, 4, 7)$, then the corresponding block upper triangular matrix will have in the diagonal 2×2-, 3×3-, and 1×1-blocks.) Rephrased in a more geometric way, $B_1(z)$ *and* $B_2(z)$ *determine the same submodule if and only if they correspond to the same flag*. Explicitly, the flag associated to $B(z) = B(1)Diag\{z^{\alpha_1}, \ldots, z^{\alpha_r}\}$ is given as follows.

Let $B(1) = (u_1, \ldots, u_r)$ be the column vectors of $B(1)$ and suppose that

$$\alpha_1 = \cdots = \alpha_{j_1} < \alpha_{j_1+1} = \cdots = \alpha_{j_2} < \cdots < \alpha_{j_s+1} = \cdots = \alpha_r,$$

then the flag associated to $B(z)$ is given by

$$\begin{aligned} &(Span\{u_1, \ldots, u_{j_1}\} \subset Span\{u_1, \ldots, u_{j_2}\} \subset \cdots \\ &\qquad \cdots \subset Span\{u_1, \ldots, u_{j_s}\} \subset \mathbb{C}^r) \in Fl_{j_1, \ldots, j_s}(\mathbb{C}^r), \end{aligned}$$

where note that the last $\mathbb{C}^r$ should be identified with $\widehat{\mathcal{V}} \in Gr_r(\mathbb{C}^n)$.

Definition 2.1.6 [admissible pair of sequences]. Recall the Hilbert polynomial $P = P(t) = (n-r)t + d + (n-r)$. Then $(\alpha_1, \ldots, \alpha_r\,;\, \beta_1, \ldots, \beta_r)$ is called an *admissible pair of sequences* with respect to $P(t)$ if it satisfies

(1) $0 \le \alpha_1 \le \ldots \le \alpha_r$, $0 \le \beta_1 \le \ldots \le \beta_r$, and
(2) $(\alpha_1 + \ldots + \alpha_r) + (\beta_1 + \ldots + \beta_r) = d$.

Theorem 2.1.7 [topology of S^1-fixed-point locus].
Let $(\alpha_1, \ldots, \alpha_r\, ;\, \beta_1, \ldots, \beta_r)$ be an admissible pair of sequences of non-negative integers, $Fl_{j_1,\ldots,j_s,r}(\mathbb{C}^n)$ and $Fl_{j'_1,\ldots,j'_{s'},r}(\mathbb{C}^n)$ be the flag manifold associated to the multiplicity of elements in $(\alpha_1, \ldots, \alpha_r)$ and $(\beta_1, \ldots, \beta_r)$ respectively, as discussed above. Let

$$Fl_{j_1,\ldots,j_s,r}(\mathbb{C}^n) \rightarrow Gr_r(\mathbb{C}^n) \quad \text{and} \quad Fl_{j'_1,\ldots,j'_{s'},r}(\mathbb{C}^n) \rightarrow Gr_r(\mathbb{C}^n)$$

be the natural projections. Then the subset $F_{\alpha_1,\ldots,\alpha_r\,;\,\beta_1,\ldots,\beta_r}$ of the S^1-fixed-point locus that is associated to $(\alpha_1, \ldots, \alpha_r\, ;\, \beta_1, \ldots, \beta_r)$ is connected and is given by the fiber product

$$Fl_{j_1,\ldots,j_s,r}(\mathbb{C}^n) \times_{Gr_r(\mathbb{C}^n)} Fl_{j'_1,\ldots,j'_{s'},r}(\mathbb{C}^n)\,.$$

2.2. The S^1-weight system of the tangent space of Quot-scheme at an S^1-fixed-point component. After recalling some related facts in the preparatory remarks, we compute the S^1-weights and their multiplicities of the tangent space of the Quot-scheme $Quot(\mathcal{E}^n)$ at an S^1-fixed-point.

Preparatory remarks.
Recall (cf. [Ch], [H-L], and [Kol]) that the tangent space of Quot-scheme at a point is given by

$$T_{(\mathcal{E}^n \rightarrow \mathcal{E}^n/\mathcal{V})}\, Quot_P(\mathcal{E}^n) \;\simeq\; Hom_{\mathcal{O}_C}(\mathcal{V}\,,\,\mathcal{E}^n/\mathcal{V})\,.$$

When $(\mathcal{E}^n \rightarrow \mathcal{E}^n/\mathcal{V})$ is an S^1-fixed-point, S^1 acts both on $\mathcal{V}$ and $\mathcal{E}^n/\mathcal{V}$. The S^1-action on $T_{(\mathcal{E}^n \rightarrow \mathcal{E}^n/\mathcal{V})}\, Quot_P(\mathcal{E}^n)$ is translated to the S^1-action on $Hom_{\mathcal{O}_C}(\mathcal{V}\,,\,\mathcal{E}^n/\mathcal{V})$ by conjugations: $f \mapsto t\cdot f\cdot t^{-1}$ for $t \in S^1$.

Recall the inclusion of S^1-invariant subsheaves $\mathcal{V} \subset \widehat{\mathcal{V}}$ in $\mathcal{E}^n$. One thus has a natural morphism $\mathcal{E}^n/\mathcal{V} \rightarrow \mathcal{E}^n/\widehat{\mathcal{V}}$. Since $\widehat{\mathcal{V}}$ is a constant rank-r subbundle in $\mathcal{E}^n$, $\mathcal{E}^n/\widehat{\mathcal{V}}$ is a rank-$(n-r)$ trivial bundle on C. Since

$$(\mathcal{E}^n/\mathcal{V})|_{C-\{0,\infty\}} \stackrel{\sim}{\rightarrow} (\mathcal{E}^n/\widehat{\mathcal{V}})|_{C-\{0,\infty\}}$$

from the restriction of the above morphism and the restriction to the stalks

$$(\mathcal{E}^n/\mathcal{V})_0 \rightarrow (\mathcal{E}^n/\widehat{\mathcal{V}})_0 \qquad (\text{ resp. } (\mathcal{E}^n/\widehat{\mathcal{V}})_\infty \rightarrow (\mathcal{E}^n/\widehat{\mathcal{V}})_\infty)$$

at 0 (resp. ∞) given by

$$\left(\mathbb{C}[z]^{\oplus n}/(\mathbb{C}[z]z^{\alpha_1} \oplus \cdots \oplus \mathbb{C}[z]z^{\alpha_r})\right) \otimes_{\mathcal{O}_C(U_0)} \mathcal{O}_{C,0} \;\longrightarrow\; \mathbb{C}[z]^{\oplus(n-r)} \otimes_{\mathcal{O}_C(U_0)} \mathcal{O}_{C,0}$$

(resp.

$$\left(\mathbb{C}[z]^{\oplus n}/(\mathbb{C}[z]z^{\beta_1} \oplus \cdots \oplus \mathbb{C}[z]z^{\beta_r})\right) \otimes_{\mathcal{O}_C(U_\infty)} \mathcal{O}_{C,\infty} \;\longrightarrow\; \mathbb{C}[z]^{\oplus(n-r)} \otimes_{\mathcal{O}_C(U_\infty)} \mathcal{O}_{C,\infty}\;)$$

are surjective, the morphism $\mathcal{E}^n/\mathcal{V} \rightarrow \mathcal{E}^n/\widehat{\mathcal{V}}$ is surjective and one has the following split exact sequence of torsion-part/locally-free-part decomposition

$$0 \longrightarrow \widehat{\mathcal{V}}/\mathcal{V} \longrightarrow \mathcal{E}^n/\mathcal{V} \longrightarrow \mathcal{E}^n/\widehat{\mathcal{V}} \longrightarrow 0\,.$$

Since any constant rank-$(n-r)$ subbundle in $\mathcal{E}^n$ that is transverse to $\widehat{\mathcal{V}}$ is S^1-invariant and is mapped isomorphically to $\mathcal{E}^n/\widehat{\mathcal{V}}$, the above decomposition is also S^1-equivariant.

The S^1-action on $Hom_{\mathcal{O}_C}(\mathcal{V}, \mathcal{E}^n/\mathcal{V})$ when $(\mathcal{E}^n \to \mathcal{E}^n/\mathcal{V})$ is an S^1-fixed-point.

The above discussion gives an S^1-invariant decomposition of the tangent space to the Quot-scheme at an S^1-fixed-point :

$$\begin{aligned} Hom_{\mathcal{O}_C}(\mathcal{V}, \mathcal{E}^n/\mathcal{V}) &= Hom_{\mathcal{O}_C}(\mathcal{V}, \ddot{\mathcal{V}}/\mathcal{V} \oplus \mathcal{E}^n/\widehat{\mathcal{V}}) \\ &= Hom_{\mathcal{O}_C}(\mathcal{V}, \mathcal{F}_0) \oplus Hom_{\mathcal{O}_C}(\mathcal{V}, \mathcal{F}_\infty) \oplus Hom_{\mathcal{O}_C}(\mathcal{V}, \mathcal{E}^n/\widehat{\mathcal{V}}), \end{aligned}$$

where $\mathcal{F}_0$ (resp. $\mathcal{F}_\infty$) is the torsion subsheaf of $\mathcal{E}^n/\mathcal{V}$ supported at 0 (resp. ∞). We shall now study the three summands in the decomposition and their S^1-weight system, denoted by Wt_1, Wt_2, and Wt_3 respectively. Due to the tediousness of the discussion, we itemize the argument below.

- *The summands $Hom_{\mathcal{O}_C}(\mathcal{V}, \mathcal{F}_0)$ and $Hom_{\mathcal{O}_C}(\mathcal{V}, \mathcal{F}_\infty)$:*
  (1) These two components can be calculated via the restriction of the former to U_0 and the latter to U_∞. The problem is reduced then to the study of the group of homomorphisms of $\mathbb{C}[z]$-modules and the S^1-action on it. Explicitly,

  $$Hom_{\mathcal{O}_C}(\mathcal{V}, \mathcal{F}_0) = Hom_{\mathbb{C}[z]}(\oplus_{j=1}^r \mathbb{C}[z] \cdot z^{\alpha_j}, \oplus_{i=1}^r \mathbb{C}[z] \cdot \bar{e}_{0\,i}),$$

  where $z^{\alpha_i} \cdot \bar{e}_{0\,i} = 0$ for $i = 1, \ldots, r$, and

  $$Hom_{\mathcal{O}_C}(\mathcal{V}, \mathcal{F}_\infty) = Hom_{\mathbb{C}[w]}(\oplus_{j=1}^r \mathbb{C}[w] \cdot w^{\beta_j}, \oplus_{i=1}^r \mathbb{C}[w] \cdot \bar{e}_{\infty\,i}),$$

  where $w^{\beta_i} \cdot \bar{e}_{\infty\,i} = 0$ for $i = 1, \ldots, r$.

  (2) *Computation of the weight systems Wt_1 and Wt_2:*

  (2.1) Realize an element in $\oplus_{i=1}^r \mathbb{C}[z] \cdot \bar{e}_{0\,i}$ as a column vector and let

  $$f(z) = (f_{ij}(z))_{i,j} \in Hom_{\mathcal{O}_C}(\mathcal{V}, \mathcal{F}_0)$$

  with respect to the local bases $(z^{\alpha_1}, \ldots, z^{\alpha^r})$ and $(\bar{e}_{01}, \ldots, \bar{e}_{0r})$ for $\mathcal{V}$ and $\mathcal{F}_0$ respectively. Then *deg* $f_{ij}(z) < \alpha_i$ and (cf. the similar computation for the weight system Wt_3 below),

  $$(t \cdot f)(z) = (t^{\alpha_j} f_{ij}(t^{-1}z))_{i,j}, \quad t \in S^1.$$

  Thus, the rank-1 S^1-eigen-spaces in $Hom_{\mathcal{O}_C}(\mathcal{V}, \mathcal{F}_0)$ can be chosen to be generated by

  $$\varepsilon_{ij}[\mu_{ij}^0] = \varepsilon_{ij}[\mu_{ij}^0](z) := (f_{kl}(z))_{k,l},$$

  $$\text{where} \quad f_{kl}(z) = \begin{cases} 0 & \text{if } (k,l) \neq (i,j), \\ z^{\alpha_j - \mu_{ij}^0} & \text{if } (k,l) = (i,j). \end{cases}$$

  whose S^1-weight is μ_{ij}^0 that satisfies

  $$\alpha_j - \alpha_i < \mu_{ij}^0 \leq \alpha_j.$$

  From this, one has

  $$Wt_1 = \coprod_{i,j=1}^r ((\alpha_j - \alpha_i, \alpha_j] \cap \mathbb{Z})$$

(2.2) Rewrite $(\alpha_1, \ldots, \alpha_r)$ as

$$\begin{array}{ccccccc} 0 & \leq & a_1\,(=\alpha_1) & < & \cdots & < & a_k\,(=\alpha_r)\,, \\ & & m_1 & & \cdots & & m_k \end{array}$$

with the multiplicity indicated. For an interval $I \subset \mathbb{R}$, let χ_I be the characteristic function $\chi_I(x) = 1$, if $x \in I$, and $= 0$, otherwise. Let $\chi^A = \sum_{i=1}^k m_i\,\chi_{(-a_i\,,\,0]}$ and define χ^A_m by $\chi^A_m(x) = \sum_{j=1}^k m_j\,\chi^A(x-a_j)$. Then the multiplicity for $\mu \in Wt_1$ is given by $\chi^A_m(\mu)$.

(2.3) Realize an element in $\oplus_{i=1}^r \mathbb{C}[w]\cdot\overline{e}_{\infty\, i}$ as a column vector and let

$$g(w) \;=\; (g_{ij}(w))_{i,j} \;\in\; Hom_{\mathcal{O}_C}(\mathcal{V}\,,\,\mathcal{F}_\infty\,)\,.$$

Then $deg\ g_{ij}(w) < \beta_i$ and (cf. the similar computation for the weight system Wt_3 below),

$$(t\cdot g)(w) \;=\; (\,t^{-\beta_j}\,g_{ij}(tw)\,)_{i,j}\,,\quad t\in S^1\,.$$

Thus, the rank-1 S^1-eigen-spaces in $Hom_{\mathcal{O}_C}(\mathcal{V}\,,\,\mathcal{F}_\infty\,)$ can be chosen to be generated by

$$\varepsilon_{ij}[\mu^\infty_{ij}] \;=\; \varepsilon_{ij}[\mu^\infty_{ij}](z) \;:=\; (g_{kl}(z))_{k,l}\,,$$

$$\text{where}\quad g_{kl}(w) \;=\; \begin{cases} 0 & \text{if } (k,l)\neq(i,j), \\ w^{\beta_j+\mu^\infty_{ij}} & \text{if } (k,l)=(i,j). \end{cases}$$

whose S^1-weight is μ^∞_{ij} that satisfies

$$-\beta_j \;\leq\; \mu^\infty_{ij} \;<\; \beta_i-\beta_j\,.$$

From this, one has

$$Wt_2 \;=\; \bigsqcup_{i,\,j=1}^{r}\;(\,[\,-\beta_j\,,\,\beta_i-\beta_j\,)\cap\mathbb{Z}\,)$$

with the same rule of counting multiplicity as for Wt_1.

(2.4) Rewrite $(\beta_1, \ldots, \beta_r)$ as

$$\begin{array}{ccccccc} 0 & \leq & b_1\,(=\beta_1) & < & \cdots & < & b_l\,(=\beta_r) \\ & & n_1 & & \cdots & & n_l \end{array}$$

with the multiplicity indicated. Let $\chi^B = \sum_{i=1}^l n_i\,\chi_{[\,0\,,\,\beta_i\,)}$ and define χ^B_m by $\chi^B_m(x) = \sum_{j=1}^l n_j\,\chi^A(x+b_j)$. Then the multiplicity for $\mu\in Wt_2$ is given by $\chi^B_m(\mu)$.

- *The summand* $Hom_{\mathcal{O}_C}(\mathcal{V}\,,\,\mathcal{E}^n/\widehat{\mathcal{V}}\,)$:

(1) Since $\mathcal{E}^n/\widehat{\mathcal{V}}$ is represented by a rank-$(n-r)$ constant subbundle in $\mathcal{E}^n$ transverse to $\widehat{\mathcal{V}}$, it can be further decomposed into a direct sum of constant line subbundles in $\mathcal{E}^n$. Since all the bundles involved are constant, the decomposition of the quotient $\mathcal{E}^n/\widehat{\mathcal{V}} = \mathcal{O}_C^{\oplus(n-r)}$ is S^1-invariant. Recall that S^1 acts on $\mathcal{E}^n$ and hence on $\mathcal{E}^n/\widehat{\mathcal{V}}$ via the trivial linearization. With respect to this decomposition, one has

$$Hom_{\mathcal{O}_C}(\,\mathcal{V}\,,\,\mathcal{E}^n/\widehat{\mathcal{V}}\,) \;=\; Hom_{\mathcal{O}_C}(\,\mathcal{V}\,,\,\mathcal{O}_C^{\oplus(n-r)}\,) \;=\; H^0(C,\mathcal{V}^\vee)^{\oplus(n-r)}\,.$$

Remark. A connected component of the fixed-point locus can be stratified by subsets labelled by the isomorphism classes of vector bundles associated to the S^1-invariant subsheaves $\mathcal{V}$ in $\mathcal{E}^n$. There can be more than one strata for a connected component.

(2) *Computation of the weight system Wt_3.*

(2.1) Recall the S^1-invariant decompositions

$$Hom_{\mathcal{O}_C}(\mathcal{V}\,,\,\mathcal{E}^n/\widehat{\mathcal{V}}\,)$$
$$= Hom_{\mathcal{O}_C}(\mathcal{V}\,,\,\mathcal{O}_C^{\oplus(n-r)}\,) \;=\; Hom_{\mathcal{O}_C}(\mathcal{V}\,,\,\mathcal{O}_C\,)^{\oplus(n-r)}\,.$$

The existence of such S^1-invariant decomposition implies that the sought-for S^1-weight system for $Hom_{\mathcal{O}_C}(\mathcal{V}\,,\,\mathcal{E}^n/\widehat{\mathcal{V}}\,)$ consists of $(n-r)$-many copies of the S^1-weight system for $Hom_{\mathcal{O}_C}(\mathcal{V}\,,\,\mathcal{O}_C\,)$.

(2.2) Let $f \in Hom_{\mathcal{O}_C}(\mathcal{V}\,,\,\mathcal{O}_C\,)$. Then, as a morphism of sheaves on C, f is described by a pair

$$(f_0, f_\infty)$$
$$\in Hom_{\mathcal{O}_C(U_0)}(\,\mathcal{V}(U_0)\,,\,\mathcal{O}_C(U_0)\,) \times Hom_{\mathcal{O}_C(U_\infty)}(\,\mathcal{V}(U_\infty)\,,\,\mathcal{O}_C(U_\infty)\,)$$
$$= Hom_{\mathbb{C}[z]}(\,\oplus_{i=1}^r \mathbb{C}[z]\cdot z^{\alpha_i}\,,\,\mathbb{C}[z]\,) \times Hom_{\mathbb{C}[w]}(\,\oplus_{i=1}^r \mathbb{C}[w]\cdot w^{\beta_i}\,,\,\mathbb{C}[w]\,)$$

such that $f_0|_{U_0\cap U_\infty} = f_\infty|_{U_0\cap U_\infty}$.

(2.3) Recall the proof of Theorem 5.3 in [Ko], which says in our case that *the weight system of the tangent bundle at an S^1-fixed-point depends only on the connected component of the fixed-point locus.* Thus, *to compute the weight system one can choose the S^1-invariant subsheaf $\mathcal{V}$ in $\mathcal{E}^n$ such that the two diagonalized local pieces on affine charts U_0 and U_∞ match (i.e. $\mathcal{V}$ becomes the direct sum of appropriate ideal sheaves in constant line subbundles in $\mathcal{E}^n$)*. From the previous discussions, there are many - even continuous families of - such $\mathcal{V}$. However, as will be clear from the explicit expression that the weight system obtained is indeed independent of which such $\mathcal{V}$ is chosen for the computation, as long as they belong to the same fixed-point component. This gives a consistency check of the method.

(2.4) Let $\mathcal{V}$ be an S^1-invariant subsheaf of $\mathcal{E}^n$ such that the two local diagonalizations match and suppose that α_i is matched with $\beta_{i'}$, $i = 1, \ldots, r$. Then $\mathcal{V}$ is decomposed into a direct sum $\oplus_{i=1}^r \mathcal{I}_{\alpha_i,\beta_{i'}}$, where $\mathcal{I}_{\alpha_i,\beta_{i'}}$ is a subsheaf in a constant line subbundle $\simeq \mathcal{O}_C$ in $\mathcal{E}^n$ with the local data as a sheaf of $\mathcal{O}_C$-module:

on U_0		on $U_0\cap U_\infty$		on $U_0\cap U_\infty$		on U_∞
$\mathbb{C}[z]\cdot z^{\alpha_i}$	$\rightarrow$	$\mathbb{C}[z,z^{-1}]$	$\stackrel{z\leftrightarrow 1/w}{\longleftrightarrow}$	$\mathbb{C}[w,w^{-1}]$	$\leftrightarrow$	$\mathbb{C}[w]\cdot w^{\beta_{i'}}$.

The vector bundle associated to $\mathcal{V}$ is isomorphic to $\oplus_{i=1}^r \mathcal{O}(-\alpha_i - \beta_{i'})$ and $Hom_{\mathcal{O}_C}(\,\mathcal{V}\,,\,\mathcal{O}_C\,)$ is further decomposed into an S^1-invariant direct sum

$$Hom_{\mathcal{O}_C}(\,\mathcal{V}\,,\,\mathcal{O}_C\,) \;=\; \oplus_{i=1}^r\, Hom_{\mathcal{O}_C}(\,\mathcal{I}_{\alpha_i,\beta_{i'}}\,,\,\mathcal{O}_C\,)\,.$$

$f \in Hom_{\mathcal{O}_C}(\mathcal{I}_{\alpha,\beta}\,,\,\mathcal{O}_C\,)$ is given by a pair

$$(f_0, f_\infty) \;\in\; Hom_{\mathbb{C}[z]}(\mathbb{C}[z]\cdot z^\alpha\,,\,\mathbb{C}[z]) \times Hom_{\mathbb{C}[w]}(\mathbb{C}[w]\cdot w^\beta\,,\,\mathbb{C}[w])\quad,$$
$$(\,z^\alpha \stackrel{f_0}{\longmapsto} h_0(z)\,,\, w^\beta \stackrel{f_\infty}{\longmapsto} h_\infty(w)\,)$$

such that the following matching condition holds

$$z^{-\alpha}h_0(z) \;=\; w^{-\beta}h_\infty(w) \quad \text{under } z \rightarrow 1/w\,.$$

Consequently,

$$\begin{aligned} & Hom_{\mathcal{O}_C}(\mathcal{I}_{\alpha,\beta}\,,\,\mathcal{O}_C\,) \\ & = \{\,(h_0(z)\,,\,h_\infty(w))\,|\, deg\, h_0(z) \le \alpha+\beta \text{ and } h_\infty(w) = w^{\alpha+\beta}h_0(1/w)\,\} \\ & \simeq\; H^0(C,\mathcal{I}^\vee) \;=\; H^0(C,\mathcal{O}_C(\alpha+\beta))\,. \end{aligned}$$

(2.6) The S^1-action on $Hom_{\mathcal{O}_C}(\mathcal{I}_{\alpha,\beta}\,,\,\mathcal{O}_C\,)$ is given by $f \mapsto t\cdot f$, where $t\cdot f$ is the composition of the following conjugation of $f = (\,h_0(z)\,,\,h_\infty(w)\,)$:

on U_0:

$$\begin{aligned} s_0(z)z^\alpha & \stackrel{t^{-1}}{\longrightarrow} && s_0(tz)(tz)^\alpha = t^\alpha s_0(tz)\cdot z^\alpha \\ & \stackrel{f}{\longrightarrow} && t^\alpha s_0(tz)\,h_0(z) \\ & \stackrel{t}{\longrightarrow} && t^\alpha s_0(tt^{-1}z)h_0(t^{-1}z) = t^\alpha s_0(z)h_0(t^{-1}z) \end{aligned}$$

on U_∞:

$$\begin{aligned} s_\infty(w)w^\beta & \stackrel{t^{-1}}{\longrightarrow} && s_\infty(t^{-1}w)(t^{-1}w)^\beta = t^{-\beta}s_\infty(t^{-1}w)\cdot w^\beta \\ & \stackrel{f}{\longrightarrow} && t^{-\beta}s_\infty(t^{-1}w)\,h_\infty(w) \\ & \stackrel{t}{\longrightarrow} && t^{-\beta}s_\infty(t^{-1}tw)h_\infty(tw) = t^{-\beta}s_\infty(w)h_\infty(tw)\,. \end{aligned}$$

One can check directly that if $(f_0,\, f_\infty)$ satisfies the matching condition, then so does $(t\cdot f)_0$, $(t\cdot f)_\infty)$. Consequently,

$$f = (\,h_0(z)\,,\,h_\infty(w)\,) \;\stackrel{t}{\longrightarrow}\; t\cdot f = (\,t^\alpha\,h_0(t^{-1}z)\,,\,t^{-\beta}\,h_\infty(tw)\,)$$

on $Hom_{\mathcal{O}_C}(\mathcal{I}_{\alpha,\beta}\,,\,\mathcal{O}_C\,)$. If f is an invariant direction of the S^1-action on $Hom_{\mathcal{O}_C}(\mathcal{I}_{\alpha,\beta}\,,\,\mathcal{O}_C\,)$, then $t\cdot f = t^\mu f$ for some $\mu \in \mathbb{Z}$. From the above expression, this means that

$$(\,t^\alpha\,h_0(t^{-1}z)\,,\,t^{-\beta}\,h_\infty(tw)\,) \;=\; (\,t^\mu\,h_0(z)\,,\,t^\mu\,h_\infty(w)\,) \quad\text{for all } t\,,$$

which implies that

$$f \;=\; f_\mu \;:=\; (\,h_0(z)\,,\,h_\infty(w)\,) \;=\; (\,c\,z^{\alpha-\mu}\,,\,c\,w^{\beta+\mu}\,)\,.$$

From this, one concludes that

$$-\beta \;\le\; \mu \;\le\; \alpha\,, \qquad \mu \in \mathbb{Z}\,;$$

with the associated weight subspace spanned by f_μ.

(2.7) Resume the indices (i, i') for $\mathcal{I}_{\alpha_i, \beta_{i'}}$. Then

Lemma [weight subsystem Wt_3].
(1) *Let Wt_3' be the system of weights of the S^1-action on $Hom_{\mathcal{O}_C}(\mathcal{V}, \mathcal{O}_C)$. Then the weight system Wt_3 for the S^1-action on*

$$Hom_{\mathcal{O}_C}(\mathcal{V}, \mathcal{E}^n/\widehat{\mathcal{V}})$$

is given by $Wt_3 = (n-r)\, Wt_3'$, i.e. same set of integers μ as in Wt_3' but with multilicity $m_\mu = (n-r)\, m_\mu'$,
(2) *Wt_3' is given by*

$$Wt_3' = \bigsqcup_{i=1}^{r} ([-\beta_{i'}, \alpha_i] \cap \mathbb{Z}).$$

Recall $(\alpha_1, \ldots, \alpha_r; \beta_1, \ldots, \beta_r)$ rewritten as

$$\begin{array}{ccccccc} 0 & \le & a_1 & < & \cdots & < & a_k\, (= \alpha_r) \\ & & m_1 & & \cdots & & m_k \end{array} \quad ,$$

$$\begin{array}{ccccccc} 0 & \le & b_1 & < & \cdots & < & b_l\, (= \beta_r) \\ & & n_1 & & \cdots & & n_l \end{array}$$

with the multiplicity indicated. Then any $\nu \in [-\beta_r, \alpha_r] \cap \mathbb{Z}$ is in Wt_3'. Its multiplicity m_μ' in Wt_3' is given by

$$m_\mu' = \begin{cases} n_l + \cdots + n_j & \text{if } -b_j \le \mu < -b_{j-1}\,. \\ r & \text{if } -b_1 \le \mu \le a_1\,, \\ m_k + \cdots + m_j & \text{if } a_{j-1} < \mu \le a_j\,, \end{cases}$$

From this expression, it is clear that Wt_3 depends only on $(\alpha_1, \ldots, \alpha_r; \beta_1, \ldots \beta_r)$ and hence only on the connected component of the S^1-fixed-point locus, as it should.

Proof of Lemma. Consider the two sets of lattice points in $\mathbb{Z} \oplus \mathbb{Z} \subset \mathbb{R}^2$:

$$\begin{aligned} A &= \{(\alpha_i, r-i+1) \mid i = 1, \ldots, r\}, \\ B &= \{(-\beta_i, r-i+1) \mid i = 1, \ldots, r\}, \end{aligned}$$

and the r-many line segments connecting $(-\beta_{i'}, r-i'+1)$ and $(\alpha_i, r-i+1)$. Let π be the projection of $\mathbb{R}^2$ to the horizontal axis $L \supset \mathbb{Z}$. Then, for an integer $\mu \in \mathbb{Z} \subset L$, the multiplicity m_μ of μ in Wt_3' is the same as the number of the line segments above whose projection into L contain μ. Thus, $m_\mu > 0$ if and only if $\mu \in [-\beta_r, \alpha_r]$. To read off m_μ, one combs the collection of line segments so that each line segment becomes a three-edged-path with the first and the third edge horizontal and the middle one vertical and contained in the vertial axis, cf. FIGURE 2-2-1. From this,

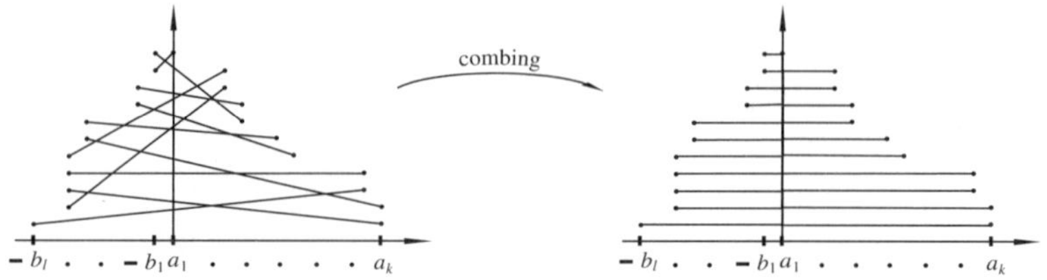

FIGURE 2-2-1. The mutiplicity of μ and the combing of the line segments.

To summarize:

Theorem 2.2.1 **[S^1-weight].** *The S^1-weights on the tangent space*

$$T_{(\mathcal{E}^n \to \mathcal{E}^n/\mathcal{V})} Quot_P(\mathcal{E}^n) \simeq Hom_{\mathcal{O}_C}(\mathcal{V}, \mathcal{E}^n/\mathcal{V})$$

of Quot-scheme $Quot(\mathcal{E}^n)$ at an S^1-fixed-point $(\mathcal{E}^n \to \mathcal{E}^n/\mathcal{V})$ are the disjoint union of Wt_1, Wt_2, and Wt_3, as given above.

2.3. Combinatorics of the S^1-weight system and the multiplicity of 0. A generating function for the multiplicity of weights in Wt_3 is immediate, following same argument as in the counting of the states at various levels in conformal field theory, e.g. [G-S-W]. An example is given by the following formal function

$$\prod_{j=0}^{\infty} \frac{1}{1 - q_0^{n-r} \cdots q_j^{n-r}\, s^j\, t}\,.$$

It remains unclear to us whether the weight systems Wt_1 and Wt_2 also have elegant generating functions; nevertheless they can be obtained from the following manipulations.

- *The weight subsystem Wt_1:*
  (1) Consider the formal expansion

$$\left. \left(\prod_{j=0}^{\infty} \frac{1}{1 - q_{-j} \cdots q_0\, s^j\, t\, v_j} \right) \right|_{q_0=1} = \sum_{k,\,l,\,P} A^{(1)}_{k,\,l,\,P}(\mathbf{q})\, A^{(2)}_P(\mathbf{v})\, s^k\, t^l\,,$$

  where $\mathbf{q}$ and $\mathbf{v}$ represent collectively the two sets of variables q_i and v_i respectively. Note that both $A^{(1)}_{k,\,l,\,P}(\mathbf{q})$ and $A^{(2)}_P(\mathbf{v})$ are monomials.
  (2) Do the substitutions

$$A^{(2)}_P(\mathbf{v}) \xrightarrow{\;v_j \to A^{(1)}_{k,\,l,\,P}(\mathbf{q}_{i \to i+j+1})\;} \widehat{A^{(2)}_P}(\mathbf{q})\,,$$

  where $\mathbf{q}_{i\to i+j+1}$ means that q_i is replaced by q_{i+j+1} for all i. The result $\widehat{A^{(2)}_P}(\mathbf{q})$ is a monomial in $\mathbf{q}$ and the multiplicity of $j \in \mathbb{Z}$ is n_j if $q_j^{n_j}$ appears as a primary factor of $\widehat{A^{(2)}_P}(\mathbf{q})$.

(Cf. See Example 2.3.2 below.)

- *The weight subsystem* Wt_2 :

(1) Consider the formal expansion

$$\left.\left(\prod_{j=0}^{\infty}\frac{1}{1-q_0\cdots q_j\, s^j\, t\, v_j}\right)\right|_{q_0=1} = \sum_{k,\,l,\,P} B^{(1)}_{k,\,l,\,P}(\mathbf{q})\, B^{(2)}_P(\mathbf{v})\, s^k\, t^l\,.$$

(2) Do the substitutions

$$B^{(2)}_P(\mathbf{v}) \xrightarrow{\ v_j\to B^{(1)}_{k,\,l,\,P}(\mathbf{q}_{i\to i-j-1})\ } \widehat{B^{(2)}_P}(\mathbf{q})\,,$$

where $\mathbf{q}_{i\to i-j-1}$ means that q_i is replaced by q_{i-j-1} for all i. The result $\widehat{B^{(2)}_P}(\mathbf{q})$ is a monomial in $\mathbf{q}$ and the multiplicity of $j\in\mathbb{Z}$ is n_j if $q_j^{n_j}$ appears as a primary factor of $\widehat{B^{(2)}_P}(\mathbf{q})$.

Remark 2.3.1.

(1) The powers k and l and the monomials $A^{(1)}_{k,\,l,\,P}(\mathbf{q})$ and $A^{(2)}_P(\mathbf{v})$ are related as follows. $A^{(2)}_P(\mathbf{v})$ is the monomial that encodes the partition P of k into the summmation l-many non-negative integers. Corresponding to P is a conjugate partition $\widehat{P}$. The Young diagram associated to P is conjugate to that associated to $\widehat{P}$. The monomial $A^{(1)}_{k,\,l,\,P}(\mathbf{q})$ is determined by the partition $\widehat{P}$. Similarly for $B^{(1)}_{k,\,l,\,P}(\mathbf{q})$ and $B^{(2)}_P(\mathbf{v})$. (Cf. Example 2.3.2 below.)

(2) Item (1) above implies that for a distinguished S^1-fixed-point component $F_{\alpha_1,\,\ldots,\,\alpha_r\,;\,0,\,\ldots,\,0}$, the subweight system Wt_1 is generated completely by the Young diagram associated to $d=\alpha_1+\cdots+\alpha_r$ as a partition of d by a "partial tensor" with the conjugate Young diagram, as illustrated in FIGURE 2-3-1.

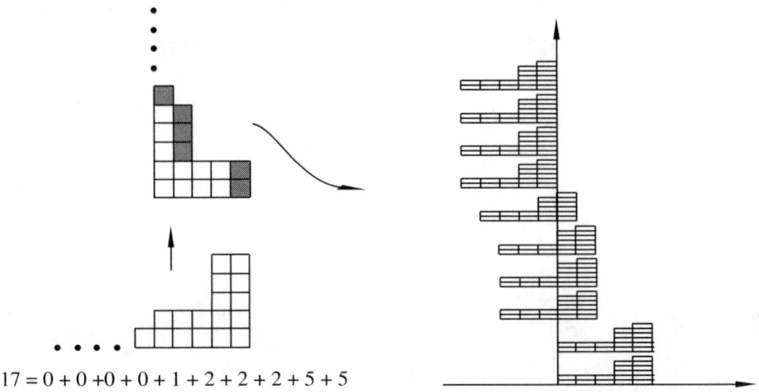

FIGURE 2-3-1. Generation of Wt_1 from a single Young

Example 2.3.2 [weight computation for Wt_1]. Consider the expansion

$$\left(\prod_{j=0}^{\infty} \frac{1}{1 - q_{-j} \cdots q_0\, s^j\, t\, v_j} \right) \Bigg|_{q_0=1} = \sum_{k,\, l,\, P} A^{(1)}_{k,\, l,\, P}(\mathbf{q})\, A^{(2)}_{P}(\mathbf{v})\, s^k\, t^l\, .$$

Consider, for example, the case $r = 10$ and $d = 17$. Then, to determine the weight subsystem Wt_1 for the normal bundle to the component $F_{0,0,0,0,1,2,2,2,5,5\,;\,0,\,\cdots\,,\,0}$ in E_0, one only needs to look at the (unique) term in the expansion with $A^{(2)}_P(\mathbf{v}) = v_0^4\, v_1\, v_2^3\, v_5^2$, corresponding to the partition $P\,:\, 17 = 0+0+0+0+1+2+2+2+5+5$:

$$\begin{aligned} & (\, A^{(1)}_{17,\, 10,\, P}(\mathbf{q})\, A^{(2)}_P(\mathbf{v})\, s^{17}\, t^{10}\,)|_{q_0=1} \\ & \quad = q_{-1}\, (q_{-2}\, q_{-1})^3\, (q_{-5}\, q_{-4}\, q_{-3}\, q_{-2}\, q_{-1})^2\, v_0^4\, v_1\, v_2^3\, v_5^2\, s^{17}\, t^{10} \\ & \quad = q_{-5}^2\, q_{-4}^2\, q_{-3}^2\, q_{-2}^5\, q_{-1}^6\, v_0^4\, v_1\, v_2^3\, v_5^2\, s^{17}\, t^{10}\, . \end{aligned}$$

Observe that the conjugate partition $\widehat{P}$: $17 = 0+0+0+0+0+2+2+2+5+6$ is encoded in the monomial in q's. Now do the substitiution with the rule of shifting the indices as given above:

$$\begin{aligned} & v_0^4\, v_1^1\, v_2^3\, v_5^2 \quad \longrightarrow \\ & \quad (\, q_{-4}^2\, q_{-3}^2\, q_{-2}^2\, q_{-1}^5\, q_{\mathbf{0}}^6\,)^{\mathbf{4}}\, (\, q_{-3}^2\, q_{-2}^2\, q_{-1}^2\, q_0^5\, q_{\mathbf{1}}^6\,)^{\mathbf{1}}\, (\, q_{-2}^2\, q_{-1}^2\, q_0^2\, q_1^5\, q_{\mathbf{2}}^6\,)^{\mathbf{3}}\, (\, q_1^2\, q_2^2\, q_3^2\, q_4^5\, q_{\mathbf{5}}^6\,)^{\mathbf{2}}\, |_{q_0=1} \\ & \quad = q_{-4}^8\, q_{-3}^{10}\, q_{-2}^{16}\, q_{-1}^{28}\, q_1^{25}\, q_2^{22}\, q_3^4\, q_4^{10}\, q_5^{12}\, , \end{aligned}$$

where some of the indices are boldfaced to make the pattern manifest. Let α be the generator of $H^*_{S^1}(pt)$. Then the S^1-weights in Wt_1 for the normal bundle to $F_{0,0,0,0,1,2,2,2,5,5\,;\,0,\,\cdots\,,\,0}$ is

$$8(-4\alpha),\ 10(-3\alpha),\ 16(-2\alpha),\ 28(-\alpha),\ 25(\alpha),\ 22(2\alpha),\ 4(3\alpha),\ 10(4\alpha),\ 12(5\alpha)\, .$$

□

Though it can be obtained also from Sec. 2.2, the following lemma follows immediately from the combinatorics of the weight system discussed in this subsection.

Lemma 2.3.3 [multiplicity of 0]. *The multiplicity of 0 in the S^1-weight system Wt to the restriction of the tangent bundle $T_*\,Quot_P(\mathcal{E}^n)$ to $F_{\alpha_1,\dots,\alpha_r\,;\,\beta_1,\dots,\beta_r}$ is equal to $\dim F_{\alpha_1,\dots,\alpha_r\,;\,\beta_1,\dots,\beta_r}$. Consequently, the S^1-weight system of the normal bundle to $F_{\alpha_1,\dots,\alpha_r\,;\,\beta_1,\dots,\beta_r}$ is exactly the subsystem of non-zero weights in Wt.*

Corollary 2.3.4 [e_{S^1} invertible]. *Let E be any of the S^1-fixed-point component in $Quot_P(\mathcal{E}^n)$. Then the S^1-weights of the normal bundle $\nu_E(Quot_P\mathcal{E}^n)$ to E are all nonzero. Consequently $e_{S^1}(\nu/Quot_P(\mathcal{E}^n))$ is invertible in $A^*(E)(\alpha)$, where α is a generator of the ring $H^*_{S^1}(pt)$.*

3. Mirror principle computation for Grassmannian manifolds.

3.1. The distinguished S^1-fixed-point components and the hyperplane-induced class. To make the comparison immediate, here we follow the notations in [L-L-Y1: III, Sec. 5.4]. Recall the following approach ibidem to compute $A(t)$ when there is a commutative diagram:

$$\begin{array}{ccccc} F_0 & \stackrel{e^Y}{\longrightarrow} & Y_0 & \stackrel{g}{\longleftarrow} & E_0 \\ \downarrow i & & \downarrow j & & \downarrow k \\ M_d & \stackrel{\varphi}{\longrightarrow} & W_d & \stackrel{\psi}{\longleftarrow} & \mathcal{Q}_d \end{array} ,$$

where $\mathcal{Q}_d$ is an S^1-manifold, $\psi : \mathcal{Q}_d \to W_d$ is an S^1-equivariant resolution of singularities of $\varphi(M_d)$, E_0 is the set of fixed-points in $\psi^{-1}(Y_0)$ and is called the distinguished S^1-fixed-point component, and $\varphi_*[M_d] = \psi_*[\mathcal{Q}_d]$ in $A_*^{S^1}(W_d)$.

In the current case, X is the Grassmannian manifold $Gr_r(\mathbb{C}^n)$, $\mathcal{Q}_d$ is the Quot-scheme $Quot_{P(t)=(n-r)t+(d+n-r)}(\mathcal{E}^n)$, and the linear sigma model W_d for X is the projective space $\mathbb{P}(H^0(C, \mathcal{O}_C(d)) \otimes \Lambda^r\mathbb{C}^n)$ of $\binom{n}{r}$-tuple of degree-d homogeneous polynomials on C. This is a linear sigma model for $\mathbb{P}(\Lambda^r\mathbb{C}^n)$ that is turned into a linear sigma model for X via the Plücker embedding $Gr_r(\mathbb{C}^n) \to \mathbb{P}(\Lambda^r\mathbb{C}^n)$.

An element in W_d can be written as

$$[\sum_j c_{1j} z_0^j z_1^{d-j} : \sum_j c_{2j} z_0^j z_1^{d-j} : \cdots],$$

where $[z_0 : z_1]$ is the homogeneous coordinates for C and $c_{ij} \in \mathbb{C}$ with $1 \le i \le \binom{n}{r}$ and $0 \le j \le d$. The group S^1 acts on W_d by

$$[\sum_j c_{1j} z_0^j z_1^{d-j} : \cdots] \longmapsto [\sum_j c_{1j} (tz_0)^j z_1^{d-j} : \cdots], \quad t \in S^1 .$$

There are $(d+1)$-many S^1-fixed-point components in W_d, each of which consists of points of the form $[\, c_{1j} z_0^j z_1^{d-j} : c_{2j} z_0^j z_1^{d-j} : \cdots]$ for $0 \le j \le d$ and is isomorphic to $\mathbb{P}(\Lambda^r(\mathbb{C}^n))$. From [L-L-Y1: II, III], the S_1-fixed-point component F_0 in M_d consists of degree-$(1,d)$ stable maps (C, f) into $C \times Gr_r(\mathbb{C}^n) \subset C \times \mathbb{P}(\Lambda^r\mathbb{C}^n)$ that is obtained by gluing a degree-$(1,0)$ stable map $(C_1 = \mathbb{CP}^1, f_1, \infty)$ and a degree-$(0,d)$ stable map (C_2, f_2, x) with $f_1(\infty) = f_2(x)$ at their marked point. Regard these as stable maps into the projective space $\mathbb{P}(\Lambda^r\mathbb{C}^n)$, then [L-L-Y1: I, Sec.2, Example 10 and III, Sec. 3] implies that the S^1-fixed-point component Y_j in W_d consists of point of the form $[\, c_{1,d-j} z_0^{d-j} z_1^j : c_{2,d-j} z_0^{d-j} z_1^j : \cdots]$. In particular, Y_0 consists of points of the form $[\, c_{1d} z_0^d : c_{2d} z_0^d : \cdots]$.

The map

$$\psi : \mathcal{Q}_d = Quot_{P(t)}(\mathcal{E}^n) \longrightarrow W_d = \mathbb{P}(H^0(C, \mathcal{O}_C(d)) \otimes \Lambda^r\mathbb{C}^n)$$

is given as follows. Write $C = \operatorname{Proj} \mathbb{C}[z_0, z_1]$, where $\mathbb{C}[z_0, z_1]$ is regarded as a graded ring with grading given by the total degree. Then $\mathcal{E}^n$ is the sheaf associated to the graded $\mathbb{C}[z_0, z_1]$-module $\mathfrak{M} := \mathbb{C}[z_0, z_1]^{\oplus n}$, whose grade-$d$ piece $\mathfrak{M}_d$ is given by a basis for $\mathfrak{N}_{\mathcal{V}}$. Express each e_i as a column vector with entries in $\mathbb{C}[z_0, z_1]$ and consider the matrix $A_{\mathcal{V}} = [\, e_1, \ldots, e_r \,]$. When the quotient sheaf $\mathcal{E}^n/\mathcal{V}$ has degree d, all the $r \times r$-minors of $A_{\mathcal{V}}$, if not zero, must be of degree d as well. The map ψ sends $(\mathcal{E}^n \to \mathcal{E}^n/\mathcal{V})$ then to the $\binom{n}{r}$-tuple of $r \times r$-minors of $A_{\mathcal{V}}$. (Cf. [Ha], [So], [Str], and [S-S].)

Since ψ is S^1-equivariant, it sends an S^1-fixed-point component in $\mathcal{Q}_d$ into an S^1-fixed-point component in W_d. To see which S^1-fixed-point component in $\mathcal{Q}_d$ is sent to Y_0, one only needs to check where a single point in $F_{\alpha_1, \ldots, \alpha_r; \beta_1, \ldots, \beta_r}$ is mapped to.

Lemma 3.1.1. $\psi(F_{\alpha_1, \ldots, \alpha_r; \beta_1, \ldots, \beta_r}) \subset Y_{\beta_1 + \cdots + \beta_r}$.

Proof. Recall that, for a fixed-point $(\mathcal{E}^n \to \mathcal{E}^n/\mathcal{V}) \in F_{\alpha_1, \ldots, \alpha_r; \beta_1, \ldots, \beta_r}$,

$$deg\, \mathcal{E}^n/\mathcal{V} \;=\; \alpha_1 + \cdots + \alpha_r + \beta_1 + \cdots + \beta_r \;=\; d\,.$$

Observe also that the special fixed-points in $F_{\alpha_1, \ldots, \alpha_r; \beta_1, \ldots, \beta_r}$, for which the two local diagonalizations match with $\alpha_i \to \beta_{i'}$, corresponds to a subsheaf $\mathcal{V}$ in $\mathcal{E}^n$ is isomorphic to the direct sum $\oplus_i \mathcal{I}_{\alpha_i(0) + \beta_{i'}(\infty)}$ of ideal sheaves $\mathcal{I}_{\alpha_i(0)+\beta_{i'}(\infty)}$ in $\mathcal{O}_C$ associated to the degree-d divisor/subscheme $\alpha_i(0) + \beta_{i'}(\infty)$ in C. Its associated matrix $A_{\mathcal{V}}$ can be written as

$$\begin{pmatrix} z_0^{\alpha_1} z_1^{\beta_{1'}} & & \\ & \ddots & \\ & & z_0^{\alpha_r} z_1^{\beta_{r'}} \\ 0 & \cdots & 0 \\ \vdots & \cdots & \vdots \\ 0 & \cdots & 0 \end{pmatrix} \qquad \text{with zero entries } a_{ij} \text{ for } i \neq j,$$

after a constant re-trivialization of $\mathcal{E}^n$. The $r \times r$-minors of this matrix are all zero except the one from the top $r \times r$-submatrix, whose value is $z_0^{\alpha_1 + \cdots + \alpha_r} z_1^{\beta_1 + \cdots + \beta_r}$. Thus, ψ maps such point to some

$$[\, 0 : \cdots : 0 : z_0^{\alpha_1 + \cdots + \alpha_r} z_1^{\beta_1 + \cdots + \beta_r} : 0 : \cdots : 0 \,]\,,$$

which lies in $Y_{\beta_1 + \cdots + \beta_r}$. This proves the lemma.

□

Since $0 \leq \beta_1 \leq \cdots \leq \beta_r$, one concludes that

Corollary 3.1.2 [distinguished components]. *The distinguished S^1-fixed-point locus E_0 is given by*

$$E_0 \;=\; \coprod_{\substack{0 \leq \alpha_1 \leq \cdots \leq \alpha_r \\ \alpha_1 + \cdots + \alpha_r = d}} F_{\alpha_1, \cdots, \alpha_r; 0, \ldots, 0}\,,$$

a disjoint union of flag manifolds determined by the multiplicities of entries in $(\alpha_1, \ldots, \alpha_r)$ with $\alpha_1 + \cdots + \alpha_r = d$. (Cf. Theorem 2.1.7)

On each distinguished S^1-fixed-point component $F_{\alpha_1, \cdots, \alpha_r; 0, \ldots, 0}$, there is the pulled-back hyperplane class $k^* \psi^* \kappa = g^* j^* \kappa$, where κ is the hyperplane class on

W_d. To see what it is, recall first the multiplicity numbers $m_1, \ldots, m_k$ for $0 \leq \alpha_1 \leq \cdots \leq \alpha_r$ and the following fact/definition:

Fact/Definition 3.1.3 [special Schubert cycle]. (Cf. [Fu1], also [Jo].) Recall that, over the flag manifold $Fl = Fl_{m_1, m_1+m_2, \ldots, r}(\mathbb{C}^n)$, there is a universal flag of bundles $S_1 \hookrightarrow S_2 \hookrightarrow \cdots \hookrightarrow S_{k+1} = Fl \times \mathbb{C}^n$ with *rank* $S_i = m_1 + \cdots + m_i$. Then *the intersection Chow ring $A^*(Fl)$ is generated by the Chern classes of the quotient bundles S_i/S_{i-1}, $1 \leq i \leq k+1$ and $S_0 = 0$, with relations determined by* $\prod_{i=1}^{k+1} c(S_i/S_{i-1}) = 1$. The Schubert cycles that represent these special generators are called *special Schubert cycles.*

Since $F_{\alpha_1, \cdots, \alpha_r; 0, \ldots, 0} \simeq Fl_{m_1, m_1+m_2, \ldots, r}(\mathbb{C}^n)$, this gives $A^*(F_{\alpha_1, \cdots, \alpha_r; 0, \ldots, 0})$. Recall also from Sec. 2.1 that points in $F_{\alpha_1, \cdots, \alpha_r; 0, \ldots, 0}$ can be represented by $n \times r$-matrices $B(z)$ with coefficients in $\mathbb{C}[z]$. Since the map g is the Plücher embedding and it sends $B(z)$ to the tuple of $r \times r$-minors of $B(1)$ multiplied by the factor z^d, the image $g(F_{\alpha_1, \cdots, \alpha_r; 0, \ldots, 0})$ coincides with the image of the Grassmannian manifold $Gr_r(\mathbb{C}^n)$ in Y_0 via Plücker embedding and g is indeed the fibration to the base Grassmannian manifold given in Theorem 2.1.7.

Let $S \hookrightarrow Gr_r(\mathbb{C}^n) \times \mathbb{C}^n$ be the universal rank-r bundle over $Gr_r(\mathbb{C}^n)$. Then the Plücker embedding in the direct bundle language is the section from the projectivization of the tautological bundle map $\bigwedge^r S = \det S \hookrightarrow Gr_r(\mathbb{C}^n) \times \bigwedge^r \mathbb{C}^n$ over $Gr_r(\mathbb{C}^n)$ and, hence, the hyperplane class on $\mathbb{CP}^{\binom{n}{r}-1}$ is pulled back to the Chern class $-c_1(S)$ on $Gr_r(\mathbb{C}^n)$ via the Plücker embedding. On the other hand, the embedding of $Y_0 \simeq \mathbb{CP}^{\binom{n}{r}-1}$ in $W_d \simeq \mathbb{CP}^{\binom{n}{r}d+\binom{n}{r}-1}$ has degree 1 from [L-L-Y1: I and II]. Together one concludes that:

Corollary 3.1.4 [pulled-back hyperplane class]. *Let κ be the hyperplance class in W_d. Then, with the notation in Fact/Definition 3.1.3, one has*

$$k^*\psi^*\kappa \;=\; g^*j^*\kappa \;=\; -\,c_1(S_k)$$

on the distinguished S^1-fixed-point component $F_{\alpha_1, \cdots, \alpha_r; 0, \ldots, 0}$. Since S^1 and $\mathbb{C}^\times$ act on these components trivially, these classes lift naturally as to classes on $(F_{\alpha_1, \cdots, \alpha_r; 0, \ldots, 0})_{\mathbb{C}^\times}$ and will be denoted by the same notation.

Remark 3.1.5 [pulled-back hyperplane in Chern roots]. In terms of Chern roots to be discussed in Sec. 3.3, this class is represented by $-(y_1 + \cdots + y_r) = y_{r+1} + \cdots + y_n$.

3.2. The weight subspace decomposition of the normal bundle to the

Reduction of structure group and the S^1-weight subspaces in matrix forms.

Note that the notation P in this section is for parabolic subgroups. Recall that the $GL(n,\mathbb{C})$-action on $\mathbb{C}^n$ induces a $GL(n,\mathbb{C})$-action on the set of local sections in $\mathcal{E}^n$. Thus, given a $g \in GL(n,\mathbb{C})$, one has a correspondence $\mathcal{V} \longmapsto g\cdot\mathcal{V}$ with a specified isomorphism from $\mathcal{V}$ to $g\cdot\mathcal{V}$. This induces a $GL(n,\mathbb{C})$-action on $Quot_{P(t)}(\mathcal{E}^n)$, which leaves all the S^1-fixed-point component invariant. This $GL(n,\mathbb{C})$-action on $Quot_{P(t)}(\mathcal{E}^n)$ commutes with the S^1-action discussed earlier. In this way, the normal bundle ν to a S^1-fixed-point component E is realized as a homogeneous $GL(n,\mathbb{C})$-bundle and its structure group is the stabilizer P of a point p in that component: $\nu_E\,\mathcal{Q}_d = GL(n,\mathbb{C})\times_P\mathbb{C}^R$, where R is the codimension of E in $\mathcal{Q}_d$, $\mathbb{C}^R$ is identified with the fiber of $\nu_E\,\mathcal{Q}_d$ at p with the P-action induced from $GL(n,\mathbb{C})$.

The existence of a flag manifold also as a compact quotient implies that one can choose a compact $U(n)$ in $GL(n,\mathbb{C})$ such that each S^1-fixed-point component is also a $U(n)$-orbit. Then the new stabilizer at a point becomes

$$P_0 \;=\; U(n)\cap P \;=\; U(m_1)\times\cdots\times U(m_k)\times U(n-r)$$

and

$$\nu_E\,\mathcal{Q}_d \;=\; GL(n,\mathbb{C})\times_P\mathbb{C}^R \;=\; U(n)\times_{P_0}\mathbb{C}^R\,.$$

In this way, we have reduced the structure group of $\nu_E\,\mathcal{Q}_d$ to P_0 that remains compatible with the S^1-action. Applying this to each of the distinguished S^1-fixed-point components $F_{\alpha_1,\,\ldots,\,\alpha_r\,;\,0,\,\ldots,\,0}$, we then realize $T_*\,Quot_{P(t)}(\mathcal{E}^n)|_{F_{\alpha_1,\,\ldots,\,\alpha_r\,;\,0,\,\ldots,\,0}}$ as a homogeneous $U(n)$-bundle, determined by a representation of P_0.

Given $0\le\alpha_1\le\ldots\le\alpha_r$ rewritten as

$$\begin{array}{ccccccc} 0 & \le & a_1 & < & \cdots & < & a_k\,(=\alpha_r) \\ & & m_1 & & \cdots & & m_k \end{array}$$

with the multiplicity indicated, fix a point on $F_{\alpha_1,\,\ldots,\,\alpha_r\,;\,0,\,\ldots,\,0}$ represented by the subsheaf $\mathcal{V}$ in $\mathcal{E}^n$ determined by

$$\mathcal{V}(U_0)=\mathbb{C}[z]\cdot z^{\alpha_1}\oplus\cdots\oplus\mathbb{C}[z]\cdot z^{\alpha_r}\oplus 0^{\oplus(n-r)} \quad\text{and}\quad \mathcal{V}(U_\infty)=\mathbb{C}[w]^{\oplus r}\oplus 0^{\oplus(n-r)}$$

(or equivalently, the graded submodule in $\mathfrak{M}$ generated by $(0,\,\ldots,\,0,\,z_0^{\alpha_i},\,0\,\ldots,\,0)$ for $1\le i\le r$, in the notation of Sec. 2.1). Then P is the subgroup of appropriate block upper triangular matrices in $GL(n,\mathbb{C})$. Fix a Hermitian inner product on $\mathbb{C}^n$, which renders $\mathcal{E}^n$ a trivialized Hermitain vector bundle, and let $U(n)\hookrightarrow GL(n,\mathbb{C})$ be the subgroup of $GL(n,\mathbb{C})$ with respect to this inner product. Then the induced action of $U(n)$ on $F_{\alpha_1,\,\ldots,\,\alpha_r\,;\,0,\,\ldots,0}$ is transitive with

$$P_0 = P\cap U(n) = U(m_1)\times\cdots\times U(m_k)\times U(n-r)$$

being the subgroup of $U(n)$ that consists of $m_1\times m_1,\,\ldots,\,m_k\times m_k$, $(n-r)\times(n-r)$ unitary diagonal blocks.

There is an embedding of Hom-groups

$$\begin{aligned} &Hom_{\mathcal{O}_C}(\mathcal{V},\mathcal{E}^n/\mathcal{V}) \\ &\hookrightarrow Hom_{\mathbb{C}[z]}(\mathbb{C}[z]\cdot z^{\alpha_1}\oplus\cdots\oplus\mathbb{C}[z]\cdot z^{\alpha_r},\;\mathbb{C}[z]\cdot\bar{e}_1\oplus\cdots\oplus\mathbb{C}[z]\cdot\bar{e}_r\oplus\mathbb{C}[z]^{\oplus(n-r)}), \end{aligned}$$

where the annihilator $Ann(\bar{e}_i)$ of $\bar{e}_i$ is the ideal (z^{α_i}) in $\mathbb{C}[z]$. Let m_0 be the multiplicity of 0 in $\alpha_1,\,\ldots,\,\alpha_r$. Then $\bar{e}_1 = \cdots = \bar{e}_{m_0} = 0$ and the image is a submodule of the latter that consists of matrices of polynomials with degree bounds:

$$f = [\,f_{ij}(z)\,]_{(n-m_0)\times r}\,,$$

where

$$deg\, f_{ij}(z) \le \begin{cases} \alpha_{m_0+i} - 1 & \text{for} \quad 1 \le i \le r - m_0 \quad \text{and} \quad 1 \le j \le r\,, \\ \alpha_j & \text{for} \quad r - m_0 + 1 \le i \le n - m_0 \quad \text{and} \quad 1 \le j \le r\,. \end{cases}$$

The P_0-action on $Hom_{\mathcal{O}_C}(\mathcal{V}, \mathcal{E}^n/\mathcal{V})$ is given by

$$f \longmapsto g \diamond f\,, \quad \text{for } f \in Hom_{\mathcal{O}_C}(\mathcal{V}, \mathcal{E}^n/\mathcal{V}) \text{ and } g \in P_0$$

with

$$g \diamond f := g \odot f \odot Diag\{z^{-\alpha_1}, \cdots, z^{-\alpha_r}\} \cdot g^{-1} \cdot Diag\{z^{\alpha_1}, \cdots, z^{\alpha_r}\}\,,$$

where g in the formula is the lower-right $(n - m_0) \times (n - m_0)$ submatrix of the defining matrix of g when acting on $\mathbb{C}^n$, g^{-1} is the $r \times r$ upper-left submatrix of the inverse of the defining matrix for g, the operation $\cdot$ is the usual matrix mutiplication, and the operation $\odot$ is the usual matrix multiplication followed by truncations of terms in an entry that exceeds the degree bound above. This shows explicitly that the P_0-action and the S^1-action on $Hom_{\mathcal{O}_C}(\mathcal{V}, \mathcal{E}^n/\mathcal{V})$ commute.

From the previous discussions on the S^1-weight system, each monomial in an entry (a Laurent polynomial in z) of

$$\widetilde{f} := f \odot Diag\{z^{-\alpha_1}, \cdots, z^{-\alpha_r}\}$$

gives an S^1-invariant subspace in $Hom_{\mathcal{O}_C}(\mathcal{V}, \mathcal{E}^n/\mathcal{V})$. The degree bound for an entry in $\widetilde{f}$ is given by

$$\begin{cases} -\alpha_j \le deg\, \widetilde{f}_{ij}(z) \le \alpha_{m_0+i} - \alpha_j - 1 & \text{for} \quad 1 \le i \le r - m_0 \quad \text{and} \quad 1 \le j \le r\,, \\ -\alpha_j \le deg\, \widetilde{f}_{ij}(z) \le 0 & \text{for} \quad r - m_0 + 1 \le i \le n - m_0 \quad \text{and} \quad 1 \le j \le r\,. \end{cases}$$

Thus one has a decomposition of the P_0-module by the S^1-weight subspaces, each of which is itself a P_0-module:

$$\widetilde{f} = z^{-\alpha_r}\, \widetilde{f}_{(\alpha_r)} + \cdots + z^{-1}\, \widetilde{f}_{(1)} + \widetilde{f}_{(0)} + z\, \widetilde{f}_{(-1)} + \cdots + z^{\alpha_r - \alpha_1 - 1}\, \widetilde{f}_{(\,-(\alpha_r - \alpha_1 - 1)\,)}\,,$$

where the S^1-weight for z^μ-component here is $-\mu$, (cf. the expression $z^{\alpha_j - \mu^0_{ij}}$ in the discusion of the S^1-weight system Wt_1). (Note that here we are assuming the generic situation, in which $\alpha_1 < \alpha_r$ and hence $\alpha_r - \alpha_1 - 1 \ge 0$. If $\alpha_1 = \cdots = \alpha_r$, then $\alpha_r - \alpha_1 - 1 = -1$ and $\widetilde{f} = z^{-\alpha_r}\, \widetilde{f}_{(\alpha_r)} + \cdots + z^{-1}\, \widetilde{f}_{(1)} + \widetilde{f}_{(0)}$.)

The P_0-module decomposition of S^1-weight spaces and the P_0-weight system.

- *S^1-weight-subspace decomposition.*
  (1) Recall the multiplicity m_i, $1 \le i \le k$, of the sequence $0 \le \alpha_1 \le \cdots \le \alpha_r$ and m_0 the multiplicity of 0 in the sequence. Then the matrices g, g^{-1}, f, $\widetilde{f}$ can be put into a block form. For example, the (I, J)-block for f is

(2) In terms of the block form, the decomposition of $\widetilde{f}$ into a summation of matrices with only one non-zero block gives the decomposition of $Hom_{\mathcal{O}_C}(\mathcal{V}, \mathcal{E}^n/\mathcal{V})$ into representations of P_0. Consequently, each such summand is the representation of the form $\rho_{m_I} \otimes (\rho_{m_J}^{-1})^t = \rho_{m_I} \otimes \overline{\rho}_{m_J}$, where ρ_{m_I} is the defining representation of $U(m_I)$, $(\rho^{-1})^t$ its inverse transpose, which is the same as its complex conjugate $\overline{\rho}$.

(3) This decomposition is compatible with the S^1-weight subspace decomposition. In fact, *the block form of the* S^1*-weight summand* $\widetilde{f}_{(s)}$, $-\alpha_r \leq s \leq \max\{\alpha_r - \alpha_1 - 1, 0\}$, *is determined by the Young diagram corresponding to the partition* $d = \alpha_1 + \cdots + \alpha_r$. They are all *"sparse-lower-triangular" block matrices*, (cf. FIGURE 3-2-1). These block forms are invariant under the conjugation followed by truncations of terms of degree higher than the upper degree bounds

$$\widetilde{f}_{(s)} \longmapsto g \odot \widetilde{f}_{(s)} \odot g^{-1}$$

and hence this gives a decomposition of the homogeneous bundle into the direct sum of S^1-weight homogeneous subbundles. In particular, the lower sub-triangular block form of $\widetilde{f}_{(0)}$ corresponds to the tangent bundle $T_*F_{\alpha_1, \ldots, \alpha_r; 0, \ldots, 0}$. The dimension is consistent with the computation in Lemma 2.3.3.

(4) These sparse-lower-triangular block matrices are determined by the Young diagram corresponding to the partition $d = \alpha_1 + \cdots + \alpha_r$. The rule from a Young diagram to the sparse-lower-triangular block forms can be summarized in three steps:

(4.1) Take the dual of the Young diagram and put the zero-matrix of the same dimension as $\widetilde{f}$ into the same block form. Copy these zero-matrices by multiplying by a weight factor z^ν with $-\alpha_r \leq \nu \leq \max\{\alpha_r - \alpha_1 - 1, 0\}$.

(4.2) Recall a_I and m_I at the beginning of this subsection. Multiply the dual Young diagram horizontally by the multiplicity m_I and fill the block forms with these multi-strip as indicated in FIGURE 3-2-1, beginning with the block form with z^{-a_I}-factor. This corresponds to the S^1-weight system Wt_1.

(4.3) Add all these matrices and fill in all the blocks in matrices with negative ν in z^ν such that some block above it is already filled. For the block form with factor z^0, fill in all the blocks in the rows lower than the last filled row. These additional filling corresponds to the S^1-weight system Wt_3.

(5) The block decomposition of each S^1-weight subspace into the direct sum of P_0-modules.

(5.1) *Definition of diagonal and off-diagonal blocks*: When $m_0 = 0$, the diagonal blocks follow the usual definition. When $m_0 > 0$, the diagonal blocks here are the blocks that are above and adjacent to the usual diagonal blocks (i.e. the $(I, I+1)$-blocks). All other blocks are called off-diagonal.

- The diagonal blocks corresponds to a representation $\rho_{m_I} \otimes \overline{\rho}_{m_I}$, where ρ_{m_I} is the defining representation of some $U(m_I)$.

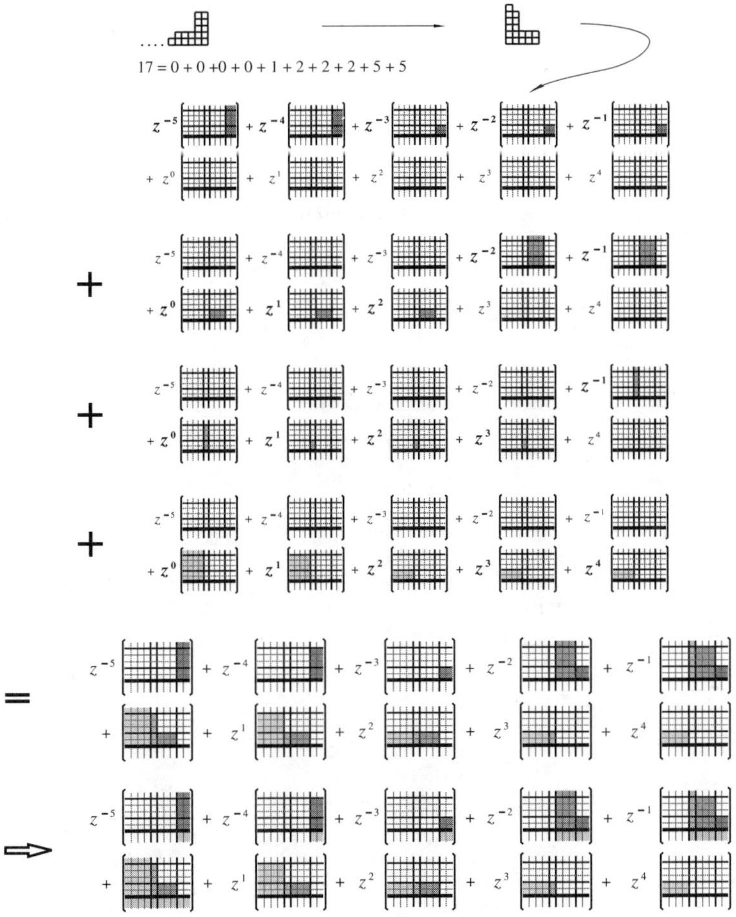

FIGURE 3-2-1. The simultaneous decomposition of $Hom_{\mathcal{O}_C}(\mathcal{V}, \mathcal{E}^n/\mathcal{V})$ by weight subspaces of S^1 and representations of P_0. Original entries in the matrix are divided by light lines while blocks are divided by dark lines. The think dark line divides the upper $(r - m_0)$ rows and the lower $(n - r)$ rows. All the unshaded blocks are zero. Observe how the block forms are all determined by the Young diagram - the conjugate Young diagram is horizontally fattened by the various multiplicities and then distribute into the block forms (cf. the blocks with the same dark shades) -.

- The off-diagonal blocks are irreducible representations $\rho_1 \otimes \rho_2^*$ of the product $U(m_{I_1}) \times U(m_{I_2})$, where ρ_j is the defining representation of $U(m_{I_j})$, $i = 1, 2$.

of $\rho_1 \otimes \overline{\rho_2}$ is given by

$$(\,\lambda_i - \lambda'_j \,|\, 1 \le i \le I_1\,,\ 1 \le j \le I_2\,)\,.$$

- *The P_0-weight system Wt_{P_0} of $Hom_{\mathcal{O}_C}(\mathcal{V}, \mathcal{E}^n/\mathcal{V})$.*
  (1) Recall the fixed maximal torus the diagonal subgroup $T = (\mathbb{C}^\times)^n$ in P_0. Let E_{ij} be a $(n - m_0) \times r$ matrix with 1 in (i,j)-entry and zero elsewhere. Then, every subspace of an S^1-weight subspace in the lower-triangular block that consists of constant multiples of some E_{ij} is a P_0-weight subspace of weight $\lambda_{i+m_0} - \lambda_j$. Consequently, the P_0-weight system can be directly read off from the collection of sparse lower-triangular block forms obtained from the Young diagram corresponding to the distinguished S^1-fixed-point component. In expression,

  $$Wt_{P_0}(\,\text{Young diagram}\,) = \bigsqcup_{(\,\text{triangular block form }\Delta\,)}\ \bigsqcup_{(\,\text{block }\square\in\Delta\,)}\ \bigsqcup_{(i,j)\in\square} (\lambda_{i+m_0} - \lambda_j)\,,$$

  where the Young diagram is the one corresponding to $F_{\alpha_1,\,\ldots,\,\alpha_r\,;\,0,\,\ldots,\,0}$ (namely the partition $d = \alpha_1 + \cdots + \alpha_r$), $\sqcup$ means the disjoint combination with multiplicity allowed, $\square \in \Delta$ means that the block is in the triangular block form Δ, and $(i,j) \in \square$ means that the (i,j)-position in the matrix for $\widetilde{f}$ lies in the block $\square$.

  (2) In summary:

  [Young diagram corresponding to a distinguished S^1-fixed-point component] $\Longrightarrow$ [Collection of sparse lower-triangular block forms associated to the S^1-weight subspaces of a fiber of the normal bundle to the distinguished S^1-fixed-point component] $\Longrightarrow$ [P_0-weight system Wt_{P_0} of each S^1-weight subspace]

Let us now turn to the computation of the $\mathbb{C}^\times$-equivariant Euler class of the normal bundle to a distinguished component in $Quot_{P(t)}(\mathcal{E}^n)$.

3.3. Structure of the induced bundle on $B\mathbb{C}^\times \times_{\mathbb{C}^\times} F_{\alpha_1,\,\ldots,\,0}$ and the $\mathbb{C}^\times$-equivariant Euler class $e_{\mathbb{C}^\times}\,\nu_{F_{\alpha_1,\,\ldots,\,0}}Quot_{P(t)}(\mathcal{E}^n)$. We compute first the Chern polynomials of the normal bundle $\nu_{F_{\alpha_1,\,\ldots,\,\alpha_r\,;\,0,\,\ldots,\,0}}Quot_{P(t)}(\mathcal{E}^n)$ of $F_{\alpha_1,\,\ldots,\,\alpha_r\,;\,0,\,\ldots,\,0}$ in $Quot_{P(t)}(\mathcal{E}^n)$ and then use this to express the $\mathbb{C}^\times$-equivariant Euler class after working out the bundle structure of the induced bundle of $\nu_{F_{\alpha_1,\,\ldots,\,\alpha_r\,;\,0,\,\ldots,\,0}}$ over $B\mathbb{C}^\times \times F_{\alpha_1,\,\ldots,\,\alpha_r\,;\,0,\,\ldots,\,0}$.

The following fact is in Borel and Hirzebruch [B-H], with slight modification to fit into our situation:

Fact 3.3.1 [Chern class and representation]. (Cf. [B-H].)

(1) *Let T be a maximal torus of $U(n)$, $\mathfrak{h}$ be the corresponding Cartan subalgebra, and $Fl(n) := U(n)/T$. There are canonical homomorphisms*

$$\{\,\text{integral linear functional on } \mathfrak{h}\,\} \;\simeq\; H^1(T,\mathbb{Z}) \;\to\; H^2(Fl(n),\mathbb{Z})\,.$$

T comes with a natural product decomposition $T = U(1)^{\times n}$ invariant under the Weyl group action. This decomposition specifies then a distinguished basis $x_1, \ldots, x_n$ for $H^1(T, \mathbb{Z})$, unique up permutations. Regard x_i also as elements in the other two groups via the above homomorphism and let $y_i = x_i$ in $H^2(Fl(n), \mathbb{Z})$. Up to permutations, y_i in $H^2(Fl(n), \mathbb{Z})$ are the first Chern class of the line bundles associated to the tautological flag bundle over $Fl(n)$.

(2) [Chern root]. *Let $P_0 = U(m_1) \times \cdots \times U(m_k) \times U(m_{k+1}) \subset U(n)$, where $m_1 + \cdots + m_k + m_{k+1} = n$, T be a maximal torus of $U(n)$ contained in P_0, and $\eta : U(n) \to B = U(n)/P_0$ be the principal P_0-bundle over B from the quotient map. Then $Fl(n)$ is a split manifold for η. Let $\zeta : Fl(n) \to B$ be the induced map from η, then $\zeta^* : H^*(B, \mathbb{Z}) \to H^*(Fl(n), \mathbb{Z})$ is injective and $\zeta^* c(\eta) = \prod_{i=1}^n (1 + y_i)$.*

(3) [naturality of Chern class]. *Let ρ be an m-dimensional unitary representation of P_0 with weights $w_j = a_{j1}x_1 + \cdots + a_{jn}x_n$, $j = 1, \ldots, m$, and $V := U(n) \times_\rho \mathbb{C}^m$ be the associated homogeneous vector bundle over $U(n)/P_0$. Then*

$$\zeta^* c(V) = \prod_{j=1}^m (1 + w_j) = \prod_{j=1}^m (1 + a_{j1}y_1 + \cdots + a_{jn}y_n)$$

is an integral polynomial function of symmetric functions in $y_1, \ldots, y_{m_1}$, in $y_{m_1+1}, \ldots, y_{m_1+m_2}$, and so on respectively. Each of these partial symmetric products of Chern roots y_i can be identified with the special Schubert cycles in the flag manifold $U(n)/P_0$.

Recall the P_0-weight system associated to the Young diagram corresponding to $F_{\alpha_1, \ldots, \alpha_r; 0, \ldots, 0}$ and Corollary 2.3.4, which says that all the S^1-weight of a fiber of the normal bundle are non-zero. Let ν be the normal bundle in consideration. Then the above fact implies that the Chern polynomial $c_\nu(t)$ of ν is given by

$$c_\nu(t) = \prod_{\left(\substack{\text{triangular block form } \Delta_w \\ \text{for non-zero } S^1\text{-weight}}\right)} \prod_{(\text{block } \square \in \Delta_w)} \prod_{(i,j) \in \square} (t + y_{i+m_0} - y_j),$$

where the first product on the right hand side of the equality ranges over all possible non-zero S^1-weights w. The result is an integral polynomial function of the special Schubert cycles in $A_*(F_{\alpha_1, \ldots, \alpha_r; 0, \ldots, 0})$, cf. Fact/Definition 3.1.3.

The S_1-action on $\nu = \nu_{F_{\alpha_1, \ldots, \alpha_r; 0, \ldots, 0}} Quot_{P(t)}(\mathcal{E}^n)$ induces a bundle

$$\mathcal{T} \to \mathbb{C}\mathrm{P}^\infty \times F_{\alpha_1, \ldots, \alpha_r; 0, \ldots, 0}.$$

Let $\nu = \oplus_w \nu_w$ be the decomposition of the normal bundle as a direct sum of S^1-weight subspace and $\mathcal{T} = \oplus_w \mathcal{T}_w$ be the induced decomposition of $\mathcal{T}$. The following Lemma is well-known.

Lemma 3.3.2 [induced bundle of S^1-weight summand]. *Let*

$$\mathbb{C}\mathrm{P}^\infty \xleftarrow{\mathrm{pr}_1} \mathbb{C}\mathrm{P}^\infty \times F_{\alpha_1, \ldots, \alpha_r; 0, \ldots, 0} \xrightarrow{\mathrm{pr}_2} F_{\alpha_1, \ldots, \alpha_r; 0, \ldots, 0}$$

Let $R(w)$ be the rank of ν_w. By the multiplicativity of Euler class and the rule under the tensor with a line bundle (cf. [Fu1]), we conclude that

Theorem 3.3.3 [Euler class]. *The S^1 equivariant Euler class of the normal bundle $\nu = \nu_{F_{\alpha_1,\ldots,\alpha_r;0,\ldots,0}} Quot_{P(t)}(\mathcal{E}^n)$ is given by*

$$e_{\mathbb{C}^\times}\nu = \prod_w e_{\mathbb{C}^\times}\nu_w = \prod_w c_{\nu_w}(-w\alpha)$$

$$= \prod_{\left(\substack{\text{triangular block form } \Delta_w \\ \text{associated to nonzero } w}\right)} \; \prod_{(\text{block}\,\square\in\Delta_w)} \; \prod_{(i,j)\in\square} (-w\alpha + y_{i+m_0} - y_j),$$

where $\alpha = c_1(\mathcal{O}_{\mathbb{CP}^\infty}(1))$, $c_{\nu_w}(t) = t^{R(w)} + c_1(\nu_w)t^{R(w)-1} + \cdots$ is the Chern polynomial of ν_w. The triangular block forms Δ_w, $-\alpha_r \le w \le max\{\alpha_r - \alpha_1 - 1, 0\}$, associated with the Young diagram corresponding to the partition $d = \alpha_1 + \cdots + \alpha_r$ are defined by Item (4) in the subheading "S^1-weight-subspace decomposition" of the heading "The P_0-module decomposition of S^1-weight spaces and the P_0-weight system" in Sec. 3.2.

Here pr_1^*, pr_2^* in the formula are omitted for simplicity of notations.

4. Illustrations by two examples.

In this section, we present two simple examples of the Mirror Principle computation that are computable by hand to illustrate the discussions in this article. With notations in Fact 3.3.1, for $Fl(3) := Fl_{1,2}(\mathbb{C}^3)$ $H^*(Fl(3),\mathbb{Z})$ is generated by y_1, y_2, $(y_3 = -(y_1 + y_2))$. The intergral of the top classes are given by

$$\int_{Fl(3)} y_1^3 = \int_{Fl(3)} y_2^3 = 0 \quad \text{and} \quad \int_{Fl(3)} y_1^2 y_2 = -\int_{Fl(3)} y_1 y_2^2 = -1,$$

following [Mo].

Example 4.1 [$Gr_2(\mathbb{C}^3)$, degree 3]. In this case, $n = 3$, $r = 2$, $n-r = 1$, and $d = 3$. There are two distinguished S^1-fixed-point components in the related component $Quot_{P(t)=t+4}(\mathcal{E}^3)$ of Quot-scheme:

- $F_{0,3\,;\,0,0} \simeq Fl_{1,2}(\mathbb{C}^3)$, $dim = 3$.

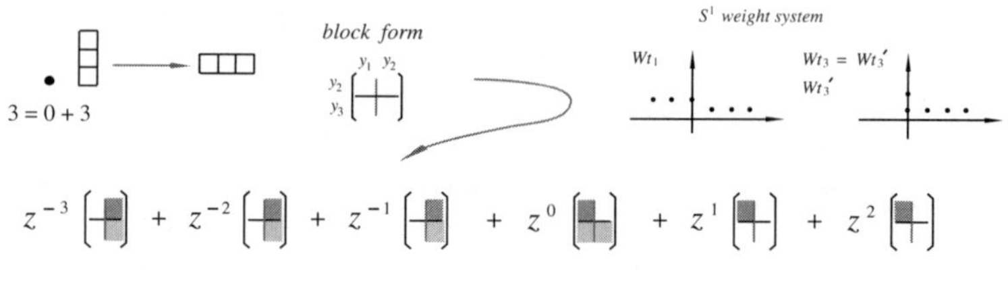

– Grouping of Chern roots: $\{y_1\,;\,y_2\,;\,y_3\}$.

– $\mathbb{C}^\times$-equivariant Euler class of normal bundle:

$$e_{\mathbb{C}^\times}(\nu) = (-3\alpha)(-3\alpha + y_3 - y_2)(-2\alpha)(-2\alpha + y_3 - y_2) \cdot (-\alpha)(-\alpha + y_3 - y_2)(\alpha + y_2 - y_1)(2\alpha + y_2 - y_1).$$

– Pulled-back hyperplane class:

$$k^*\psi^*\kappa = g^*j^*\kappa = -(y_1 + y_2).$$

– The integral over the component:

$$\int_E \frac{k^*\psi^* e^{\kappa\cdot\zeta}}{e_{\mathbb{C}^\times}(E/\mathcal{Q}_d)} = -\frac{103}{1296}\frac{1}{\alpha^{11}} - \frac{23}{108}\frac{\zeta}{\alpha^{10}} - \frac{29}{864}\frac{\zeta^2}{\alpha^9}.$$

- $F_{1,2\,;\,0,0} \simeq Fl_{1,2}(\mathbb{C}^3)$, $dim = 3$.

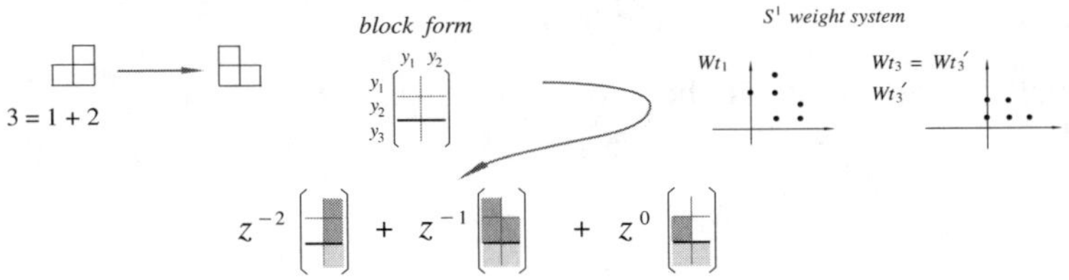

– Grouping of Chern roots: $\{y_1\,;\,y_2\,;\,y_3\}$.
– $\mathbb{C}^\times$-equivariant Euler class of normal bundle:

$$e_{\mathbb{C}^\times}(\nu) = (-2\alpha + y_1 - y_2)(-2\alpha)(-2\alpha + y_3 - y_2) \cdot (-\alpha)(-\alpha + y_2 - y_1)(-\alpha + y_3 - y_1)(-\alpha)(-\alpha + y_3 - y_2).$$

– Pulled-back hyperplane class:

$$k^*\psi^*\kappa = g^*j^*\kappa = -(y_1 + y_2).$$

– The integral over the component:

$$\int_E \frac{k^*\psi^* e^{\kappa\cdot\zeta}}{e_{\mathbb{C}^\times}(E/\mathcal{Q}_d)} = \frac{3}{16}\frac{\zeta}{\alpha^{10}} + \frac{1}{32}\frac{\zeta^2}{\alpha^9}.$$

$$\text{Total integral} = -\frac{103}{1296}\frac{1}{\alpha^{11}} - \frac{11}{432}\frac{\zeta}{\alpha^{10}} - \frac{1}{432}\frac{\zeta^2}{\alpha^9}.$$

□

Example 4.2 [$Gr_1(\mathbb{C}^3)$, **degree 3**]. In this case, $n = 3$, $r = 1$, $n - r = 2$, and $d = 3$. There is one distinguished S^1-fixed-point component in the related component $Quot_{P(t)=2t+5}(\mathcal{E}^3)$ of Quot-scheme:

– $F_{3\,;\,0} \simeq Gr_1(\mathbb{C}^3)$, $dim = 2$.

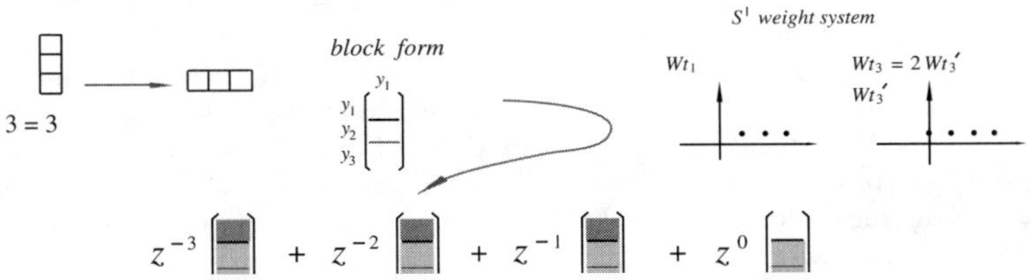

– Grouping of Chern roots: $\{ y_1 \,;\, y_2, y_3 \}$.
– $\mathbb{C}^\times$-equivariant Euler class of normal bundle:

$$e_{\mathbb{C}^\times}(\nu) = (-3\alpha)(-3\alpha+y_2-y_1)(-3\alpha+y_3-y_1)(-2\alpha)(-2\alpha+y_2-y_1) \cdot(-2\alpha+y_3-y_1)(-\alpha)(-\alpha+y_2-y_1)(-\alpha+y_3-y_1)\,.$$

– Pulled-back hyperplane class:

$$k^*\psi^*\kappa = g^*j^*\kappa = -y_1\,.$$

– The integral over the component:

$$\int_E \frac{k^*\psi^* e^{\kappa\cdot\zeta}}{e_{\mathbb{C}^\times}(E/\mathcal{Q}_d)} = -\frac{103}{1296}\frac{1}{\alpha^{11}} - \frac{11}{432}\frac{\zeta}{\alpha^{10}} - \frac{1}{432}\frac{\zeta^2}{\alpha^9}\,,$$

which is the same as the total integral in Example 4.1, as it should be since
$Gr_1(\mathbb{C}^3) = Gr_2(\mathbb{C}^3)$.

□

Remark 4.3. One can check that the integral values are correct, using the result in [L-L-Y1, I] for the computation for $\mathbb{C}\mathrm{P}^2$.

Remark 4.4. Now that we can compute the integral that is related to the intersection numbers on the moduli space of rational stable maps into Grassmannian manifolds, the A-model for Calabi-Yau complete intersections in a Grassmannian manifold can also be computed explicitly.

Remark 4.5. Two immediate questions follow from the current work:

(1) the automatization of the calculations via a computer code, following the diagrammatic rules discussed, and the computation for more examples and
(2) generalization of the discussion to flag manifolds, which involves hyper-Quot schemes.

The study of them will be reported in another work.

With these remarks, we conclude this paper.

Acknowledgements. We would like to thank Raoul Bott, Jiun-Cheng Chen, Joe Harris, Shinobo Hosono, Yi Hu, Deepee Khosla, Albrecht Klemm, Jun Li, Hui-Wen Lin, Chiu-Chu Liu, Alina Marian, Keiji Oguiso, Chin-Lung Wang, and Xiaowei Wang for valuable conversations and group meeting talks that influence the work. C.-H. L would like to thank in addition Orlando Alvarez and William Thurston for educations; J.H., Sean Keel, Mihnea Popa, and Ravi Vakil for the various courses on algebraic geometry; Dima Arinkin, Hung-Wen Chang, J.-C.C., Izzet Coskun, M.P., and Jason Starr for discussions on related objects in algebraic geometry; organizers for the inspiring conference and invitation of presentation, Nathan Dunfield for making Adobe figures latex compatible, and Ling-Miao Chou for the tremendous moral support.

References

[Ak] D.N. Akhiezer, *Lie group actions in complex analysis*, Aspects in Math. E27, Vieweg, 1995.

[A-B] M.F. Atiyah and R. Bott, *The moment map and equivariant cohomology*, Topology, **23** (1984), pp. 1 - 28.

[Ber] A. Bertram, *Quantum Schubert calculus*, *Adv. Math.* **128** (1997), pp. 289 - 305.

[BCKS1] V.V. Batyrev, I. Ciocan-Fontanine, B. Kim, D. van Straten, *Conifold transitions and mirror symmetry for calabi-Yau complete intersections in Grassmannians*, math.AG/9710022.

[BCKS2] ——, *Mirror symmetry and toric degenerations of partial flag manifolds*, math.AG/9803108.

[B-D-W] A. Bertram, G. Daskalopoulos, and R. Wentworth, *Gromov invariants for holomorphic maps from Riemann surfaces to Grassmannians*, *J. Amer. Math. Soc.* **9** (1996), pp. 529 - 571.

[B-G-G] I.N. Bernstein, I.M. Gel'fand, and S.I. Gel'fand, *Schubert cells and cohomology of the space G/P*, Russian Math. Surv. **28(3)** (1973), pp. 1 - 26.

[B-H] A. Borel and F. Hirzebruch, *Characteristic classes and homogeneous spaces, I. Amer. J. Math.* **80** (1958), pp. 458 - 538; *II. Amer. J. Math.* **81** (1959), pp. 315 - 382; *III. Amer. J. Math.* **82** (1960), pp. 491 - 504.

[CF1] I. Ciocan-Fontanine, *The quantum cohomology ring of flag varieties*, Transactions *Amer. Math. Soc.* **351** (1999), pp. 2695 - 2729.

[CF2] ——, *On quantum cohomology rings of partial flag varieties*, *Duke Math. J.* **98** (1999), pp. 485 - 524.

[Ch] L. Chen, *Poincaré polynomials of hyperquot schemes*, math.AG/0003077.

[Ei] D. Eisenbud, *Commutative algebra - with a view toward algebraic geometry*, GTM 150, 1994.

[E-H] D. Eisenbud and J. Harris, *The geometry of schemes*, GTM 197, Springer-Verlag, 2000.

[Fri] R. Friedman, *Algebraic surfaces and holomorphic vector bundles*, Springer-Verlag, 1998.

[Fu1] W. Fulton, *Intersection theory*, Ser. Mod. Surv. Math. 2, Springer-Verlag, 1984.

[Fu2] ——, *Flags, Schubert polynomials, degeneracy loci, and determinantal formulas*, *Duke Math. J.* **65** (1991), pp. 381 - 420.

[Gr] A. Grothendieck *Techniques de construction et théorèmes d'existence en géométrie algébrique IV: Les schémas de Hilbert*, Séminaire Bourbaki, (1960/1961), no. 221, pp. 1 - 28.

[G-S-W] M.B. Green, J.H. Schwarz, and E. Witten, *Superstring theory*, vol. 1 and vol. 2, Cambridge Univ. Press, 1987.

[Ha] R. Hartshorne, *Algebraic geometry*, GTM 52, Springer-Verlag, 1977.

[Hi] F. Hirzebruch, *Topological methods in algebraic geometry*, Grund. Math. Wiss. Ein. 131, Springer-Verlag, 1966.

[H-L] D. Huybrechts and M. Lehn, *The geometry of moduli spaces of sheaves*, Vieweg, 1997.

[H-M] J. Harris and I. Morrison, *Moduli of curves*, GTM 187, Springer-Verlag, 1998.

[H-V] H. Hori and C. Vafa, *Mirror symmetry*, hep-th/0002222.

[Ja] N. Jacobson, *Basic algebra I*, W.H. Freeman and Company, 1974.

[Jo] J.P. Jouanolou, *Cohomologie de quelques schméas classiques et théorie cohomologique des classes de Chern*, SGA 5, 1965-66, pp. 282 - 350, Lect. Notes Math. 589, Springer-Verlag, 1977.

[Ke] G.R. Kempf, *Algebraic varieties*, London Math. Soc. Lect. Note Ser. 172, Cambridge Univ. Press, 1993.

[Kim] B. Kim, *Gromov-Witten invariants for flag manifolds*, Ph.D. dissertation, University of California at Berkeley, 1996.

[Ko] S. Kobayashi, *Transformation groups in differential geometry*, Springer-Verlag, 1972.

[Kol] J. Kollár, *Rational curves on algebraic varieties*, Ser. Mod. Surv. Math. 32, Springer-Verlag, 1996.

[Kos] B. Kostant, *Lie algebra cohomology and generalized Schubert cells*, Ann. Math. **77**

[K-M] M. Kontsevich and Yu. Manin, *Gromov-Witten classes, quantum cohomology, and enumerative geometry*, Commun. Math. Phys. **164** (1994), pp. 525 - 562.

[L-L-Y1] B. Lian, K. Liu, and S.-T. Yau, *Mirror principle, I*, Asian J. Math. **1** (1997), pp. 729 - 763; *II*, Asian J. Math. **3** (1999), pp. 109 - 146; *III*, Asian J. Math. **3** (1999), pp. 771 - 800; *IV*, `math.AG/0007104`.

[L-L-Y2] ———, *A survey of mirror principle*, `math.AG/0010064`.

[L-T1] J. Li and G. Tian, *Quantum cohomology of homogeneous varieties*, J. Alg. Geom. **6** (1997), pp. 267 - 305.

[L-T2] ———, *Virtual moduli cycle and Gromov-Witten invariants of algebraic varieties*, J. Amer. Math. Soc. **11** (1998), pp. 119 - 174.

[Mo] D. Monk, *The geometry of flag manifolds*, Proc. London Math. Soc. **9** (1959), pp. 253 - 286.

[Mu1] D. Mumford, *Lectures on curves on an algebraic surface*, Ann. Math. Study 59, Princeton Univ. Press, 1966.

[Mu2] ———, *The red book of varieties and schemes*, Lect. Notes Math. 1358, Springer-Verlag, 1988.

[Rø] E.A. Rødland, *The Pfaffian Calabi-Yau, its mirror, and their link to the Grassmannian $G(2,7)$*, `math.AG/9801092`.

[Sa] H. Samelson, *Notes on Lie algebra*, Springer-Verlag, 1990.

[So] F. Sottile, *Real rational curves in Grassmannians*, J. Amer. Math. Soc. **13** (2000), pp. 333 - 341.

[Str] S.A. Strømme, *On parametrized rational curves in Grassmann varieties*, in *Space curves*, F. Ghione, C. Peskine, and E. Sernesi eds, pp. 251 - 272, Lect. Notes Math. 1266, Springer-Verlag, 1987.

[S-S] F. Sottile and B. Sturmfels, *A Sagbi basis for the quantum Grassmannian*, `math.AG/9908016`.

[Wi1] E. Witten, *Phases of $N = 2$ theories in two dimensions*, Nucl. Phys. **B403** (1993), pp. 159 - 222.

[Wi2] ———, *The Verlinde algebra and the cohomology of the Grassmannian*, in *Geometry, topology, and physics for Raoul Bott*, S.-T. Yau ed., pp. 357 - 422, International Press, 1994.

Department of Mathematics, Brandeis University, Waltham, MA 02154
E-mail address: `lian@brandeis.edu`

Department of Mathematics, Harvard University, Cambridge, MA 02138
E-mail address: `chienliu@math.harvard.edu`

Department of Mathematics, University of California at Los Angelas, Los Angelas, CA 90095
E-mail address: `liu@math.ucla.edu`

Department of Mathematics, Harvard University, Cambridge, MA 02138
E-mail address: `yau@math.harvard.edu`

Contemporary Mathematics
Volume **322**, 2003

The Semi-infinity of Floer (co)homologies

Weiping Li

1. Introduction

Floer's beautiful proof of the Arnold conjecture for monotone symplectic manifolds in [**16, 18**] plays a very important role in the study of symplectic topology. A by-product of the proof is the definition of the celebrated symplectic Floer (co)homology. The Arnold conjecture in [**1**] states that the number of fixed points of an exact symplectic diffeomorphism on a symplectic manifold can be bounded below by the sum of Betti numbers of the symplectic manifold if all fixed points are nondegenerate. The first breakthrough toward the Arnold conjecture is Conley and Zehnder's solution [**9**] for tori by the finite-dimensional reduction of the Morse theorems. An entirely different approach by Gromov [**23**] was the analysis of pseudoholomorphic curves. Floer in [**18**] combined the ideas in [**9, 23**] and Witten's description of the Morse complex [**16**] to develop a relative index and a generalized Morse complex in the infinite dimensional setup of the loop space. This leads to the definition of the symplectic Floer homology. The same idea was quickly adapted into gauge theory to define the instanton Floer (co)homology for integral homology 3-spheres in [**17**] by using Taubes' results in [**48**].

In the 1990's, people tried to understand multiply covered J-holomorphic curves with negative first Chern class in order to extend the Floer homology for any compact closed symplectic manifold. Several groups of researchers (see [**21, 34, 45, 46, 44**] for details and historical accounts) overcame this difficulty by introducing stable J-maps. This answers the (nondegenerate version) Arnold conjecture affirmatively for any compact closed symplectic manifold. There are some applications of the symplectic Floer homology in the study of symplectic topology. For instance, quantum cohomology, counting rational curves and Gromov-Witten invariants are related to the symplectic Floer (co)homology. Kontsevich [**26**] initiated the homological algebra properties of the symplectic Floer homology and made his homological mirror conjecture by finding certain cohomological operations in the Floer homology (see [**22**] for more). The Floer homology as a homology theory is different from usual homology.

In this paper, we are interested in the symplectic/instanton Floer (co)homology with certain semi-infinite properties. In [**7**], Cohen, Jones and Segal proposed a semi-infinite Floer homotopy theory through the semi-infinity of the spectral flow

---

1991 *Mathematics Subject Classification.* Primary 57R99, 58F05; Secondary 58E05, 57R57.

analysis (polarization). Their theory has been verified in examples [**7, 25**]. In [**2**] p298, Atiyah stated that *In the product we then have four types of cycle*

finite $\times$ *finite*, *cofinite* $\times$ *cofinite*,

cofinite $\times$ *finite*, *finite* $\times$ *cofinite*.

The first two give ordinary homology and cohomology respectively. The other two are quite different from these and give in an obvious sense "middle-dimensional" homology, one "positive" and one "negative." The well-defined cap-product on the instanton Floer cohomology shows that the semi-infinite Floer cohomology cannot be reduced to ordinary homology and cohomology by a factorization (see [**2**]). The author has elaborated this comment and proved it rigorously for the instanton Floer homology in [**29**]. In a formal way, the instanton Floer homology can be viewed as the "middle-dimensional" homology of the infinite dimensional space $\mathcal{B}_Y$ of gauge equivalence classes of $SU(2)$-connections on $Y \times SU(2)$, where Y is an integral homology 3-sphere. Let $Z(X_i)(i = 1, 2)$ be the set of boundary values of instantons over the smooth 4-manifold X_i. The $Z(X_i)$ represents a cycle of homology class $\Phi(X_i)$ which is independent of metrics. The pairing $\langle \Phi(X_1), \Phi(X_2) \rangle$ can be identified with the intersection $Z(X_1) \cap Z(X_2)$ in $\mathcal{B}_Y$ (see [**10**]). The formalization alone might suggest that there should be a product from $H^*(\mathcal{B}_Y) \otimes HF_*(Y)$ to $HF_*(Y)$, as multiplying a finite dimensional cohomology class by an infinite dimensional class to get another infinite dimensional class (see [**2**] p.298). Such a cap-product structure on the instanton Floer homology shows that the Floer homology groups are, from a topological point of view, something essentially new [**2**] p298. Dually, this is equivalent to a cup-product structure on the instanton Floer cohomology $H^*(\mathcal{B}_Y) \otimes HF^*(Y) \to HF^*(Y)$ (see [**2**]).

Note that unlike the instanton case, the cap/cup-product structures on the symplectic Floer cohomology induce a module structure. This reflects that the symplectic Floer cohomology can be related to the ordinary cohomology of the underlying symplectic manifolds for exact Hamiltonian diffeomorphisms. One can also extend this to symplectic diffeomorphisms which are symplectically isotopic to the identity [**30, 32**]. In order to demonstrate the semi-infinity of the symplectic Floer cohomology, some new ideas are needed. We approach this by extending the symplectic Floer cohomology to the infinite-dimensional symplectic manifold ΩG (based loop groups). For the well-defined symplectic Floer cohomology $HF^*_{\mathrm{sym}}(\Omega G, \Lambda_\omega)$, we build the isomorphism between $HF^*_{\mathrm{sym}}(\Omega G, \Lambda_\omega)$ and the semi-infinite cohomology of its associated loop algebra defined in [**12, 20, 50**].

The present paper is mainly a survey article. The paper is organized as follows: Section 2 briefly reviews (i) Morse theory and its ramifications, (ii) Novikov homology on finite-dimensional manifolds and (iii) Feign's semi-infinite cohomology of infinite-dimensional Lie algebras. These constructions are used in later sections; Section 3 we survey our module structure result for symplectic Floer cohomology, and show that there is an isomorphism between $HF^*_{\mathrm{sym}}(\Omega G, \Lambda_\omega)$ and the semi-infinite cohomology of its associated loop algebra; Section 4 recalls the cap-product structure on the instanton Floer homology, gives an instanton Floer homology with Novikov ring coefficients, and extends the methods in §3.1 to obtain a module structure on the monopole homology of an integral homology 3-sphere.

2. Morse Theory and its ramifications

In [**38**], Morse first studied finite dimensional manifolds with critical points and the connecting orbits between them. For a function $f : M \to \mathbb{R}$, a critical point is a point $p \in M$ at which $df(p) = 0$ or $\frac{\partial f}{\partial x_i}(p) = 0$ for a local chart $(x_1, \cdots, x_n)$ of p, the Hessian $H_p(f) = \frac{\partial^2 f}{\partial x_i \partial x_j}(p)$ at the critical point p has two intrinsic invariants, the nullity $n(p)$ of the symmetric matrix $H_p(f)$ and the number of negative eigenvalues (index of p) $\lambda(p)$. A nondegenerate function is a function with nullity zero for every critical point.

THEOREM 2.1. *(Morse homotopy theorems) (1) Suppose that a nondegenerate function f has no critical points in the range $[a, b]$. Let $M_t = \{x \in M : f(x) \leq t\}$ be the "half space". Then there exists a diffeomorphism of M_a with M_b: $M_a \cong M_b$.*

(2) Suppose that a nondegenerate function f has only one critical point p with $\lambda(p) = k$ and $a < f(p) < b$ in the range $[a, b]$. Then M_b is diffeomorphic to M_a with a thickened k-cell attached: $M_b \cong M_a \cup (e^k \times e^{n-k})$.

(3) There exists a polynomial $Q(t)$ with nonnegative coefficients such that

$$\mathcal{M}_t(M, f) - P_t(M) = (1 + t)Q(t),$$

where $\mathcal{M}_t(M, f) = \sum_{p\in M, df(p)=0} t^{\lambda(p)}$ is the Morse polynomial and $P_t(M)$ is the Poincaré polynomial of M.

See [**5, 37, 38**] for more explicit explanations and references on the Morse theory.

The homological version of Morse theory can be formulated as follows. Define the chain group $C_i(M, f) = \mathbb{Z}\langle x : df(x) = 0, \lambda(x) = i\rangle$, where $\langle\cdot\rangle$ means free generate. The boundary map $\partial_i : C_i(M, f) \to C_{i-1}(M, f)$ is defined by

$$\partial_i(x) = \sum_{df(y)=0, \lambda(x)-\lambda(y)=1} \#\mathcal{M}(x, y) \cdot y,$$

where $\mathcal{M}(x, y) = \{u : \frac{\partial u}{\partial t} = -\nabla f(u), u(-\infty) = x, u(+\infty) = y\}$ is the moduli space of trajectory flow lines of f, and the orientation for each critical point is fixed from the orientation of $T_x M$, the orientation of y induced from the orientation of the moduli space gives ± 1 according to the match or unmatch with the fixed orientation of $T_y M$. Then the following result is attributed to Conley, Milnor, Smale, Thom and Witten.

THEOREM 2.2. *(Morse Homology Theorem) For any $i \in \mathbb{Z}$, (1) $\partial_{i-1} \circ \partial_i = 0$; (2) $H_i(M, f) = \ker \partial_i / Im \partial_{i+1}$ is isomorphic to the usual homology $H_i(M; \mathbb{Z})$.*

REMARK 2.3. The symplectic/instanton Floer homology is essentially constructed as above [**16**] for infinite dimensional spaces M such as loop spaces or equivalence classes of connections and f as the symplectic action a_H or the Chern-Simons functional cs. The spectral flow plays the role of a (relative) index and the moduli space of trajectory flow lines becomes the space of J-holomorphic curves or instantons (see [**16, 17, 18**]).

In [**42**], Pitcher generalized Morse theory from a Morse function $f : M \to \mathbb{R}$ to a lifted Morse theory from a circle-valued function $f : M \to S^1$. The idea is to

To mention a few ramifications of Morse theory, Thom explored the cobordism theory in the 1940's; Bott [**5**] extended it to the critical submanifolds to obtain the famous Bott periodicity in the 1950's; Smale strengthened the handlebody theory to prove the generalized Poincaré conjecture in 1950's; Palais extended Morse theory into infinite-dimensional Banach manifolds in the 1960's; Conley in [**8**] used isolated invariant sets and isolating blocks to define homotopy index and study dynamical systems and bifurcation theory in the 1970's; In the 1980's, Goresky and MacPherson studied the natural extension of Morse theory to stratified topological spaces (intersection homology theory), Witten used physical insights to encode Hodge theory with Morse theory in the 1980's, and Floer used Witten's idea in the infinite-dimensional case to prove the Arnold conjecture in the 1980's; Cohen, Jones and Segal [**7**] proposed the infinite dimensional Morse homotopy theory through the flow category construction in the 1990's. There are many other extensions and generalizations of Morse theory. It is not easy to give a complete list of references. We will restrict ourselves to those references which are close to our discussion.

2.1. Novikov's approach for closed 1-forms. Novikov [**47**] around 1981 studied multi-valued Morse function arising from Hamiltonian systems. The analogous Morse theory for closed 1-forms is constructed over a proper covering space with Laurent polynomial coefficients (called the Novikov ring) and is measures the semi-infinite trajectories from a level bounded above (or below). Floer constructed an analogous theory for closed 1-forms on an infinite-dimensional space in order to study the Arnold conjecture [**18, 19**] which combines the ideas of Novikov and Witten. Hofer and Salamon [**24**] combined Floer's construction and Novikov's construction to define the symplectic Floer cohomology of weakly monotone symplectic manifolds. Now such a generalization has been pushed to cover all compact, closed symplectic manifolds of finite dimensions by many people.

We recall the Novikov-Morse theory for closed 1-forms on finite-dimensional manifolds. Let M be a closed manifold and ω be a closed 1-form ($d\omega = 0$) on M. Define $I_\omega : \pi_1(M) \to \mathbb{R}$ by $I_\omega(a_i) = \langle \omega, a_i \rangle$ for a basis $a_i \in H_1(M;\mathbb{Z})$. We choose loops $a_1, \cdots, a_k, a_{k+1}, \cdots, a_n$ with $\langle \omega, a_i \rangle = r_i, 1 \leq i \leq k$ and $\langle \omega, a_i \rangle = 0$ otherwise. The real numbers $\{r_i : 1 \leq i \leq k\}$ are linearly independent over $\mathbb{Z}$ for generic ω. Suppose $\{r_i : 1 \leq i \leq k\}$ defines a point in $\mathbb{R}P^{k-1}$, we say ω has degree of irrationality $k - 1$. For instance, if $k = 0$, then $\omega = df$ is an exact 1-form; if $k = 1$, then there is a scalar $\lambda \neq 0$ such that $\lambda\omega$ defines $\psi : M \to S^1$ by $\psi(x) = \exp(2\pi i \int_{x_0}^{x} \lambda\omega)$. This gives a *circle valued Morse function.* For $k \geq 1$, there is an infinite foldcovering $\pi : \hat{M} \to M$ such that $\pi^*\omega = df$ for $f : \hat{M} \to \mathbb{R}$. The function f is called a many-valued Morse function after suitable generic choices. The space $\hat{M}$ is a $H_1(M;\mathbb{Z})/\ker I_\omega$-covering space.

For a generic Riemannian metric g_M on M, the gradient flow of f with respect to $\pi^*(g_M) = g_{\hat{M}}$ is of Morse-Smale type. Define $C_i(M, \omega; g_M)$ to be the freely generated Λ_ω-module generated by the critical points of f with index i, where Λ_ω is the Novikov ring associated with the closed 1-form ω. In general, the Novikov ring associated to an integral domain F (denoted by $\Lambda(H_1(M;\mathbb{Z})/\ker I_\omega; I_\omega; F)$) is the F-module of all functions $H_1(M;\mathbb{Z})/\ker I_\omega \to F$ (as $a \mapsto \lambda_a$) such that the set $\{a \in H_1(M;\mathbb{Z})/\ker I_\omega : \lambda_a \neq 0, I_\omega(a) \leq c\}$ is finite for every constant $c \in \mathbb{R}$, and the product is given by convolution $(\lambda_1 * \lambda_2)_a = \sum_i (\lambda_1)_{a_i} (\lambda_2)_{a_i^{-1}a}$. If $F = \mathbb{Z}$ and $I_\omega = 0$, then the module Λ_ω is a group ring. For the injective map $I_\omega : H_1(M;\mathbb{Z})/\ker I_\omega \to$

$\mathbb{R}$, $\Lambda(H_1(M;\mathbb{Z})/\ker I_\omega; I_\omega; F)$ is an integral domain for an integral domain F by Sikorav's result. Our Novikov ring Λ_ω is $\Lambda(H_1(M;\mathbb{Z})/\ker I_\omega; I_\omega; \mathbb{Z})$. There are at most finitely many trajectories of $-\nabla f$ connecting two critical points $\hat{x}, \hat{y}$ in $\hat{M}$ with $i(\hat{x}) - i(\hat{y}) = 1$. One can define the boundary map $\partial_i : C_i(M, \omega; g_M) \to C_{i-1}(M, \omega; g_M)$ by counting the signed number of the 1-dimensional moduli space $\mathcal{M}(\hat{x}, \hat{y}) = \{u : \frac{\partial u}{\partial t} = -\nabla f(u), u(-\infty) = \hat{x}, u(\infty) = \hat{y}\}$. The rank of $C_i(M, \omega; g_M)$ is infinite dimensional over $\mathbb{Z}$, but is *finite* over Λ_ω. The boundary map ∂_* can be extended naturally to an Λ_ω-module homomorphism. Hence the homology of the complex $(C_*(M, \omega; g_M), \partial_*)$ is called the Novikov homology and denoted by $H_*(M, \omega; g_M)$.

THEOREM 2.4. *The Novikov homology $H_*(M, \omega; g_M)$ is independent of the choices of Riemannian metric g_M for which the gradient flow of f is of Morse-Smale type and the lifting function f with $df = \pi^*\omega$. Furthermore $H_*(M, \omega; g_M) \cong H_*(M, r\omega; g_M)$ for $\omega \in H^1(M;\mathbb{R})$ and $r \neq 0$ (see* **[13, 14, 41, 47]**) *for more).*

The Novikov ring plays a semi-infinite role in the construction of the Novikov homology for closed 1-forms in the sense of selecting the right generators among infiniteily many critical points of $f : \hat{M} \to \mathbb{R}$. Let $b_i(M, \omega)$ be the rank of $H_i(M, \omega; g_M)$ over Λ_ω, and $t_i(M, \omega)$ be the minimal number of generators of the torsion subgroup $Tor_{\Lambda_\omega} H_i(M, \omega; g_M)$. The classical Morse inequalities are

$$m_i(f) \geq b_i(M) + t_i(M) + t_{i-1}(M), \tag{2.1}$$

where $m_i(f)$ is the rank of $C_i(M, f)$ (free Abelian group generated by those critical points of f with Morse index i) over $\mathbb{Z}$, and $b_i(M) = rank_{\mathbb{Z}} H_i(M;\mathbb{Z})$ and $t_i(M)$ is the minimal number of generators in $Tor(H_i(M;\mathbb{Z}))$. The similar inequality for the Novikov homology

$$m_i(f, \omega) \geq b_i(M, \omega) + t_i(M, \omega) + t_{i-1}(M, \omega)$$

also holds. For simply connected manifolds M of dimension ≥ 6, Smale showed that there exist Morse functions on M such that the inequalities (2.1) turn into equalities. For equality in the Novikov homology, Farber proved the similar result **[13]**.

2.2. Feign's approach for infinite-dimensional algebras. Feigin **[12]** in 1984 considered "semi-infinite" wedge products in the infinite-dimensional Lie algebra and defined a differential on such products to form a well-defined cochain complex without geometric topology interpretation. The cohomology of such a complex is called a semi-infinite cohomology. This is completely new from the regular cohomology of Lie algebrae. This construction was further developed by Frenkel, Garland and Zuckerman **[20]** and the semi-infinite cohomology has many powerful applications to string theory, conformal field theory, 2D quantum gravity and many other physical worlds (see **[20, 50]** for more references). We briefly recall the cohomology of Lie algebra in order to present Feign's construction **[12, 20, 50]** of the $\frac{\infty}{2}$-cohomology of infinite dimensional algebras.

The Cohomology of Lie algebrae: Let $\mathfrak{g} = Der\mathcal{A}$ (tangent vector field) be a module, and $\mathcal{A}$ be a commutative associative algebra. The Zariski r-forms are an isomorphism in general. For $w \in (\wedge^r_{\mathcal{A}}\mathfrak{g})^*$, we have $w(X_1, \cdots, X_r) \in \mathcal{A}$ for $X_i \in Der\mathcal{A} = \mathfrak{g}$. The followings are the ingredients for the cohomology of Lie algebrae. (1) Interior-product ι_X: $(\iota_X w)(X_2, \cdots, X_r) = w(X, X_2, \cdots, X_r)$; (2) Lie-derivative $\mathcal{L}_X$: $(\mathcal{L}_X w)(X_1, \cdots, X_r) = X(w(X_1, \cdots, X_r)) - \sum_{i=1}^r w(X_1, \cdots, [X, X_i], \cdots, X_r)$; (3) $[\mathcal{L}_X, \mathcal{L}_Y] = \mathcal{L}_{[X,Y]}$, $[\mathcal{L}_X, \iota_Y] = \iota_{[X,Y]}$, $\{\iota_X, \iota_Y\} = 0$, $\iota_X^2 = 0$; (4) Let $\mathcal{C}^\bullet = \oplus_{r=1}^\infty (\wedge^r_{\mathcal{A}}\mathfrak{g})^*$. There exists a unique $d : \mathcal{C}^r \to \mathcal{C}^{r+1}$ with $\mathcal{L}_X = \iota_X \circ d + d \circ \iota_X$ and $d(fw) = df \wedge w + fdw$ for all $X \in \mathfrak{g}$, $f \in \mathcal{A}, w \in \mathcal{C}^r$; (5) the differential is given by

$$dw(X_0, X_1, \cdots, X_r) = \frac{1}{r+1}\Big(\sum_{i=0}^r (-1)^i X_i(w(X_0, \cdots, \hat{X}_i, \cdots, X_r))$$

$$+ \sum_{0 \le i < j \le r} (-1)^{i+j} w([X_i, X_j], X_0, \cdots, \hat{X}_i, \cdots, \hat{X}_j, \cdots, X_r)\Big);$$

(6) $d \circ d = 0$. Let $T\mathfrak{g}$ be a tensor algebra of $\mathfrak{g}$ where $T\mathfrak{g} = \oplus_{n=0}^\infty T_n\mathfrak{g}$, where $T_n\mathfrak{g} = \mathfrak{g} \otimes \cdots \otimes \mathfrak{g}$, and $T_0\mathfrak{g} = K$ the base field. There is a two-sided ideal, I, generated by $x \otimes y - y \otimes x - [x, y]$ for $x, y \in \mathfrak{g}$. The universal enveloping algebra $U\mathfrak{g} = T\mathfrak{g}/I$. Let V be the underlying vector space of $\mathfrak{g}$. $EV = \oplus_{n=0}^\infty E_nV$ is the exterior algebra on V, $E_nV = \Lambda^n V$ and $E_0V = K$. Define $C_n = U\mathfrak{g} \otimes E_nV$ with $C_0 = U\mathfrak{g}$ and $\varepsilon : C_0 \to K$ augmentation. The homomorphisms $d_n : C_n \to C_{n-1}$ are the $\mathfrak{g}$-module maps

$$d_n(\langle x_1, \cdots, x_n\rangle) = \sum_{i=1}^n (-1)^{i+1} x_i \langle x_1, \cdots, \hat{x}_i, \cdots, x_n\rangle$$

$$+ \sum_{1 \le i < j \le n} (-1)^{i+j} \langle [x_i, x_j], x_1, \cdots, \hat{x}_i, \cdots, \hat{x}_j, \cdots, x_n\rangle.$$

Thus (C_*, d_*) is a $\mathfrak{g}$-free resolution of K. Form $Hom_{U\mathfrak{g}}(C_*, V)$ and a co-differential $\delta^* = Hom_{U\mathfrak{g}}(d_*, V)$ to get a cochain complex. Define

$$Ext^*_{U\mathfrak{g}}(K, V) = H^*(Hom_{U\mathfrak{g}}(C_*, V), \delta^*).$$

By the Cartan-Eilenberg theorem $Ext^*_{U\mathfrak{g}}(A, B) \cong H^*(\mathfrak{g}, Hom_K(A, B))$, we have

$$H^*(\mathfrak{g}, V) \cong H^*(\mathfrak{g}, Hom_K(K, V)) = Ext^*_{U\mathfrak{g}}(K, V).$$

The Semi-infinite cohomology of infinite-dimensional algebras: The theory of semi-infinite cohomology of graded Lie algebras was first introduced by Feign [**12**], developed by Frenkel, Garland and Zuckerman [**20**] and explored with homological algebra (two-sided derived functor) property by Voronov [**50**]. The semi-infinite cohomology is also called BRST (Bechi-Rouet-Stora-Tyutin) cohomology in the mathematical physics literature.

Let $\mathfrak{g} = \oplus_{n\in\mathbb{Z}}\mathfrak{g}_n$ be a graded Lie algebra over K, i.e., $[\mathfrak{g}_m, \mathfrak{g}_n] \subset \mathfrak{g}_{m+n}$ such that $\dim_K \mathfrak{g}_n < \infty$ for each $n, m \in \mathbb{Z}$; and satisfy the minimal properties that any self-respecting differential should possess. Denote $\mathcal{N}_\pm = \oplus_{\pm n>0}\mathfrak{g}_n$. Fix a basis $\{e_i\}_{i\in\mathbb{Z}}$ for $\mathfrak{g}$ such that if $e_i \in \mathfrak{g}_n$ then either $e_{i+1} \in \mathfrak{g}_n$ or $e_{i+1} \in \mathfrak{g}_{n+1}$, and let $\mathfrak{g}^* = \oplus_{n\in\mathbb{Z}}\mathfrak{g}_n^*$ be the restricted dual of $\mathfrak{g}$ with $\mathfrak{g}_n^* = Hom_K(\mathfrak{g}_{-n}, K)$ and $\{e^i\}$ the dual basis in $\mathfrak{g}^*$. By the Poincaré-Birkhoff-Weyl theorem, the universal enveloping algebra of $\mathfrak{g}$ is $\mathcal{U}\mathfrak{g} \cong \mathcal{U}\mathcal{N}_{\le 0} \otimes \mathcal{U}\mathcal{N}_+$ as vector spaces. Denote sdeg the grading in $\mathcal{U}\mathcal{N}_+$, as opposed to deg in $\mathfrak{g}$. A homogeneous element $\sum_{k=1}^r u_kv_k \in \mathcal{U}\mathfrak{g}$ with $u_k \in \mathcal{U}\mathcal{N}_{\le 0}$ and $v_k \in \mathcal{U}\mathcal{N}_+$ has $\deg(u_kv_k) = \deg(u_lv_l)$ for all k and l. If an expression $\sum_{k=1}^\infty u_kv_k$ ($u_k \in \mathcal{U}\mathcal{N}_{\le 0}$ and $v_k \in \mathcal{U}\mathcal{N}_+$ with $\deg v_k = sdeg(u_kv_k) \to \infty$

as $k \to \infty$) is allowed, one obtains a canonical completion $\mathcal{U}^\infty\mathfrak{g}$ of $\mathcal{U}\mathfrak{g}$ with respect to deg.

Let $s(\mathfrak{g}) = \mathfrak{g} \oplus \mathfrak{g}^* \oplus K_c$ be the super Heisenberg Lie algebra with center K_c in which elements of $\mathfrak{g}$ and $\mathfrak{g}^*$ will be specified by symbols ι and ε to emphasize that $\mathfrak{g}$ and $\mathfrak{g}^*$ are now the adjoint and co-adjoint representations. The $*$-degrees for elements of $\mathfrak{g}, \mathfrak{g}^*$ and K_c are $-1, 1$ and 0 respectively. Denote both commutators and anti-commutators by the super-bracket $[,]$. For $x, y \in \mathfrak{g}$ and $x^*, y^* \in \mathfrak{g}^*$, $[\iota(x), \iota(y)] = [\varepsilon(x^*), \varepsilon(y^*)] = 0$, $[\varepsilon(x^*), \iota(y)] = \langle x^*, y \rangle$. A secondary $\mathbb{Z}$-grading is given by $\deg \iota(x) = n$, $x \in \mathfrak{g}_n$; $\deg \varepsilon(x^*) = -n$, $x^* \in \mathfrak{g}^*$; and $\deg c = 0$. Now we complete the universal enveloping algebra of $s(\mathfrak{g})$ with respect to deg. Let $C^\infty\mathcal{G} = \mathcal{U}^\infty s(\mathfrak{g})/(c = 1)$ be the completed Clifford algebra on $\mathfrak{g}$. By the Poincaré-Birkhoff-Weyl theorem, $C^\infty\mathcal{G} \cong (\Lambda\mathfrak{g} \otimes \Lambda\mathfrak{g}^*)^\infty$ as vector spaces.

The semi-infinite form $\Lambda^{\infty/2+*}\mathfrak{g}^*$ is the $\mathcal{U}s(\mathfrak{g})/(c = 1)$-module induced from the one-dimensional trivial representation of the subalgebra $\Lambda(\mathcal{N}_{\geq 0} \oplus \mathcal{N}_-^*)$ whose elements are called annihilation operators. The space $\Lambda^{\infty/2+*}\mathfrak{g}^*$ can be viewed as the exterior algebra $\Lambda^\bullet\mathfrak{g}^*$ as a module of $C^\infty\mathcal{G}$. This module is the span of the semi-infinite forms (see Lemma in [**12**] and [**20**] §1.1):

$$\nu = e^{i_1} \wedge e^{i_2} \wedge \cdots \wedge e^{i_k} \wedge \cdots, \tag{2.2}$$

with the property that there exists $N(\nu) \in \mathbb{Z}$ such that $i_1 > i_2 > \cdots > i_k > \cdots$, and $i_{k+1} = i_k - 1$ for $k > N(\nu)$. While an ordinary exterior algebra $\Lambda^*\mathfrak{g}^*$ has the structure of a $\mathfrak{g}$-module, $\Lambda^{\infty/2+*}\mathfrak{g}^*$ in general admits only the structure of a $\tilde{\mathfrak{g}}$-module, where $\tilde{\mathfrak{g}}$ is a one-dimensional central extension of $\mathfrak{g}$. For $x \in \mathfrak{g}$ and $x^* \in \mathfrak{g}^*$ and ν in (2.2), we define

$$\varepsilon(x^*)\nu = x^*\nu, \quad \iota(x)\nu = \sum_{k \geq 1} (-1)^{k-1} \langle x, e^{i_k} \rangle e^{i_1} \wedge e^{i_2} \wedge \cdots \wedge \widehat{e^{i_k}} \wedge \cdots.$$

Let ad_x and ad_x^* ($x \in \mathfrak{g}$) denote the adjoint and the coadjoint action of $\mathfrak{g}$. Define an action of $\mathfrak{g}$ on $\Lambda^{\infty/2+*}\mathfrak{g}^*$ by an inner derivation $\rho(x)\nu = \sum_{k \geq 1} e^{i_1} \wedge \cdots \wedge (ad_x^* \cdot e^{i_k}) \wedge e^{i_{k+1}} \wedge \cdots$. By the definition of $\Lambda^{\infty/2+*}\mathfrak{g}^*$, one has

$$[\rho(x), \iota(y)] = \iota(ad_x y), \quad [\rho(x), \varepsilon(y^*)] = \varepsilon(ad_x^* y^*). \tag{2.3}$$

A "vacuum" semi-infinite form in $\Lambda^{\infty/2+*}\mathfrak{g}^*$ is an element $\nu_0 \in \Lambda^{\infty/2+*}\mathfrak{g}^*$ such that $[x, y]\nu_0$ is proportional to ν_0 for any $x \in \mathfrak{g}_n$ and $y \in \mathfrak{g}_{-n}$ with $n \neq 0$. For a given vacuum ν_0 and a fixed element $\beta \in \mathfrak{g}_0^*$ trivial on $[\mathfrak{g}_0, \mathfrak{g}_0]$, we define for $x \in \mathfrak{g}_0$,

$$\rho(x)\nu_0 = \langle \beta, x \rangle \nu_0. \tag{2.4}$$

The action ρ can be uniquely extended to $\Lambda^{\infty/2+*}\mathfrak{g}$ by requiring relations (2.3) and (2.4). A representation of $\mathfrak{g}$ is defined by

$$\rho(x) = \sum_{i \in \mathbb{Z}} : \varepsilon(ad_x^* e^i)\iota(e_i) : + \langle \beta, x \rangle, \quad x \in \mathfrak{g},$$

where $: \cdot :$ is the Wick normal ordering operation which means a factor with positive degree should appear on the right, and if one has to change the given order one had better multiply that term by -1 (see §1.1 in [**20**]):

Note that for $H^2(\mathfrak{g}) = 0$ (e.g. $\mathfrak{g}$ is Virasoro, $L\mathfrak{g} \oplus K_c$, affine Kac-Moody algebra) and any ν_0, there exists a β such that ρ is a representation of $\mathfrak{g}$ on $\Lambda^{\infty/2+*}\mathfrak{g}^*$ (see [**50**] §2.4).

Let $\tilde{s}(\mathfrak{g})$ be the Lie superalgebra as the direct sum $\tilde{\mathfrak{g}} \oplus s(\mathfrak{g})$, where $\tilde{\mathfrak{g}}$ is the central extension of $\mathfrak{g}$ defined by $\rho([x,y]) - [\rho(x), \rho(y)]$ with $*$-deg 0. Note that $*$-deg is the super degree, $*$-deg$\iota(x) = -1$ and $*$-deg$\varepsilon(x^*) = 1$, $*$-deg$\nu_0 \in \mathbb{Z}$ has the freedom. Define

$$\Lambda^{\infty/2+m}\mathfrak{g}^* = \{\nu \in \Lambda^{\infty/2+*}\mathfrak{g}^* : *\text{-deg}\nu = m\},$$

is a $\mathfrak{g}$-module for every $m \in \mathbb{Z}$, and it carries a $\mathbb{Z}$-graded structure. We fix $*$-deg $\nu_0 \in \mathbb{Z}$, for $x \in \mathfrak{g}_n, x^* \in \mathfrak{g}_n^*$ with $\deg \iota(x) = n$ and $\deg \varepsilon(x^*) = -n$,

$$\Lambda^{\infty/2+m}\mathfrak{g}^* = \oplus_{n\in\mathbb{Z}}\Lambda^{\infty/2+m}\mathfrak{g}_n^*,$$

where $\Lambda^{\infty/2+m}\mathfrak{g}_n^* = \{\nu \in \Lambda^{\infty/2+m}\mathfrak{g}^* : \deg \nu = n\}$. It is easy to see that there exists an integer n_0 such that $\dim \Lambda^{\infty/2+m}\mathfrak{g}_n^* = 0$ for $n > n_0$ and $\dim \Lambda^{\infty/2+m}\mathfrak{g}_n^* < \infty$. Elements of $\tilde{\mathfrak{g}}$ will be denoted by $\overline{x}$, and will have secondary grading induced from $\mathfrak{g}$ with $\deg \overline{c} = 0$. Note that $[\overline{x}, \overline{y}] = \overline{[x,y]_{\tilde{\mathfrak{g}}}}$. By the Poincaré-Birkhoff-Weyl theorem,

$$\mathcal{U}^\infty \tilde{s}(\mathfrak{g})/(c = 1, \overline{c} = 1) \cong (\mathcal{U}\tilde{\mathfrak{g}} \otimes \Lambda\mathfrak{g} \otimes \Lambda\mathfrak{g}^*)^\infty.$$

THEOREM 2.5. *If M is any smooth $\tilde{\mathfrak{g}}$-module on which the central element acts by 1, there is a unique element*

$$d = \sum_i \overline{e_i}\varepsilon(e^i) + \sum_{i<j} : \iota([e_i, e_j])\varepsilon(e^i)\varepsilon(e^j) : + \varepsilon(\beta).$$

(1) *The element d is a well-defined differential on the $\mathcal{U}^\infty\tilde{s}(\mathfrak{g})/(c = 1, \overline{c} = 1)$-module $M \otimes \Lambda^{\infty/2+*}\mathfrak{g}^*$. I.e., $d^2 = 0$.*
(2) *The cohomology $H^*(M \otimes \Lambda^{\infty/2+*}\mathfrak{g}^*, d)$, denoted by $H^{\infty/2+*}(\mathfrak{g}, \omega; M)$, is called the semi-infinite cohomology of $\mathfrak{g}$ with coefficients in the graded $\mathfrak{g}$-module M, where ω is to emphasize the corresponding central extension $\tilde{\mathfrak{g}}$ of $\mathfrak{g}$. (See also* [**12, 20, 50**]*)*

□

We consider a $\mathbb{Z}$-graded $\mathfrak{g}$-module M such that the action mapping $\mathcal{U}(\mathfrak{g}) \times M \to M$ is continuous, where the universal enveloping algebra $\mathcal{U}(\mathfrak{g})$ is endowed with the topology defined by the filtration which is associated to the grading of $\mathcal{U}(\mathfrak{g})$ induced from that on $\mathfrak{g}$ and M is endowed with the discrete topology. Let $\mathcal{O}_\mathfrak{g}$ be a category of such $\mathfrak{g}$-modules with $\mathfrak{g}$-homomorphisms. It is an Abelian category with arbitrary direct limit. For our purpose, we have that $\mathfrak{g}$-module $\Lambda^{\infty/2+*}$ of semi-infinite forms is an object of the category $\mathcal{O}_\mathfrak{g}$.

DEFINITION 2.6. (1) A $\mathfrak{g}$-module S is semijective if the functor $Hom_{\mathcal{N}_+}(\cdot, S)$ is exact on the category of $\mathcal{N}_+$-modules ($\mathcal{N}_+$-injective), and the functor $Hom_\mathfrak{g}(S, \cdot)$ on the category $\mathfrak{g}$-modules is exact on $\mathcal{N}_+$-split sequences (projective relative to $\mathcal{N}_+$).

(2) A semijective resolution $S^\bullet$ of a complex $K^\bullet$ of $\mathfrak{g}$-modules is an isomorphism $S^\bullet \to K^\bullet$ in the derived category, i.e., there exists a pair of quasi-isomorphisms in the category of complexes of $\mathfrak{g}$-modules in $\mathcal{O}_\mathfrak{g}$, $S^\bullet \to I^\bullet \leftarrow K^\bullet$, where $S^\bullet$ is semijective.

In [**50**] §3, Voronov showed that (i) there exists a standard semijective module SS such that there are natural $\mathfrak{g}$-module isomorphisms

$$Hom_{\mathcal{N}_{\leq 0}}(\mathcal{U}(\mathfrak{g}),\mathcal{U}(\mathcal{N}_{\leq 0}))^{(-\beta)} \cong SS \cong \mathcal{U}(\mathcal{N}_+)^* \otimes_{\mathcal{N}_+} \mathcal{U}(\mathfrak{g}),$$

where $(-\beta)$ is the twisted action of the subalgebra $\mathcal{N}_{\leq 0}$ by the 1-cocycle $-\beta$, and SS is considered with right and left regular representations respectively; (ii) for any complex $K^\bullet$ of $\mathfrak{g}$-modules in $\mathcal{O}_\mathfrak{g}$, there exists a semijective resolution. (iii) for any $\mathfrak{g}$-module $M \in \mathcal{O}_\mathfrak{g}$, there exists a semijective resolution of M:

$$M \otimes S^\bullet = M \otimes C^{\infty/2+\bullet}(\mathfrak{g}, SS) = M \otimes SS \otimes \Lambda^{\infty/2+\bullet}\mathfrak{g}^*,$$

by Theorem 3.2, Theorem 3.4 and Theorem 3.7 of [**50**].

3. Floer cohomology from symplectic topology

3.1. The symplectic Floer cohomology. Let (P,ω) be a compact or tamed symplectic manifold with a closed non-degenerate 2-form ω. The 2-form ω defines a 2-dimensional cohomology class $[\omega] \in H^2(P,\mathbb{R})$. By choosing an almost complex structure J on (P,ω) such that $\omega(\cdot, J\cdot)$ defines a Riemannian metric, we have an integer-valued second cohomology class $c_1(P) \in H^2(P,\mathbb{Z})$ (the first Chern class). These two cohomology classes define two homomorphisms $I_\omega : \pi_2(P) \to \mathbb{R}$ and $I_{c_1} : \pi_2(P) \to \mathbb{Z}$. We call the symplectic manifold (P,ω) *monotone* if $I_\omega = \alpha I_{c_1}$, for some $\alpha \geq 0$. Let $H : \mathbb{R} \times P \to \mathbb{R}$, $H(s,x) = H_s(x)$ be a smooth time-dependent Hamiltonian function with $H_s = H_{s+1} \circ \phi$ and ψ_s be the symplectic diffeomorphism generated by H. We have $\psi_{s+1} \circ \phi_H = \phi \circ \psi_s$ with $\phi_H = \psi_1^{-1} \circ \phi$. Define the space of smooth contractible paths

$$\Omega_\phi = \{\gamma : \mathbb{R} \to P \mid \gamma(s+1) = \phi(\gamma(s)), [\phi^{-1} \circ \gamma] = 0 \in \pi_1(P)\}.$$

The (perturbed) symplectic action functional $a_H : \Omega_\phi \to \mathbb{R}/2\alpha N\mathbb{Z}$ is defined as a functional whose differential is given by

$$da_H(\gamma)\xi = \int_0^1 \omega(\dot{\gamma} + J\nabla H_s(\gamma), \xi)ds, \tag{3.1}$$

where N is the minimal Chern number defined by $I_{c_1}(\pi_2(P)) = N\mathbb{Z}$ with $N \geq 0$. So the critical points of a_H are in one-to-one correspondence with the elements in the fixed-point set $\mathrm{Fix}\,(\phi_H)$. We make use of an infinite cyclic cover $\tilde{\Omega}_\phi$ of Ω_ϕ so that the symplectic action $\tilde{a}_H$ on $\tilde{\Omega}_\phi$ and the Maslov index function μ on $\tilde{Z}_\phi$ are defined as $\tilde{a}_H : \tilde{\Omega}_\phi \to \mathbb{R}$ and $\mu : \tilde{Z}_\phi \to \mathbb{Z}$. The space Ω_ϕ is path connected and $\pi_1(\Omega_\phi, \alpha_0)$ is independent of the base point $\alpha_0 \in \Omega_\phi$. Take $z_0 \in \mathrm{Fix}\,(\phi_H) \subset \Omega_\phi$. We denote $\mathrm{Fix}\,(\phi_H)$ by $\mathrm{Fix}\,\phi$ for simplicity.

Given $x \in \mathrm{Fix}\,\phi \subset \Omega_\phi$, let $x^{(r)} \in \tilde{\mathrm{Fix}}\,\phi \subset \tilde{\Omega}_\phi$ be the unique lift of x such that $\tilde{a}_H(x^{(r)}) \in (r, r+2\alpha N)$. Note that if $\alpha N = 0$ then one already has a $\mathbb{Z}$-graded symplectic Floer homology as in [**17**], so we work on the case $\alpha N \neq 0$. Let $\mu^{(r)}(x) = \mu(x^{(r)}, z_0) \in \mathbb{Z}$ and define the $\mathbb{Z}$-graded symplectic Floer cochain group

$$C_n^{(r)}(\phi; P, J) = \mathbb{Z}\{x \in \mathrm{Fix}\,\phi \mid \mu^{(r)}(x) = n \in \mathbb{Z}\}, \tag{3.2}$$

the base point z_0. An integral Floer coboundary map $\partial^{(r)} : C^{(r)}_{n-1}(\phi; P, J) \to C^{(r)}_n(\phi; P, J)$ is

$$\partial^{(r)} x = \sum_{y \in C^{(r)}_n(\phi; P, J)} \#\hat{\mathcal{M}}(x, y) \cdot y,$$

where $\hat{\mathcal{M}}(x, y) = \mathcal{M}(x, y)/\mathbb{R}$ is the zero-dimensional quotient with respect to the action of $\mathbb{R}$ by translation and $\mathcal{M}(x, y)$ is the union of one-dimensional components of moduli spaces of J-holomorphic curves. The number $\#\hat{\mathcal{M}}(x, y)$ counts the points with sign in [**19**]. Now $(C^{(r)}_n(\phi; P, J), \partial^{(r)}_n)_{n \in \mathbb{Z}}$ is indeed a $\mathbb{Z}$-graded symplectic Floer cochain complex (see [**30**]). We call its cohomology *$\mathbb{Z}$-graded symplectic Floer cohomology*, denoted by

$$I^{(r)}_*(\phi; P, J) = H^*(C^{(r)}_*(\phi; P, J), \partial^{(r)}), \quad * \in \mathbb{Z}.$$

For $r \in \mathbb{R}_\phi, j \in \mathbb{Z}_{2N}$ and $n \equiv j \pmod{2N}$, define a free $\mathbb{Z}$-module

$$F^{(r)}_n C^j(\phi; P, J) = \sum_{k \geq 0} C^{(r)}_{n+(2N)k}(\phi; P, J),$$

which gives a natural decreasing filtration on $C^*(\phi; P, J)$ $(* \in \mathbb{Z}_{2N})$.

THEOREM 3.1. *(i) There is a spectral sequence $(E^k_{n,j}, d^k)$ with*

$$E^1_{n,j}(\phi; P, J) \cong I^{(r)}_n(\phi; P, J), \quad n \equiv j \pmod{(2N)},$$

and the higher differential d^k maps $E^k_{n,j}(\phi; P, J)$ into $E^k_{n+(2N)k+1,j+1}(\phi; P, J)$, and

$$E^\infty_{n,j}(\phi; P, J) \cong F^{(r)}_n HF^j(\phi; P, J)/F^{(r)}_{n+(2N)} HF^j(\phi; P, J),$$

*where $F^{(r)}_n HF^j(\phi; P, J) = \ker(I^{(r)}_j(\phi; P, J) \to F^{(r)}_{n-2N} I^{(r)}_j(\phi; P, J))$ (see [**32**]).*

*(ii) The spectral sequence $(E^k_{n,j}, d^k)$ converges to $HF^*_{sym}(\phi, P)$ (the $\mathbb{Z}_{2N}$-graded symplectic Floer cohomology in [**18**]).*

(iii) For any continuation $(J^\lambda, \phi^\lambda) \in \mathcal{P}_{1,\varepsilon/2}$ which is regular at the ends, there exists an isomorphism

$$E^1_{n,j}(\phi; P, J^0) \cong E^1_{n,j}(\phi; P, J^1).$$

(iv) For each $k \geq 1$, $E^k_{n,j}(\phi; P, J)$ are symplectic invariants under continuous deformations of $(J^\lambda, \phi^\lambda)$ within the set of continuations.

In [**30**], we define the cup product with the cohomology class $a \in H^*(\Omega_\phi)$ on the $\mathbb{Z}$-graded symplectic Floer cochain groups:

$$a \cup : C^{(r)}_n(\phi, P, J) \to C^{(r)}_{n+p}(\phi, P, J) \tag{3.3}$$

$$x \mapsto \sum_{-\mu^{(r)}(x) + \mu^{(r)}(y) = p} \#(\mathcal{M}_J(x, y) \cap PD(a)) \cdot y,$$

where $\#(\mathcal{M}_J(x, y) \cap PD(a))$ is the algebraic number of $PD(a)$ intersecting $\mathcal{M}_J(x, y)$ (see also [**18**]). Any holomorphic bubbling required energy at least $2N$ forces the corresponding moduli space $\mathcal{M}_J(x, y)$ to have different lifting for x and y. So the moduli space with fixed lifting $x^{(r)}$ and $y^{(r)}$ contains no holomorphic bubbling spheres. We can rule out the codimension-one boundary component in the intersection $\mathcal{M}_J(x, y) \cap PD(a)$. This proves that $\mathcal{M}_J(x, y)$ is a genuine cycle which can be evaluated on any cohomology class $a \in H^*(\Omega_\phi)$ by a purely algebraic topology method. Thus the cup-product in (3.3) is well-defined in Proposition 4.1 and Proposition 4.3 of [**30**], and we have our module structures in the following Theorem 5.7 of [**30**].

THEOREM 3.2. (1) *$a\cup$ commutes with the higher differentials $d^k (k \geq 1)$ for the spectral sequence $E^k_{*,*}(\phi, P)$.*

(2) *All $E^k_{*,*}(\phi, P)$ has an $H^*(\Omega_\phi)$-module structure. In particular, the $E^\infty_{*,*}$ term ($HF^*_{sym}(\phi, P)$) has an $H^*(\Omega_\phi)$-module structure.*

Theorem 3.2 answers affirmatively the Floer conjectures in [**18**].

3.2. Floer homotopy theory. In [**7**], Cohen, Jones and Segal developed the semi-infinite structures on the loop space and defined the flow category. Their goal is to associate a pro-spectrum of the (compactified) flow category such that the homology of this spectrum is exactly the Floer homology. They verified this principle for $\mathbb{C}P^k$ and Hurtubise [**25**] verified this for $G_{n,n+k}(\mathbb{C})$ the complex Grassmannian manifold of n-planes in $\mathbb{C}^{n+k}$.

Define the flow category $\mathcal{C}_f(M)$ of $f : M \to \mathbb{R}$ to be a topological category whose objects are the critical points (or submanifolds) of f and whose morphisms are the unperturbed piecewise gradient flow lines of f. Let $B\mathcal{C}_f(M)$ be the geometric realization of $\mathcal{C}_f(M)$. For finite-dimensional M and a Morse function f, $\mathcal{C}_f(M)$ is homotopy-equivalent to M by Morse theory (see [**5, 7, 37, 38**]). In general, one needs to specify certain compactifications (such as Floer-Uhlenbeck's compactification in [**17, 18**]) for the flow category on infinite-dimensional manifolds. Using the gradient flow, one can define a partial order $\leq$ in $Ob(\mathcal{C}_f(M))$. Fixing an element $C_0 \in Ob(\mathcal{C}_f(M))$, denote $\overline{W(C_1, C_2)}$ by the (specified) compactification of the gradient flow lines between C_1 and C_2.

Let $I_B = \{C \in Ob(\mathcal{C}_f(M)) : C_0 \leq C, \overline{W(C, B)} \text{ is a manifold}\}$ and

$$I = \{B \in Ob(\mathcal{C}_f(M)) : \overline{W(C_0, B)} \text{ is a manifold}\}.$$

Let ν_C be the normal bundle of $B\mathcal{C}_f(M)|_B^{C_0} \hookrightarrow B\mathcal{C}_f(M)|_B^C$. Then *the Floer homotopy type $\mathcal{FH}(M, f)$ is the cofinal system*

$$\varinjlim_{B\in I} \{(B\mathcal{C}_f(M)|_B^{C_0})^{-\nu_C}\}_{C\in I_B}$$

defined in [**7, 25**]. The maps in the cofinal system are inclusion maps. For $M = \tilde{\mathcal{L}}G_{n,n+k}(\mathbb{C})$ and $f = a_\omega$ the symplectic action, the flow category $\mathcal{C}_{a_\omega}(\tilde{\mathcal{L}}G_{n,n+k}(\mathbb{C}))$ has

$$\mathcal{FH}(\tilde{\mathcal{L}}G_{n,n+k}(\mathbb{C}), a_\omega) : \; G_n(\mathbb{C}^\infty) \leftarrow G_n(\mathbb{C}^\infty)^{-(n+k)\gamma_n} \leftarrow G_n(\mathbb{C}^\infty)^{-2(n+k)\gamma_n} \leftarrow \cdots,$$

where $\tilde{\mathcal{L}}G_{n,n+k}(\mathbb{C})$ is the universal cover of the free loop space of $G_{n,n+k}(\mathbb{C})$ and γ_n is the tautological n-plane bundle over $G_n(\mathbb{C}^\infty)$. By applying H^* and using the Thom isomorphism, one obtains a direct system $\{H^{*+2in(n+k)}(G_n(\mathbb{C}^\infty)), \cup c_n^{n+k}\}_{i\geq 0}$ and its limit $H^*(G_n(\mathbb{C}^\infty))[c_n^{-(n+k)}]$. Note that $H^*(G_n(\mathbb{C}^\infty))[c_n^{-(n+k)}]$ is a cohomology group over $\mathbb{Z}$ with period $2(n+k)$, where $(n+k)$ is the minimal first Chern number of $G_{n,n+k}(\mathbb{C})$. For $n = 1$, the cohomology ring $H^*(G_1(\mathbb{C}^\infty))[c_1^{-(1+k)}]$ is the some classical groups G, the based loop groups ΩG have infinite-dimensional Grassmannian models.

For any simply connected, compact and semisimple Lie group G (through the discussion of this section), we have $\pi_1(\Omega G) = 0$, $\pi_2(\Omega G) = \pi_3(G)$ and ΩG is a strong symplectic monotone manifold of infinite-dimension. In fact, ΩG is an infinite-dimensional Kähler manifold. For our G, the Lie algebra extension $\mathbb{R} \to \tilde{\Omega}\mathfrak{g} \to \Omega\mathfrak{g}$ defined by a cocycle ω corresponds to a group extension

$$U(1) \to \tilde{\Omega}_G \to \Omega G$$

if and only if $\omega/2\pi$ represents an integral cohomology class of ΩG. If $\mathfrak{g}$ is a simple Lie algebra, the minimal Chern number N gives the basic central extension of ΩG, which is a universal $U(1)$ bundle over ΩG. For simply connected, compact, semisimple Lie group G, N is the minimal Coxeter number (≥ 2) of G. Let $H : \Omega G \to \mathbf{R}$ be the so-called energy function defined by $H(\gamma) = \frac{1}{4\pi}\int_0^{2\pi} \|\gamma^{-1}(\theta)\gamma'(\theta)\|^2 d\theta$. For the based loop group $(\Omega G, \omega)$, the energy function H is Hamiltonian. The Hamiltonian vector field X_H on ΩG which corresponds to H is the generator of the flow defined by rotating the loops $e^{it} \cdot \gamma(\theta) = \gamma(t+\theta)\gamma^{-1}(\theta)$ (see [**33**] Section 2). Let $\tilde{\mathcal{P}}(H)$ be the set of those pair $[x, u] \in L(\Omega G)$, where $x \in \mathcal{P}(H)$,

$$\mathcal{P}(H) = \{x(\theta) \in \Omega G \mid \frac{dx}{dt} = X_H(x_t(\theta))\}.$$

These are the critical points of the symplectic action $a_H : L(\Omega G) \to \mathbb{R}/(2\pi N)\mathbb{Z}$ defined by

$$a_H([x, u]) = \int_D u^*\omega + \int_0^{2\pi} H(\theta, x(\theta))d\theta. \tag{3.4}$$

The critical set of a_H is the set of periodic solutions of the rotation flow, and can be also identified with $\mathcal{R}_+^*(\tilde{\Omega}_G) = \mathcal{R}_+^*(U(1) \rtimes \tilde{\Omega}_G) = L_\omega$ by Proposition 2.6, Theorem 2.7 and Lemma 2.8 of [**33**].

Denote by $\tilde{\mathcal{P}}_k(H)$ the subset of all $[\tilde{x}, \tilde{u}] \in \tilde{L}(\Omega G)$ with $\mu([\tilde{x}, \tilde{u}]) = k$, and the k-th cochain group $C^k(H)$ by all the formal sums $\sum \xi_{[\tilde{x},\tilde{u}]}\cdot[\tilde{x}, \tilde{u}]$, for $[\tilde{x}, \tilde{u}] \in \tilde{\mathcal{P}}_k(H)$ and $\xi_{[\tilde{x},\tilde{u}]} \in \mathbb{Z}$ such that the set $\{[\tilde{x}, \tilde{u}] | \xi_{[\tilde{x},\tilde{u}]} \neq 0, \tilde{a}_H([\tilde{x}, \tilde{u}]) \leq c\}$ is finite for all $c \in \mathbb{R}$. The choice of the ring (or field) which $\xi_{[\tilde{x},\tilde{u}]}$ belongs to depends on the orientation issue of the moduli spaces $\mathcal{M}_J([\tilde{x}, \tilde{u}], [\tilde{y}, \tilde{v}])$ of the J-holomorphic curves. For our Kähler (monotone symplectic) manifold $(\Omega G, \omega)$, the coherent orientation follows from the same argument of [**19**] (see also §2.7 of [**33**]). As usual, Λ_ω is the Novikov ring associated with $\phi_\omega : \pi_2(\Omega G) \to \mathbb{R}$. Hence we obtain a well-defined symplectic Floer cohomology of $(\Omega G, \omega)$.

THEOREM 3.3. *The k-th cochain group $C^k(H)$ is a torsion-free module over the Novikov ring Λ_ω. The map δ^k defined by $\delta^k([\tilde{x}, \tilde{u}]) = \sum_{\mu([\tilde{y},\tilde{v}])=k+1} n([\tilde{x}, \tilde{u}], [\tilde{y}, \tilde{v}]) \cdot [\tilde{y}, \tilde{v}]$ satisfies $\delta^{k+1}\circ\delta^k = 0$ for all k. The cohomology $HF^k(\Omega G, \omega) = \ker \delta^k / Im\delta^{k-1}$ is a sympelctic invariant of $(\Omega G, \omega)$.*

For any $\tilde{\Omega}\mathfrak{g}$-module M with central elements acting by 1, there is a well-defined semi-infinite cohomology of $\tilde{\Omega}\mathfrak{g}$ with coefficients M.

THEOREM 3.4. *For a simply connected, compact and semisimple Lie group G, there is an isomorphism between the sympelctic Floer cohomology of ΩG with* *coefficients* Λ_ω *and the semi-infinite cohomology of the* $\mathbb{Z}$*-graded Lie algebra* $\tilde{\Omega}\mathfrak{g}$ *with coefficients* Λ_ω *in the sense of Feign* [**12, 20**]*:*

$$HF^*(\Omega G;\Lambda_\omega) \cong H^{\infty/2+*}(\tilde{\Omega}\mathfrak{g};\Lambda_\omega),$$

where Λ_ω *is the Novikov ring corresponding to the function* $I_\omega : \pi_2(\Omega G) \to \mathbb{R}$ *with a symplectic form* ω *chosen from a central extension* $\tilde{\Omega}\mathfrak{g}$ *of* $\Omega\mathfrak{g}$*.*

Proof: We first have that $C^k(H)$ is a $\tilde{\Omega}\mathfrak{g}$-module through the identification of the critical points of a_H with the irreducible representations of $\tilde{\Omega}_G$ and the Novikov ring Λ_ω is a $\tilde{\Omega}\mathfrak{g}$-module through the action on $\pi_2(\Omega G) = H_2(\Omega G) = H^2(\Omega\mathfrak{g})^*$ (see [**33**]). Then we formulate a double complex

$$C^{p,q} = C^p(\Omega G, \Lambda_\omega) \otimes SS \otimes \Lambda^{\infty/2+q}(\tilde{\Omega}\mathfrak{g})^*,$$

where $C^p(\Omega G, \Lambda_\omega) = C^p(H)$, p is the Maslov index, q is the grading in the semi-infinite forms $\Lambda^{\infty/2+q}(\tilde{\Omega}\mathfrak{g})^*$, and SS is the standard semijective $\tilde{\Omega}\mathfrak{g}$-module in [**50**] §3. Note that the double complex $C^{p,q}$ is a $\tilde{\Omega}\mathfrak{g}$-module in the category $\mathcal{O}_{\tilde{\Omega}\mathfrak{g}}$ and $C^{p,\bullet}$ is a semijective resolution of $C^p(H)$. The double complex $C^{p,q}$ can be thought as the Floer cochain groups $\mathcal{P}_k(H)$ with coefficients $\Lambda_\omega \otimes SS \otimes \Lambda^{\infty/2+q}(\tilde{\Omega}\mathfrak{g})^*$ or the semijective complex $C^{p,\bullet}$ of the $\tilde{\Omega}\mathfrak{g}$-module $C^p(H)$. The double complex has horizontal operator δ (the Floer coboundary operator) and the vertical operator d (the semi-infinite differential operator). Note that $d\delta = \delta d$ by the definitions of d and δ since their definitions are independent of the coefficients. This gives us a total complex Tot which is defined by

$$(Tot)^n = \oplus_{p+q=n} C^p(\Omega G, \Lambda_\omega) \otimes SS \otimes \Lambda^{\infty/2+q}(\tilde{\Omega}\mathfrak{g})^*, \quad D = \delta + (-1)^p d. \tag{3.5}$$

There are spectral sequences $\{{}_IE_r^{*,*}, {}_Id_r\}$ and $\{{}_{II}E_r^{*,*}, {}_{II}d_r\}$ with

$${}_IE_2^{*,*} \cong H_\delta^{p,q}H_d(Tot); \quad {}_{II}E_2^{*,*} \cong H_d^{p,q}H_\delta(Tot). \tag{3.6}$$

Both spectral sequences converge to $H^*(Tot, D)$. We calculate the E_1 term of the two spectral sequences

$${}_IE_1^{*,*} = H_d(Tot) = H^{\infty/2+*}(\tilde{\Omega}\mathfrak{g}, C^*(H));$$

$${}_{II}E_1^{*,*} = H_\delta(Tot) = HF^*(\Omega G, \Lambda_\omega \otimes S^*),$$

where $S^* = SS \otimes \Lambda^{\infty/2+*}(\tilde{\Omega}\mathfrak{g})^*$ is the standard semijective complex. By using the semijective resolution, we have

$${}_{II}E_2^{p,q} = \begin{cases} HF^p(\Omega G, \Lambda_\omega) & q = 0 \\ 0 & q \neq 0 \end{cases}$$

Using vanishing results in [**20, 50**] and positive energy for irreducible representations, we can compute the first spectral sequence

$${}_IE_2^{p,q} = \begin{cases} 0 & p \neq 0 \\ H^{\infty/2+q}(\tilde{\Omega}\mathfrak{g}, \Lambda_\omega) & p = 0 \end{cases}$$

The result follows from the identification:

$$H^n(Tot, D) = \oplus_{p+q=n}\, {}_IE_\infty^{p,q}$$

$$
\begin{aligned}
H^n(Tot, D) &= \oplus_{p+q=n}\, {}_{II}E_{\infty}^{p,q} \\
&= \oplus_{p+q=n}\, {}_{II}E_2^{p,q} = \oplus_{p+0=n}\, {}_{II}E_2^{p,0} \\
&= HF^n(\Omega G, \Lambda_\omega).
\end{aligned} \tag{3.8}
$$

□

Theorem 3.4 can be thought as a natural extension of the Chevally-Eilenberg isomorphism $H^*(G, \mathbb{R}) \cong H^*(\mathfrak{g}, \mathbb{R})$ and the Pressley-Segal isomorphism (see [**43**]) $H^*(\tilde{\Omega}_G; E) \cong H^*(\tilde{\Omega}\mathfrak{g}; E)$, where $H^*(\tilde{\Omega}_G; E)$ is the de Rham cohomology of $\tilde{\Omega}_G$ with values in E, and $H^*(\tilde{\Omega}\mathfrak{g}; E)$ is the continuous Lie algebra cohomology of a central extension $\tilde{\Omega}\mathfrak{g}$ of $\Omega\mathfrak{g}$. *But the most important aspect of the isomorphism in Theorem 3.4 is that it provides the homological algebra interpretation of the symplectic Floer cohomology and the geometric topology interpretation of the semi-infinite cohomology of infinite-dimensional Lie algebras.* See [**33**] for more discussions.

4. Floer homology from gauge theory

4.1. The instanton Floer homology. Let Y be an oriented closed three dimensional smooth manifold with $H_1(Y, \mathbb{Z}) = 0$, and let $P \to Y$ be a smooth principal $SU(2)$-bundle (always trivial). Fix a trivialization $Y \times SU(2)$ of P and let θ be the associated trivial connection. Denote the Sobolev L_k^p space of connections on P by $\mathcal{A}_Y$. $\mathcal{A}_Y$ is acted upon by the gauge group $\mathcal{G} = L_{k+1}^p(\Omega^0(Y, adP))$ of bundle automorphisms of P, and the orbit space $\mathcal{B}_Y = \mathcal{A}_Y/\mathcal{G}$ is well-defined when $k+1 > 3/p$. The universal cover $\tilde{\mathcal{B}}_Y$ of $\mathcal{B}_Y$ defined as $\tilde{\mathcal{B}}_Y = \mathcal{A}_Y/\mathcal{G}_0$, where $\mathcal{G}_0 \subset \mathcal{G}$ is the group of degree zero gauge transformations. The Chern-Simons functional $cs : \mathcal{A}_Y \to \mathbb{R}$ is defined as

$$cs(a) = \frac{1}{8\pi^2}\int_Y tr(a \wedge da + \frac{2}{3} a \wedge a \wedge a).$$

and satisfies $cs(g \cdot a) = cs(a) + \deg(g)$ for gauge transformations $g : Y \to SU(2)$. Thus cs is well-defined on $\tilde{\mathcal{B}}(P) = \mathcal{A}(P)/\{g \in \mathcal{G} : \deg(g) = 0\}$ and it descends to a function $cs : \mathcal{B}(P) \to \mathbb{R}/\mathbb{Z}$ which plays the role of a Morse function in defining Floer homology. The critical set of cs consists of the flat connections $\mathcal{R}(\mathcal{B}(P)) = \{a \in \mathcal{B}(P) |\ F_a = 0\}$ (see [**10, 17, 15**]). It is well-known that the elements of $\mathcal{R}(\mathcal{B}(P))$ are in 1-1 correspondence with those of

$$\mathcal{R}(Y) = Hom(\pi_1(Y), SU(2))/adSU(2),$$

the $SU(2)$-representations of $\pi_1(Y)$ modulo conjugacy. Let $\mathcal{R}^*(Y) = \mathcal{R}(Y) \backslash \{\theta\}$ be the space of irreducible $SU(2)$ representations. Define the weighted Sobolev space $L_{k,\delta}^p$ on sections ξ of a bundle over $Y \times \mathbb{R}$ to be the space of ξ for which $e_\delta \cdot \xi$ is in L_k^p, where $e_\delta(y, t) = e^{\delta|t|}$ for $|t| \geq 1$. For some $\delta > 0$ and any $SO(3)$ anti-self-dual connection A on the bundle over $Y \times \mathbb{R}$, the anti-self-duality operator

$$d_A^* \oplus d_A^+ : L_{k+1,\delta}^p(\Omega^1(Y \times \mathbb{R}, adP)) \to L_{k,\delta}^p((\Omega^0 \oplus \Omega_+^2)(Y \times \mathbb{R}, adP))$$

is Fredholm. The spectral flow is

$$SF(a_\alpha, a_\beta) \equiv Index(d_A^* \oplus d_A^+)(a_\alpha, a_\beta) \pmod{2p_1(A)},$$

where $p_1(A)$ is the Pontryagin form. The $Index(d_A^* \oplus d_A^+)(a_\alpha, a_\beta)$ is given by

$$Index(d_A^* \oplus d_A^+)(a_\alpha, a_\beta) = -2\int_{Y \times \mathbb{R}} p_1(A) - \frac{h_\beta - \rho_\beta(0)}{2} - \frac{h_\alpha + \rho_\alpha(0)}{2}, \tag{4.1}$$

where h_α is the sum of the dimensions of $H^i(Y, ada_\alpha)(i = 0, 1)$, and $\rho_\alpha(0)$ is the ρ-invariant of the signature operator $*d_{a_\alpha} - d_{a_\alpha}*$ over Y restricted to even forms (see [**15, 17**]). We use the trivial connection and a fixed tangent vector to the trivial connection to fix $\mu(\theta) = 0$. Floer's chain group $C_j(Y)$ is defined to be the free module generated by irreducible flat connections $a_\alpha \in f^{-1}(0)$ with $\mu(a_\alpha) \equiv Index(d_A^* \oplus d_A^+)(a_\alpha, \theta) \equiv j \pmod 8$. Floer's crucial observation is that trajectories of the vector field f, i.e., the flow lines of

$$\frac{\partial a}{\partial t} + f(a(t)) = 0, \quad \text{or} \quad \frac{\partial a}{\partial t} = \star F(a(t)),$$

can be identified with instantons A on $Y \times \mathbb{R}$ and $A|_{Y \times \{t\}} = a(t)$. Let $\mathcal{M}(a_\alpha, a_\beta)$ be the space of (perturbed) anti-self-dual connections A with asymptotic flat connections a_α and a_β. Let $\mathcal{M}^1(a_\alpha, a_\beta)$ be the union of all the 1-dimensional components in $\mathcal{M}(a_\alpha, a_\beta)$. Then $\mathcal{M}^1(a_\alpha, a_\beta)$ has a canonical orientation together with a proper, free $\mathbb{R}$-action induced by the translations on $Y \times \mathbb{R}$. The differential $\partial_Y : C_j(Y) \to C_{j-1}(Y)$ of the Floer chain complex is defined by

$$\partial_Y a_\alpha = \sum_{a_\beta \in C_{j-1}(Y)} \#\hat{\mathcal{M}}(a_\alpha, a_\beta) a_\beta,$$

where $\#\hat{\mathcal{M}}(a_\alpha, a_\beta)$ is the algebraic number of points, and the sign is given by the spectral flow. Floer has shown that $\partial^2 = 0$, hence the homology of this complex $\{C_j(Y), \partial_Y\}_{j \in \mathbb{Z}_8}$ is the (instanton) Floer homology $HF_*(Y)(* \in \mathbb{Z}_8)$. It is shown that $HF_*(Y)$ is independent of the choice of metric σ on Y and of regular perturbations by the Theorem 2 in [**17**].

For $\mu(a) \equiv j \pmod 8), b = \theta$ and r a regular value of the Chern-Simons functional, the spectral flow $SF(a^{(r)})$ is independent of the choice of trivial connection θ used to define Chern-Simons functional on $\mathcal{A}_\delta(a, \theta)$. Define $C_n^{(r)}(Y) = \mathbb{Z}\{a \in \mathcal{R}_Y^* \mid SF(a^{(r)}) = n \in \mathbb{Z}\}$ and a finite-length decreasing filtration $F_n^{(r)} C_j(Y) = \sum_{k \geq 0} C_{n+8k}^{(r)}(Y)$ of $C_j(Y)(j \in \mathbb{Z}_8)$ in [**15**]. Fintushel and Stern [**15**] obtained a spectral sequence which converges to the $\mathbb{Z}_8$-graded instanton Floer homology:

THEOREM 4.1. *(i) There is a spectral sequence $(E_{n,j}^k, d^k)$ (periodic in j with period 8) with*

$$E_{n,j}^1(Y) \cong I_n^{(r)}(Y), \qquad d^k : E_{n,j}^k(Y) \to E_{n+8k-1,j-1}^k(Y),$$

$$E_{n,j}^\infty(Y) = F_n^{(r)} I_j(Y) / F_{n+8}^{(r)} I_j(Y), \qquad (j \in \mathbb{Z}_8, n \in \mathbb{Z}, n \equiv j \pmod 8)).$$

Furthermore, the groups $E_{n,j}^k(Y)$ are topological invariants.

(ii) Any period of the spectral sequence $(E_{n,j}^k(Y), d^k)$ converges to $E_{*,j}^\infty(Y)(j \in \mathbb{Z}_8)$ which is isomorphic to the bigraded module associated to the filtration $F_n^{(r)}$ of $I_j(Y)$ $(j \in \mathbb{Z}_8)$:

$$HF_j(Y) \cong \bigoplus_{k \in \mathbb{Z}} E_{n+8k,j}^\infty(Y).$$

In order to obtain the cap-product structure, we have to overcome the main difficulty arising from the *non-compactness* of the moduli space $\mathcal{M}_{Y \times \mathbb{R}}(a, b)$ with flat connections a and b over Y. For a cohomology class $\omega \in H^*(\mathcal{B}_Y, \mathbb{Q})$ with space into codimension one (or ≥ 1) pieces. The $\mathbb{R}$-action is not serious; We can work on the space $\mathcal{M}_{Y\times\mathbb{R}}(a,b)/\mathbb{R} = \hat{\mathcal{M}}_{Y\times\mathbb{R}}(a,b)$. The non-compactness of the bubbling type is problematic. We use the $\mathbb{Z}$-graded instanton Floer homology defined by Fintushel-Stern [**15**] to fix Chern-Simons values in one of the energy bands $(r, r+1)$, where r is a regular value of the Chern-Simons functional. Thus any bubbling requires energy at least 1 forces the corresponding moduli space to have different lifting for the generators a and b. So the moduli space with fixed lifting $a^{(r)}$ and $b^{(r)}$ contains no bubbling limit points. The evaluation of a cohomology class in $H^*(\mathcal{B}_{Y\times\mathbb{R}};\mathbb{Q})$ on the compactification does not make sense. This is why we restrict ourself to the rational cohomology group $H^*(\mathcal{B}_Y,\mathbb{Q})$. The rational cohomology group $H^*(\mathcal{B}_Y,\mathbb{Q})$ is generated by the μ-map construction. Thus we can represent each cohomology class by its Poincaré dual as a divisor from the homology cycle of the 3-manifold Y itself. Thus adapting the 4-dimensional techniques (dimension counting replaced by spectral flow counting), we can show that the set $P.D(\omega)\cap \hat{\mathcal{M}}_{Y\times\mathbb{R}}(a,b)$ is compact by ruling out the three bad parts.

Theorem [Theorem 4.3 of [**29**]] *There is a well-defined cap-product action of* $H^*(\mathcal{B}_Y;\mathbb{Q})$ *on the Fintushel-Stern spectral sequence* $(E^k_{n,j}(Y), d^k)$ *given in Theorem 4.1.*

In particular, Theorem 4.3 of [**29**] shows that there is a well-defined cap-product (cup-product) structure on the instanton Floer homology (cohomology). This answers one of Atiyah's questions in [**2**] as mentioned in §1.

4.2. The instanton Floer homology with Novikov ring coefficients. For integral homology 3-spheres Y, the Chern-Simons functional $cs : \mathcal{B}_Y \to \mathbb{R}/\mathbb{Z}$ has a lift (still denoted by cs) $cs : \tilde{\mathcal{B}}_Y \to \mathbb{R}$ from the pullback of the universal cover of $\mathbb{R}/\mathbb{Z}$. Note that the covering transformation group of $\tilde{\mathcal{B}}_Y \to \mathcal{B}_Y$ is $\mathcal{G}/\mathcal{G}_0 = SU(2)$. Note that there is a monotonicity between the Chern-Simons functional and the spectral flow (see [**15, 17**]). Thus $\ker cs = \ker SF \cong \mathcal{G}_0$ the subgroup of the gauge group with degree zero. Hence the Novikov ring $\Lambda_{cs} = \Lambda(\mathcal{G}/\mathcal{G}_0; cs;\mathbb{Z}) = \Lambda(SU(2); cs;\mathbb{Z})$ is an Abelian group of all functions $\lambda : \mathcal{G}/\mathcal{G}_0 \to \mathbb{Z}$ such that $\{g \in SU(2) : \lambda_g \neq 0, deg(g) \leq k\}$ is finite for every integer $k\in\mathbb{Z}$. We denote $(a,g)\in\tilde{\mathcal{B}}_Y$ for the lift $g\cdot a$ of $a\in\mathcal{B}_Y$ corresponding to the transformation element $g\in\mathcal{G}/\mathcal{G}_0$. Define $C_j(Y;\Lambda_{cs})$ to be the collection of formal sums $\sum \lambda_{(a_\alpha,g_\alpha)}\cdot(a_\alpha,g_\alpha)$ for $g_\alpha\cdot a_\alpha \in \tilde{\mathcal{R}}^*(Y)\subset\tilde{\mathcal{B}}_Y$ and $\lambda_{(a_\alpha,g_\alpha)}\in\mathbb{Z}$ such that $\{(a_\alpha,g_\alpha)\in\tilde{\mathcal{B}}_Y : \lambda_{(a_\alpha,g_\alpha)}\neq 0, cs(g_\alpha\cdot a_\alpha)\leq c\}$ is finite for all $c\in\mathbb{R}$ with $SF(a_\alpha,g_\alpha) = SF(g_\alpha\cdot a_\alpha,\theta) = j\in\mathbb{Z}$. Note that there is one representative for each element $a_\alpha\in\mathcal{R}^*(Y)$.

Proposition 4.2. *(i) The chain group* $C_j(Y;\Lambda_{cs})$ *is a torsion-free module over the Novikov ring* Λ_{cs} *and* $\sum_{j\in\mathbb{Z}} rank_{\Lambda_{cs}} C_j(Y;\Lambda_{cs})$ *is the cardinality of the set* $\mathcal{R}^*(Y)$.

(ii) $\partial_{Y,j}(a_\alpha,g_\alpha) = \sum_{(b_\beta,g_\beta)\in C_{j-1}(Y;\Lambda_{cs})} \#\tilde{\mathcal{M}}(g_\alpha\cdot a_\alpha, g_\beta\cdot a_\beta)\cdot(b_\beta,g_\beta)$ *gives a well-defined* Λ_{cs}*-homomophism* $\partial_{Y,j} : C_j(Y;\Lambda_{cs}) \to C_{j-1}(Y;\Lambda_{cs})$. *We have* $\partial_{Y,j}\circ\partial_{Y,j-1} = 0$.

Proof: (i) follows from the definition. (ii) The moduli space $\tilde{\mathcal{M}}(g_\alpha\cdot a_\alpha, g_\beta\cdot a_\beta)$ used in the definition of $\partial_{Y,j}$ is the time-translation invariant quotient of the trajectory flow lines in $\tilde{\mathcal{B}}_Y$ with asymptotics $g_\alpha\cdot a_\alpha$ and $g_\beta\cdot a_\beta$. For fixed $cs(g_\alpha\cdot a_\alpha)$, the Chern-Simons functional value of $g_\beta\cdot a_\beta$ is bounded whenever $SF(g_\alpha\cdot a_\alpha) - SF(g_\beta\cdot a_\beta)$ is fixed by the monotonicity in [**15**]. If $SF(g_\alpha\cdot a_\alpha) - SF(g_\beta\cdot a_\beta) = 1$, then the moduli space $\tilde{\mathcal{M}}(g_\alpha \cdot a_\alpha, g_\beta \cdot a_\beta)$ consists of finitely many points so that only finite many $g_\beta \cdot a_\beta$ occur in $\partial_{Y,j}(a_\alpha, g_\alpha)$ (see [**4**]). The regularity and compactness in [**4, 15, 17**] show that $\partial_{Y,j} \circ \partial_{Y,j-1} = 0$ by the same argument in [**4, 15, 17**]. □

THEOREM 4.3. *There is a well-defined homology with Novikov ring coefficients, $HF_j(Y; \Lambda_{cs}) = \ker \partial_{Y,j} / Im\, \partial_{Y,j+1}$, as a topological invariant of the integral homology 3-sphere Y, and*

$$\sum_{j\in\mathbb{Z}}(-1)^j rank_{\Lambda_{cs}} HF_j(Y;\Lambda_{cs}) = \sum_{i\in\mathbb{Z}_8}(-1)^i rank_{\mathbb{Z}} HF_i(Y).$$

Proof: The independence of perturbations and Riemannian metrics follow from the same argument in [**4, 15, 17**]. The Euler characteristic of the instanton Floer homology over the Novikov ring Λ_{cs} is the same as the $\mathbb{Z}_8$-graded instanton Floer homology over $\mathbb{Z}$ (twice the negative Casson invariant of Y). □

It would be interesting to find a relation between the $\mathbb{Z}$-graded instanton homology $I_*^{(r)}(Y)$ defined by Fintushel and Stern in [**15**] (see also Theorem 4.1) and the $\mathbb{Z}$-graded instanton homology $HF_*(Y; \Lambda_{cs})$ over the Novikov ring Λ_{cs}.

4.3. The monopole homology over integral homology 3-spheres. In 1994, Seiberg and Witten (see [**51**]) introduced a new (simpler) kind of differential-geometric equation. In a very short time after the equation was introduced, some long-standing problems were solved, new and unexpected results were discovered. For instance, Kronheimer and Mrowka [**27**] proved the Thom conjecture affirmatively, and Auckley proved the three-dimensional version of the Thom conjecture [**3**]. Taubes [**49**] showed that there are more constraints on symplectic structures and the beautiful equality $SW = Gr$. See [**11**] for a survey of the Seiberg-Witten theory.

Using the dimension-reduction principle, one expects Floer-type homology of 3-manifolds via the Seiberg-Witten equation (see [**40**] for a symplectic approach). Indeed Kronheimer and Mrowka [**27**] analyzed the Seiberg-Witten-Floer theory for $\Sigma \times S^1$, where Σ is a closed oriented surface. Marcolli studied the Seiberg-Witten-Floer homology for 3-manifolds with first Betti number positive in [**35**]. For a connected compact oriented 3-manifold with positive first Betti number and zero Euler characteristic, Meng and Taubes [**36**] showed that a (average) version of Seiberg-Witten invariant is the same as the Milnor torsion. The interesting class of 3-manifolds consisting of integral (rational) homology 3-spheres lacks a well-posed analytic/topological theory. Although various authors attempted to resolve the problem of defining a "Seiberg-Witten-Floer" theory, the new phenomenon of harmonic-spinor jumps and the dependence of Riemannian metrics is not addressed clearly. The metric-dependence (also related to the harmonic-spinors) issue is quickly realized by many experts in this field (see [**11, 39**]). In [**39**], the irreducible Seiberg-Witten-Floer homology of the Seifert spaces is shown to be dependent on the metric and the choice of a connection on the tangent bundle (as a reference η_0 in [**31**]).

In [**31**], the present author constructed a monopole homology from the Seiberg-Witten equation in the same way as an instanton Floer homology from the Self-Duality equation in Donaldson-Floer theory [**17, 10**]. The key point is that by

The unique $U(1)$-reducible Θ gives a spectral flow $I_\eta(\Theta;\eta_0)$ as a Maslov index in [**6**] Part III. The spectral flow $I_\eta(\Theta;\eta_0) = \mu_\eta(\Theta) - \mu_{\eta_0}(\Theta)$ with respect to a reference η_0 fixes a class of admissible perturbations consisting of Riemannian metrics and 1-forms. The spectral flow $I_\eta(\Theta;\eta_0)$ is not a topological invariant, and is dependent upon the Riemannian metric. Without fixing a class of Riemannian metrics with the same $I_\eta(\Theta;\eta_0)$, one cannot obtain well-defined notions such as spectral flow of irreducible Seiberg-Witten solutions on Y, and the gluing formula as well as the relative Seiberg-Witten invariant. We emphasized the Riemann-metric dependence and understand the role of such a fixing spectral flow of $(\Theta;\eta_0)$ in [**31**]. Transversality, decay estimates, gluing and splitting result are also proved in [**31**] as well.

THEOREM 4.4. *(Theorem A of* [**31**]*)*
(1) For an integral homology 3-sphere Y and any admissible perturbation η, there is a well-defined $\mathbb{Z}$-graded monopole homology $MH_*(Y, I_\eta(\Theta;\eta_0))$ constructed from the Seiberg-Witten equation over $Y \times \mathbb{R}$.
(2) For any two admissible perturbations η_1 and η_2, there is a group homomorphism Ψ_* between $MH_*(Y, I_{\eta_1}(\Theta;\eta_0))$ and $MH_*(Y, I_{\eta_2}(\Theta;\eta_0))$.
(3) If $I_{\eta_1}(\Theta;\eta_0) = I_{\eta_2}(\Theta;\eta_0)$, then the homomorphism Ψ_* is an isomorphism.

Our fixed-class $I_\eta(\Theta;\eta_0)$ of Riemannian metrics gains control of the birth and death of irreducible solutions of the Seiberg-Witten equation on the integral homology 3-sphere Y. Changing the reference η_0 into η_0' corresponds to an overall degree-shifting by $\mu_{\eta_0'}(\Theta) - \mu_{\eta_0}(\Theta)$ for the monopole homologies. This provides a chamber-like structure for the monopole homology of integral homology 3-spheres. The assigned function

$$MH_{\mathrm{SWF}} : \{I_\eta(\Theta;\eta_0)\} \to \{MH_*(Y, I_\eta(\Theta;\eta_0))\}$$

is a topological invariant (as Seiberg-Witten-Floer Theory).

For an integral homology 3-sphere Y, the cohomology classes in the group $H^*(\mathcal{B}_{Y\times\mathbb{R}}(c_1,c_2);\mathbb{Z})$ are independent of $c_1, c_2 \in \mathcal{R}^*_{SW}(Y)$. We denoted it by the notation $H^*(\mathcal{B}_{Y\times\mathbb{R}};\mathbb{Z})$. Let $a \in H^p(\mathcal{B}_{Y\times\mathbb{R}};\mathbb{Z})$ and define the map from the monopole n-th chain group $MC_n(Y,\eta)$ to $MC_{n-p}(Y,\eta)$ given by

$$(a\cup)(c_1) = \sum_{c_2 \in \mathcal{R}^{n-p}_{SW}(Y,\eta)} \#(\mathcal{M}_{Y\times\mathbb{R}}(c_1,c_2) \cap PD(a)) \cdot c_2,$$

where the $\#(\mathcal{M}_{Y\times\mathbb{R}}(c_1,c_2) \cap PD(a))$ is the algebraic number of PD(a) intersecting with the Seiberg-Witten monopole moduli space $\mathcal{M}_{Y\times\mathbb{R}}(c_1,c_2)$ in $\mathcal{B}_{Y\times\mathbb{R}}(c_1,c_2)$, and c_i is the (deformed) Seiberg-Witten solution on Y. Note that the action is well-defined since the moduli space $\mathcal{M}_{Y\times\mathbb{R}}(c_1,c_2)$ is compact up to the transitivity.

PROPOSITION 4.5. *The compactification of $\mathcal{M}_{Y\times\mathbb{R}}(c_0,c_{n+1})$ can be described as*

$$\overline{\mathcal{M}_{Y\times\mathbb{R}}(c_0,c_{n+1})} = \cup(\times_{i=1}^{n+1}\mathcal{M}_{Y\times\mathbb{R}}(c_{i-1},c_i)),$$

the union over all sequence $c_0, c_1, \cdots, c_{n+1} \in \mathcal{R}^*_{SW}(Y,\eta)$ such that each moduli space $\mathcal{M}_{Y\times\mathbb{R}}(c_{i-1},c_i)$ is nonempty for all $1 \le i \le n+1$.

For any sequence $c_0, c_1, \cdots, c_{n+1} \in \mathcal{R}^*_{SW}(Y,\eta)$, there is a gluing map

$$G : \times_{i=1}^{n+1}\mathcal{M}^{bal}_{Y\times\mathbb{R}}(c_{i-1},c_i) \times \Delta^{n+1} \to \overline{\mathcal{M}_{Y\times\mathbb{R}}(c_0,c_{n+1})},$$

where $\Delta^{n+1} = \{(\lambda_0, \cdots, \lambda_n) \in [-\infty,\infty]^{n+1} : 1 + \lambda_{i-1} < \lambda_i, 1 \le i \le n\}$.

(1) *The image of* G *is a neighborhood of* $\times_{i=1}^{n+1}\mathcal{M}^{bal}_{Y\times\mathbb{R}}(c_{i-1},c_i)$ *in the compactification with chain solutions.*
(2) *The restriction of* G *to* $\times_{i=1}^{n+1}\mathcal{M}^{bal}_{Y\times\mathbb{R}}(c_{i-1},c_i)\times Int(\Delta^{n+1})$ *is an orientation preserving diffeomorphism onto its image.*

Proposition 4.5 (as Proposition 6.9 in [**31**]) shows that the moduli space used in the cohomology action shares the same property as the trajectory flow lines of finite-dimensional Morse functions. Thus one can obtain a $H^*(\mathcal{B}_{Y\times\mathbb{R}};\mathbb{Z})$-module structure on the monopole homology of the integral homology 3-sphere Y.

THEOREM 4.6. *For fixed* $I_\eta(\Theta;\eta_0)$,
(1) $a\cup$ *commutes with the boundary maps in the monopole homology.*
(2) $(a\cup)\circ(b\cup)=(a\cup b)\cup$ *on the monopole homology. Therefore* $MH_*(Y;I_\eta(\Theta))$ *admits an* $H^*(\mathcal{B}_{Y\times\mathbb{R}};\mathbb{Z})$*-module structure.*

Proof: (1) follows from the generalized Stokes theorem and Proposition 4.5 (see [**29, 30**]). The composition $(a\cup)\circ(b\cup)(c_1)$ can be evaluated by the following

$$\langle a\times b, \cup_{c_2,c_3}\mathcal{M}_{Y\times\mathbb{R}}(c_1,c_2)\times\mathcal{M}_{Y\times\mathbb{R}}(c_2,c_3)\rangle.$$

On the other hand, we have

$$(a\cup b)(c_1)=\langle\Delta^*(a\times b),\mathcal{M}_{Y\times\mathbb{R}}(c_1,c_3)\rangle=\langle a\times b,\Delta_*\mathcal{M}_{Y\times\mathbb{R}}(c_1,c_3)\rangle,$$

where Δ_* is the unique chain map up to chain homotopy induced from the diagonal map $\Delta:\mathcal{B}_{Y\times\mathbb{R}}\to\mathcal{B}_{Y\times\mathbb{R}}\times\mathcal{B}_{Y\times\mathbb{R}}$. For any c_2 with $\mu(c_2)\neq\mu(c_1)-\deg(a)$, dimension counting gives

$$\langle a\times b,\mathcal{M}_{Y\times\mathbb{R}}(c_1,c_2)\times\mathcal{M}_{Y\times\mathbb{R}}(c_2,c_3)\rangle=0.$$

Hence the term $\langle a\times b,\Delta_*\mathcal{M}_{Y\times\mathbb{R}}(c_1,c_3)\rangle$ equals to

$$\langle a\times b,\cup_{\mu(c_2)=\mu(c_1)-\deg(a)}\mathcal{M}_{Y\times\mathbb{R}}(c_1,c_2)\times\mathcal{M}_{Y\times\mathbb{R}}(c_2,c_3)\rangle,$$

which is the composition $(a\cup)\circ(b\cup)(c_1)$. □

Hence the Seiberg-Witten-Floer theory $\{I_\eta(\Theta;\eta_0)\}\to\{MH_*(Y;I_\eta(\Theta;\eta_0))\}$ also admits the $H^*(\mathcal{B}_{Y\times\mathbb{R}};\mathbb{Z})$-module structure.

References

[1] V. I. Arnold, Mathematical methods of Classical Mechanics, GTM60, Springer-Verlag, 1978.
[2] M. Atiyah, *New invariants of 3 and 4 dimensional manifolds, in 'The mathematical Heritage of Hermann Weyl,'* Proceedings Symposia Pure Math **48**, 285-299 (1988).
[3] D. Auckly, *The Thurston Norm and three-dimensional Seiberg-Witten Theory*, Osaka J. Math., **33** (1996), 737-750.
[4] D. Austin and P. Braam, *Equivariant Floer theory and gluing Donaldson polynomials*, Topology, **35**, No. 1(1996), 167-200.
[5] R. Bott, Collected Works, Volume 1, 4, edited by R. MacPherson, Birkäuser, 1995.
[6] S. Cappell, R. Lee and E. Miller, *Self-adjoint elliptic operators and manifold decompositions*, Comm. Pure Appl. Math. I, 49(1996), no.8, 825-866; II, 49 (1996), no.9, 869-909; III, 52(1999), no.5, 543-611.
[7] R. Cohen, J, Jones and G. Segal, *Floer's infinite dimensional Morse theory and homotopy theory*, The Floer Memorial Volume PM **133** (1995), 297-325.
[8] C. Conley, *Isolated invariant sets and the Morse index*, CBMS Regional Conference Series in Mathematics, 38. AMS, Providence, R.I., 1978.
[9] C. Conley and E. Zehnder, *The Birkhoff-Lewis fixed point theorem and a conjecture by V. I.*

[11] S. Donaldson, *The Seiberg-Witten Equations and 4-Manifolds Topology*, Bull, AMS, Vol **33** (1996), No. 1, 45-70.
[12] B. Feign, *Semi-infinite cohomology of Kac-Moody and Virasoro algebras*, Russ. Math. Surv **39**, No. 2, 155-156(1984).
[13] M. Farber, *Sharpness of the Novikov inequalitiesr* (Russian), Funktsional. Anal. i Prilozhen. 19 (1985), no. 1, 49-59, 96.
[14] M. Farber, *Topology of closed 1-forms and their critical points*, Topology 40 (2001), no. 2, 235-258.
[15] R. Fintushel and R. Stern, *Integer graded instanton homology groups for homology three spheres*, Topology, **31**: 589-604, 1992.
[16] A. Floer, *Witten's complex and infinite dimensional Morse theory*, J. Diff. Geom., **30** (1989), 207-221.
[17] A. Floer, *An instanton invariant of 3-manifolds*, Comm. Math. Phys. **118** (1988), 215 - 240.
[18] A. Floer, *Symplectic fixed points and holomorphic spheres*, Comm. Math. Phys. **120** (1989), 575-611.
[19] A. Floer and H. Hofer, *Coherent orientations for periodic orbit problems in symplectic geometry*, Math. Z., **212** (1993), 13-38.
[20] I. Frenkel, H. Garland and G. Zuckerman, *Semi-infinite cohomology and string theory*, Proc. Natl. Acad. Sci. USA **83**, 8442-8446(1986).
[21] K. Fukaya and K. Ono, *Arnold conjecture and Gromov-Witten invariant*, Topology 38 (1999), no. 5, 933-1048.
[22] K. Fukaya, Y. Oh, Ohta and K. Ono, *Lagrangin intersection Floer theory - anomaly and obstructon*, Kyoto preprint 2000.
[23] M. Gromov, *Pseudoholomorphic curves in symplectic manifolds*, Invent. Math. **82**, 307-347, (1985).
[24] H. Hofer and D. Salamon, *Floer homology and Novikov ring* , The Floer Memorial Volume, Progress in Math. Vol. **133** (1995), 483-524.
[25] D. Hurtubise, *The flow category of the action functional on $\mathcal{L}G_{N,N+K}(C)$*, Illinois J. Math., Vol. **44**, No. 1 (2000), 33-50.
[26] M. Kontsevich, *Homological algebra of Mirror symmetry*, Proc. ICM 1994 (Zürich, Switzerland), Vol 1, 120-139.
[27] P. Kronheimer and T. Mrowka, *The genus of embedded surfaces in the projective plane*, Math. Res. Letters **1** (1994), 797-808.
[28] Lê Hông Vân and K. Ono, *Symplectic fixed points, the Calabi invariant and Novikov homology*, Topology, Vol. **34**, No. 1, 155-176(1995).
[29] W. Li, *Cap-product structures on the Fintushel-Stern spectral sequence*, Math. Proc. Cambridge Philosophical Soc. Part 2, Volume **131** (2001), 265-278.
[30] W. Li, *A Module structure on the symplectic Floer cohomology*, Communications in Mathematical Physics, Vol **211**, No. 1 (2000), 137-151.
[31] W. Li, *A Monopole Homology of Integral Homology 3-spheres*, to appear in Turkish J. Math.
[32] W. Li, *Künneth formulae and cross products for the symplectic Floer cohomology*, Topology and its Application, Volume 110, Issue 3, 30 (2001), 211-236.
[33] W. Li, *The semi-infinite cohomology and the Floer cohomology*, OSU preprint (submitted).
[34] G. Liu and G. Tian, *Floer homology and Arnold conjecture*, J. Diff. Geom., **49** (1998), 1-74.
[35] M. Marcolli, *Seiberg-Witten-Floer homology and Heegaard splitting*, Intern. Jour. of Maths., Vol **7**, No. 5 (1996) 671-696.
[36] G. Meng and C. Taubes, *SW = Milnor torsion*, Math. Res. Lett **3** (1996), 661-674.
[37] J. Milnor, Morse Theory, Princeton University Press, 1963.
[38] M. Morse, *Relations between the critical points of a real function of n independent variables*, Trans. AMS, **27** (1925), 345-396.
[39] T. Mrowka, P. Ozsvath and B. Yu, *Seiberg-Witten monopoles on Seifert fibered spaces*, Comm. Annl. Geom., **5** (4) (1997), 685-791.
[40] P. Ozsvath and Z. Szabo, *Holomorphic disks and three-manifold invariants: properties and applications*, math.SG/0105202.
[41] A. Pazhitnov, *Morse theory of closed 1-forms*, Algebraic topology Poznań 1989, 98-110, Lecture Notes in Math., 1474, Springer, Berlin, 199
[42] E. Pitcher, *Critical points of a map to a circle*, Proc. Nat. Acad. Sci. U. S. A. 25, (1939). 428-431.

[43] A. Pressley and G. Segal, Loop groups. Oxford Math.Mono, Clarendon Press, Oxford 1986.

[44] Y. Ruan, *Virtual neighborhoods and the monopole equations*, Topics in symplectic 4-manifolds (Irvine, CA, 1996), 101-116, First Int. Press Lect. Ser., I, Internat. Press, Cambridge, MA, 1998.

[45] D. Salamon, *Lectures on Floer homology* Symplectic geometry and topology (Park City, UT, 1997), 143-229, IAS/Park City Math. Ser., 7, AMS, Providence, RI, 1999.

[46] B. Siebert, *Symplectic Gromov-Witten invariants*, London Math. Soc. Lecture Note Ser., 264, (1999), 375-424.

[47] S. Novikov, *Multivalued functions and functionals, An analogue of Morse theory*, Soviet Math. Dokl. **24** (1981), 31-35.

[48] C. Taubes, *Casson's invariant and gauge theory*, J. Diff. Geom. **31** (1990), 547-599.

[49] C. Taubes, Seiberg-Witten and Gromov Invariants for symplectic 4-manifolds, First Int. Press Lecture Series, Vol. **2**, 2000.

[50] A. Voronov, *Semi-infinite homological algebra*, Invent. Math., **113**, 103-146(1993).

[51] E. Witten, *Monopoles and 4-manifolds*, Math. Res. Letters **1** (1994), 769-796.

Department of Mathematics, Oklahoma State University
Stillwater, Oklahoma 74078-0613
E-mail address: wli@math.okstate.edu

Contemporary Mathematics
Volume **322**, 2003

Automorphism Sheaves, Spectral Covers, and the Kostant and Steinberg Sections

Robert Friedman and John W. Morgan

Introduction

Throughout this paper, G denotes a simple and simply connected linear algebraic group over $\mathbb{C}$ of rank r and H is a Cartan subgroup, with Lie algebras $\mathfrak{g} = \operatorname{Lie} G$ and $\mathfrak{h} = \operatorname{Lie} H$. Let R be the root system of the pair (G, H), W the Weyl group, and $\Lambda \subseteq \mathfrak{h}$ the coroot lattice. Fix once and for all a positive Weyl chamber, i.e. a set of simple roots Δ. The geometric invariant theory quotient of $\mathfrak{g}$ by the adjoint action of G is identified with $\mathfrak{h}/W$, which by a theorem of Chevalley is isomorphic as a variety to affine r-space $\mathbb{A}^r$. Let $\mathfrak{g}_{\mathrm{reg}} \subseteq \mathfrak{g}$ be the Zariski open and dense subset of $\mathfrak{g}$ consisting of regular elements, i.e. of elements X whose centralizer (in $\mathfrak{g}$ or G) has the minimal dimension r. If X is regular, then the centralizers of X in $\mathfrak{g}$ and in G, i.e. the stabilizer of X in G under the adjoint representation, are abelian. One can show that each fiber of the adjoint quotient morphism $\mathfrak{g} \to \mathfrak{h}/W$ contains a unique orbit of regular elements, so that the restriction of the adjoint quotient morphism to $\mathfrak{g}_{\mathrm{reg}}$ induces an isomorphism $\mathfrak{g}_{\mathrm{reg}}/G \to \mathfrak{h}/W$. In 1963, Kostant [**14**] proved that the morphism $\mathfrak{g}_{\mathrm{reg}} \to \mathfrak{h}/W$ is smooth and constructed a section $\sigma\colon \mathfrak{h}/W \to \mathfrak{g}_{\mathrm{reg}}$ of the adjoint quotient morphism, generalizing the existence of rational canonical forms of a matrix. Actually, the construction yields a family of such sections, defined as follows. Let $X \in \mathfrak{g}_{\mathrm{reg}}$ be a principal nilpotent matrix and let $L \subseteq \mathfrak{g}$ be a linear complement to $\operatorname{Im} \operatorname{ad} X$. Then the affine subvariety $L + X$ is contained in $\mathfrak{g}_{\mathrm{reg}}$ and maps isomorphically to $\mathfrak{h}/W$, and so defines a section of the adjoint quotient morphism.

One can make an analogous construction for the adjoint quotient of G, i.e. the geometric invariant theory quotient of the action of G on itself by conjugation. In this case, the adjoint quotient is equal to H/W, which is again isomorphic as a variety to $\mathbb{A}^r$ [**2**]. Let G_{reg} be the Zariski open and dense subset of regular elements, where by definition $g \in G$ is regular if and only if the dimension of its

2000 *Mathematics Subject Classification.* Primary: 14F05, 17B10; Secondary: 14H52, 20G05.

Key words and phrases. Automorphism sheaf, spectral cover, Weierstrass cubic, Kostant and Steinberg sections, minuscule and quasi-minuscule weights.

centralizer is r, the minimal possible dimension. If g is regular, then its centralizer is abelian. Then again each fiber of the morphism $G \to H/W$ contains a unique conjugacy class of regular elements, so that G_{reg}/G is isomorphic to H/W. In 1965, using methods very different from those of Kostant, Steinberg [**15**] proved that the morphism $G_{\mathrm{reg}} \to H/W$ is smooth and constructed a section $\sigma\colon H/W \to G_{\mathrm{reg}}$ of the adjoint quotient morphism. Although the constructions are quite different, they are related as follows. Let $\sigma\colon H/W \to G_{\mathrm{reg}}$ be any section of the adjoint quotient morphism, and let $X \in \mathfrak{g}_{\mathrm{reg}}$ be the unique principal nilpotent matrix such that $\exp X = \sigma(e)$, where we denote by e the image of the identity in H/W. Then $\operatorname{Im} d\sigma_e$ is a linear subspace L of $\mathfrak{g}$ which is a complement to $\operatorname{Im}\operatorname{ad} X$, and so $L+X$ defines a Kostant section of the adjoint quotient morphism for $\mathfrak{g}$.

In [**9**], the authors gave another method for constructing sections of the adjoint quotient morphism, either for $\mathfrak{g}$ or for G, as well as generalizations to other contexts, which will be explained below. It is natural, then, to ask if these sections are essentially the same as those defined by Kostant and Steinberg, and more generally whether the sections of Kostant and Steinberg are in some sense unique. It follows from the work of Kostant and Steinberg that any such section must take values in the open set of regular elements. Clearly, if $\sigma\colon \mathfrak{h}/W \to \mathfrak{g}_{\mathrm{reg}}$ is such a section and $f\colon \mathfrak{h}/W \to G$ is any morphism, then $\operatorname{ad} f(\sigma)\colon \mathfrak{h}/W \to \mathfrak{g}_{\mathrm{reg}}$ is again a section, and similarly for sections of $G \to H/W$. We shall refer to two sections which are related in this way as *conjugate*. The main result of this paper is then the following:

THEOREM. *All sections of the adjoint quotient morphism $\mathfrak{g} \to \mathfrak{h}/W$ are conjugate. Similarly, all sections of the adjoint quotient morphism $G \to H/W$ are conjugate.*

These problems can be translated into a more geometric context. Let E be an irreducible plane cubic curve with a cusp singularity and a marked point $p_0 \in E_{\mathrm{reg}}$, where E_{reg} is the open subset of smooth points of E. Then there is a natural group structure on E_{reg} with origin p_0, and it is isomorphic to the additive group $\mathbb{G}_a$. Let $n\colon \widetilde{E} \cong \mathbb{P}^1 \to E$ be the normalization of E. Let (ξ, θ) be a pair consisting of a principal G-bundle on E and an isomorphism of principal bundles $\theta\colon n^*\xi \to \widetilde{E}\times G$. By [**9**, Corollary 3.1.2], such pairs are classified by elements $X \in \mathfrak{g}$, and hence principal G-bundles ξ such that $n^*\xi$ is trivial are classified by G-orbits in $\mathfrak{g}$. If the bundle ξ corresponds to the G-orbit containing X, then the group $\operatorname{Aut}\xi$ of automorphisms of the principal bundle ξ is isomorphic to the stabilizer in G of X under the adjoint representation. Thus, if ξ is a G-bundle over E such that $n^*\xi$ is trivial, we define ξ to be *regular* if $\operatorname{Aut}\xi$ has dimension r. More generally, let B be a scheme over $\mathbb{C}$ and let (Ξ, Θ) consist of a principal G-bundle over $B \times E$ and a bundle isomorphism $\Theta\colon (\operatorname{Id}\times n)^*\Xi \cong B \times \widetilde{E} \times G$. Then such pairs are classified by morphisms $\varphi\colon B \to \mathfrak{g}$. Given two such pairs (Ξ, Θ) and (Ξ', Θ') corresponding to morphisms φ and φ' from B to $\mathfrak{g}$, the bundles Ξ and Ξ' are isomorphic if and only if there exists a morphism $f\colon B \to G$ such that $\operatorname{ad} f(\varphi) = \varphi'$. Applying this construction in particular to a Kostant section $\sigma\colon \mathfrak{h}/W \to \mathfrak{g}_{\mathrm{reg}}$ defines a G-bundle $\Xi \to (\mathfrak{h}/W) \times E$ whose restriction to each slice $\{x\} \times E$ is regular. Clearly, $\mathfrak{h}/W$ is a coarse moduli space for the set of bundles which become trivial on $\widetilde{E}$, Ξ is a "universal" bundle in an appropriate sense, and our theorem becomes the statement that there is a unique such bundle up to isomorphism.

There is an analogous construction in the case of the group G. In this case, let E be an irreducible plane curve with a node and a marked point $p_0 \in E_{\mathrm{reg}}$. Then the natural group structure on E_{reg} with origin p_0 is isomorphic to $\mathbb{G}_m$. Pairs (ξ, θ) consisting of a principal G-bundle on E and an isomorphism of principal bundles $\theta\colon n^*\xi \to \widetilde{E} \times G$ are classified by an element $g \in G$, and more generally pairs (Ξ, Θ) consisting of a principal G-bundle over $B \times E$ and a bundle isomorphism $\Theta\colon (\mathrm{Id} \times n)^*\Xi \cong B \times \widetilde{E} \times G$ are classified by morphisms $\varphi\colon B \to G$. Regularity is defined as before. Using the Steinberg section now to define a G-bundle over $(H/W) \times E$, H/W is a coarse moduli space for the set of bundles which become trivial on $\widetilde{E}$, Ξ is a "universal" bundle in an appropriate sense, and our theorem is again the statement that there is a unique such bundle up to isomorphism.

These constructions can be further generalized. Aside from nodal and cuspidal cubics, one can also consider smooth cubic curves E together with an origin p_0, i.e. elliptic curves. The group law on E gives it the structure of a one-dimensional algebraic group, analogous to $\mathbb{G}_a$ and $\mathbb{G}_m$. By analogy with $\mathfrak{h}/W = (\mathbb{G}_a \otimes \Lambda)/W$ and $H/W = (\mathbb{G}_m \otimes \Lambda)/W$, one can also consider $(E \otimes \Lambda)/W = \mathcal{M}_G(E)$. By a theorem of Looijenga, $\mathcal{M}_G(E)$ is the coarse moduli space of semistable G-bundles over E, and has been extensively studied in [**7, 8**] via the so-called parabolic construction. The parabolic construction extends to the case of a nodal or cuspidal curve and defines a compactification of the adjoint quotient of G or $\mathfrak{g}$. Over the adjoint quotient itself, the parabolic construction defines a universal bundle, and this is the method used in [**9**] to define a section of the adjoint quotient morphism.

For a smooth elliptic curve E, one can try to use the methods of this paper to investigate different universal bundles Ξ over $\mathcal{M}_G(E) \times E$. However, there are some important differences. The bundle Ξ might only exist as a conformal bundle over $\mathcal{M}_G(E) \times E$, i.e. as a bundle over some quotient of G by a subgroup of the center. If G is not of type A_n or C_n, then $\mathcal{M}_G(E)$ is singular, and the bundle Ξ will not exist, even locally or in some conformal form of G, over the singular points. Finally, even when $\mathcal{M}_G(E)$ is smooth, the bundle Ξ will not be unique. However, the methods of this paper, in a relative setting, can be used in many cases to study families of regular semistable bundles over E or over an elliptic fibration.

The general method for proving our main theorem is as follows. We work entirely in the context of G-bundles. Let E be a Weierstrass cubic, i.e. a reduced irreducible plane cubic curve with a marked smooth point p_0. Define $\widetilde{\mathcal{M}} = E_{\text{reg}} \otimes \Lambda$ and $\mathcal{M} = \widetilde{\mathcal{M}}/W$. Given a universal bundle $\Xi \to \mathcal{M} \times E$, there is a corresponding automorphism sheaf $\mathcal{A}$ over $\mathcal{M}$, defined by $\mathcal{A}(U) = \mathrm{Aut}(\Xi|U \times E)$. One can show that, up to isomorphism, $\mathcal{A}$ is independent of the choice of Ξ. A standard argument shows that the set of all universal bundles is a principal homogeneous space over $H^1(\mathcal{M}; \mathcal{A})$. Thus we must show that $H^1(\mathcal{M}; \mathcal{A})$ is trivial. This is done by analyzing the sheaf $\mathcal{A}$ and relating it to various finite covers of $\mathcal{M}$ associated to Ξ as follows. Let $\rho\colon G \to GL(V)$ be a representation of G, which in practice will always be minuscule or quasi-minuscule (to be defined in Section 1). We can then form the associated vector bundle $\mathcal{V} = \Xi \times_G V$, which is a family of semistable vector bundles over $\mathcal{M} \times E$. The Fourier-Mukai correspondence functorially associates to $\mathcal{V}$ a finite cover T of $\mathcal{M}$, not necessarily reduced, and a sheaf of $\mathcal{O}_T$-modules $\mathcal{Q}$, finite and flat over $\mathcal{M}$, which determines $\mathcal{V}$. The scheme T is called the *spectral cover* of $\mathcal{M}$ associated to $\mathcal{V}$, and we are able to describe $\mathcal{A}$ in terms of the sheaf of invertible functions on T and various correspondences associated to T. Once we have this

Automorphism sheaves and spectral covers have been studied by many authors, including Donagi [**4**], Donagi-Gaitsgory [**5**], and Kanev [**12**]. Knop has shown (cf. [**13**]) that, if $\pi\colon \widetilde{\mathcal{M}} \to \mathcal{M}$ is the projection and $\underline{H}$ is the sheaf over $\widetilde{\mathcal{M}}$ associated to the constant group scheme $\widetilde{\mathcal{M}} \times H$, then $\mathcal{A} = (\pi_*\underline{H})^W = (\pi_*\mathbb{G}_{m,\widetilde{\mathcal{M}}} \otimes \Lambda)^W$. It is likely that this description could also lead to a proof of the main theorem.

Notation and conventions. All schemes in this paper will be of finite type over $\mathbb{C}$, all morphisms are morphisms of schemes, all sheaves will be in the étale topology and cohomology will always refer to étale cohomology. Minor modifications of our methods will prove the analogous results for the uniqueness of holomorphic sections of the adjoint quotient. In general, we will consider sheaves of abelian groups, usually written multiplicatively. The sheaf $\mathbb{G}_{m,T}$ refers to the sheaf (in the étale topology) of sections of the constant group scheme $T \times \mathbb{G}_m$. The corresponding sheaf in the Zariski topology will be denoted by $\mathcal{O}_T^*$. Given a coherent sheaf $\mathcal{F}$ over a scheme X, we will also use $\mathcal{F}$ to denote the associated sheaf in the étale topology. We will usually identify a vector bundle with its locally free sheaf of sections. Finally, E always denotes a Weierstrass cubic, necessarily either cuspidal, nodal, or smooth, with a marked smooth point p_0.

1. Preliminaries

1.1. Preliminaries on sheaves.

DEFINITION 1.1.1. Let X be a reduced irreducible scheme and let $\mathcal{S}$ be a sheaf of sets or abelian groups on $\mathcal{S}$.

(i) The sheaf $\mathcal{S}$ is *torsion free* if, for every étale open set $f\colon U \to X$, all sections $s, s' \in \Gamma(U, \mathcal{S})$ and every Zariski open dense subset U' of U, if $s|U' = s'|U'$, then $s = s'$. Equivalently, for every nonempty Zariski open subset V of X, if $j\colon V \to X$ is the inclusion, then the natural morphism $\mathcal{S} \to j_*j^*\mathcal{S}$ is an injection. Since $\mathcal{S}$ is a sheaf in the étale topology, for every étale open set $U \to X$, if $U' \to U$ is an étale open set with dense image and $s, s' \in \Gamma(U, \mathcal{S})$ are such that their restrictions to U' are equal, then $s = s'$. Note that every subsheaf of a torsion free sheaf is torsion free.

(ii) The sheaf $\mathcal{S}$ has the *Hartogs property* if for every étale open set $U \to X$ and every Zariski open subset $U' \subseteq U$ whose complement is of codimension at least 2, every section $s \in \Gamma(U', \mathcal{S})$ extends to a section $\hat{s} \in \Gamma(U, \mathcal{S})$. If in addition $\mathcal{S}$ is torsion free, then there is a unique such $\hat{s}$. Equivalently, $\mathcal{S}$ has the Hartogs property if for every Zariski open subset V of X such that $X - V$ has codimension at least two, if $j\colon V \to X$ is the inclusion, then $\mathcal{S} \to j_*j^*\mathcal{S}$ is surjective, and is an isomorphism if in addition $\mathcal{S}$ is torsion free.

(iii) If $\mathcal{S}$ is a torsion free sheaf of sets over X and $\mathcal{S}' \subseteq \mathcal{S}$ is a subsheaf, then $\mathcal{S}'$ is a *closed subsheaf* of $\mathcal{S}$ if for every étale open set $f\colon U \to X$, every section $s \in \Gamma(U, \mathcal{S})$ and every étale open dense subset U' of U, if $s|U'$ lies in $\Gamma(U', \mathcal{S}')$, then $s \in \Gamma(U, \mathcal{S}')$. A standard argument shows that it is enough to check this property for every Zariski open and dense subset U' of U. If $\mathcal{S}$ is a sheaf of abelian groups and $\mathcal{S}'$ is a subsheaf of abelian groups, then $\mathcal{S}'$ is a closed subsheaf of $\mathcal{S}$ if and only if the quotient sheaf $\mathcal{S}/\mathcal{S}'$ is torsion free.

For example, if V' is a subbundle of a vector bundle V, then the sheaf of sections of V' in the étale topology is a closed subsheaf of the sheaf of sections of V.

The following lemma is then clear:

LEMMA 1.1.2. (i) *Suppose that $\mathcal{S}$ is torsion free and has the Hartogs property. Let $\mathcal{S}'$ be a closed subsheaf of $\mathcal{S}$. Then $\mathcal{S}'$ is torsion free and has the Hartogs property as well.*

(ii) *Suppose that $\mathcal{S}'$ is a subsheaf of a torsion free sheaf $\mathcal{S}$, and that $\mathcal{C}$ is a closed subsheaf of $\mathcal{S}$. Then $\mathcal{C} \cap \mathcal{S}'$ is a closed subsheaf of $\mathcal{S}'$.*

(iii) *If $\mathcal{S}$ and $\mathcal{S}'$ are torsion free sheaves of abelian groups and $c\colon \mathcal{S} \to \mathcal{S}'$ is a homomorphism, then* $\operatorname{Ker} c$ *is a closed subsheaf of $\mathcal{S}$.*

□

1.2. A geometric lemma.

LEMMA 1.2.1. *Let X be a scheme and T a torus such that there exists a morphism $\pi\colon X \to T$ and a finite étale covering $\widetilde{T} \to T$ such that the fiber product $X \times_T \widetilde{T}$ is isomorphic to $\mathbb{A}^k \times \widetilde{T}$ as schemes over $\widetilde{T}$. Then*

(i) *Every morphism $f\colon X \to \mathbb{G}_m$ is induced by a morphism $T \to \mathbb{G}_m$, and hence is given up to translation by an element of $\mathbb{G}_m$ by a character of T. In particular, every morphism from an affine space to $\mathbb{G}_m$ is constant.*

(ii) $\operatorname{Pic} X$ *is trivial.*

Proof. The proof of (i) is left to the reader. To see (ii), let L be a line bundle on X and let d be the degree of the morphism $\widetilde{T} \to T$. Then the pullback $\widetilde{L}$ of L to $X \times_T \widetilde{T}$ is trivial, as is $\operatorname{Norm}(\widetilde{L}) = L^d$. Thus every element in $\operatorname{Pic} X$ is d-torsion. Let $(\operatorname{Pic} X)_d$ be the group of d-torsion line bundles of X. The Kummer exact sequence gives an exact sequence

$$0 \to H^0(X; \boldsymbol{\mu}_d) \to H^0(X; \mathbb{G}_{m,X}) \xrightarrow{z \mapsto z^d} H^0(X; \mathbb{G}_{m,X}) \to$$
$$\to H^1(X; \boldsymbol{\mu}_d) \to (\operatorname{Pic} X)_d \to 0.$$

Since X is a fiber bundle over T with fibers isomorphic to $\mathbb{A}^k$, the comparison of étale and classical cohomology implies that $H^1(X; \boldsymbol{\mu}_d) \cong (\mathbb{Z}/d\mathbb{Z})^r$, where $r = \dim T$. By Part (i), $H^0(X; \mathbb{G}_{m,X}) \cong \mathbb{C}^* \times \mathsf{X}(T)$, where $\mathsf{X}(T)$, the character group of T, is a free abelian group of rank r. Since the homomorphism $H^0(X; \mathbb{G}_{m,X}) \to H^0(X; \mathbb{G}_{m,X})$ is raising to the power d, it follows that the cokernel has order d^r. Hence the cokernel maps isomorphically onto $H^1(X; \boldsymbol{\mu}_d)$ and $(\operatorname{Pic} X)_d$ is trivial. Thus $\operatorname{Pic} X$ is trivial as well. □

1.3. Minuscule and quasi-minuscule weights. We begin by recalling some standard definitions from representation theory. The following three sets of objects may be identified: homomorphisms $\Lambda \to \mathbb{Z}$, linear maps $\mathfrak{h} \to \mathbb{C}$ which are integral on Λ, and characters $H \to \mathbb{C}^*$. We will call a homomorphism $\mu\colon \Lambda \to \mathbb{Z}$ a *weight* and will also μ to denote the obvious extension of μ to a linear map $\mathfrak{h} \to \mathbb{C}$ and occasionally the corresponding character $H \to \mathbb{C}^*$.

DEFINITION 1.3.1. Let $\rho\colon G \to GL(V)$ be an irreducible representation with

$w\mu_1 = \mu_2$. Since highest weights have multiplicity one, if ρ is quasi-minuscule, every non-zero weight of ρ has multiplicity one. Finally, ρ is *properly quasi-minuscule* if it is quasi-minuscule but not minuscule.

The following is a standard result, cf. [**2**]:

LEMMA 1.3.2. *With notation as above,*

(i) *The minuscule representations are exactly the representations whose highest weight is a fundamental weight ϖ_α for a simple root α such that, if $\widetilde{\alpha} = \sum_{\beta\in\Delta} h_\beta\beta$ is the highest root, then $h_\alpha = 1$. Thus, the number of isomorphism classes of minuscule representations is $\#Z(G) - 1$, where $Z(G)$ is the center of G.*

(ii) *For each G, there is exactly one properly quasi-minuscule representation ρ up to isomorphism. Its highest weight is the highest short root of G. Hence, if G is simply laced, ρ is the adjoint representation.* □

REMARK 1.3.3. Let $\rho\colon G \to GL(V)$ be a properly quasi-minuscule representation. Let $V = \bigoplus V_\mu$ be the decomposition of V into weight spaces, where the μ are either 0 or the short roots of G and, for $\mu \neq 0$, $\dim V_\mu = 1$. Then there exists a non-degenerate G-invariant symmetric bilinear form $B\colon V \otimes V \to \mathbb{C}$. If μ_1 and μ_2 are nonzero, then $B(V_{\mu_1}, V_{\mu_2}) \neq 0$ if and only if $\mu_1 + \mu_2 = 0$. There also exists a G-invariant 3-tensor $C\colon V \otimes V \otimes V \to \mathbb{C}$, with the property that, if μ_1, μ_2, μ_3 are nonzero, then $C(V_{\mu_1}, V_{\mu_2}, V_{\mu_3}) \neq 0$ if and only if $\mu_1 + \mu_2 + \mu_3 = 0$. For example, if G is simply laced, then ρ is the adjoint representation. In this case, we let $B(X, Y)$ be the Killing form and let $C(X, Y, Z)$ be the Cartan form $B([X, Y], Z)$. It is easy to see that B and C have the desired properties. The remaining groups are done on a case-by-case basis. For $G = Spin(2n+1)$, the representation ρ is the standard representation $G \to SO(2n+1) \subseteq GL(2n+1, \mathbb{C})$. In this case, B is the standard form and $C = 0$. (Note that two short roots of G never sum to a short root.) If $G = Sp(2n, \mathbb{C})$, let V_0 be the standard $2n$-dimensional representation of G with the invariant symplectic form $\langle\cdot,\cdot\rangle$. Then V is a direct summand of $\bigwedge^2 V_0$ complementary to the line defined by $\langle\cdot,\cdot\rangle$. The form B is then the restriction to V of the form on $\bigwedge^2 V_0$ induced by

$$B(v_1 \wedge v_2, w_1 \wedge w_2) = \langle v_1, w_1\rangle\langle v_2, w_2\rangle + \langle v_1, w_2\rangle\langle v_2, w_1\rangle.$$

Viewing C as a G-invariant homomorphism $\varphi\colon V \otimes V \to V^* \cong V$ via the form B, the tensor C corresponds to

$$\varphi(v_1\wedge v_2, w_1\wedge w_2) = \langle v_2, w_1\rangle v_1\wedge w_2 - \langle v_1, w_1\rangle v_2\wedge w_2 - \langle v_2, w_2\rangle v_1\wedge w_1 + \langle v_1, w_2\rangle v_2\wedge w_1.$$

The definitions of B and C for the groups of type G_2 and F_4 are well-known [**1**].

In all cases, $\rho(G)$, which is the adjoint form of G, is exactly the stabilizer in $GL(V)$ of the forms B and C.

While we will not use this remark, it motivates the constructions of Section 5.

2. Automorphism sheaves and G-bundles

2.1. Definition of automorphism sheaves. Let X and Y be schemes, with Y projective, and let $\Xi \to X \times Y$ be a principal G-bundle over $X \times Y$. We define a sheaf $\pi_{1*}\operatorname{Aut}\Xi$ over X as follows: For each étale open set $f\colon U \to X$, let $\Xi_U = (f \times \mathrm{Id})^*\Xi$ and let $\pi_{1*}\operatorname{Aut}\Xi(U) = \operatorname{Aut}\Xi_U$. Thus $\pi_{1*}\operatorname{Aut}\Xi(U)$ is the group of sections of $\operatorname{Ad}\Xi_U = \Xi_U \times_G G$, where G acts on itself by conjugation. If $\mathcal{V}$ is a vector bundle over $X \times Y$, the sheaf $\pi_{1*}\operatorname{Aut}\mathcal{V}$ is defined similarly.

There is also the related functor $\mathbf{A}$ from schemes over X to groups defined similarly for each $f\colon X' \to X$ by $\mathbf{A}(X') = \mathrm{Aut}(f \times \mathrm{Id})^*\Xi$. In this case, one can show the following:

PROPOSITION 2.1.1. *Let E be a singular Weierstrass cubic. Suppose that $\Xi \to X \times E$ is a principal G-bundle such that, for every $x \in X$, $\Xi|\{x\} \times E$ pulls back to the trivial bundle on $\widetilde{E}$ and is regular. Then $\mathbf{A}$ is representable by a commutative affine group scheme which is smooth over X.* □

Since we will not use Proposition 2.1.1 in this paper, we omit the proof.

We now discuss some general properties of $\pi_{1*}\,\mathrm{Aut}\,\Xi$ and $\pi_{1*}\,\mathrm{Aut}\,\mathcal{V}$. First we have the following:

LEMMA 2.1.2. *Let X be normal and let $\mathcal{V}$ be a locally free sheaf over $X \times E$. Then $\pi_{1*}\mathcal{V}$ has the Hartogs property.*

Proof. It clearly suffices to show that $\mathcal{O}_{X\times E}$ has the Hartogs property. This is an easy consequence of the fact that $X \times \widetilde{E}$ is normal and that, locally, any section of $\mathcal{O}_{X\times \widetilde{E}}$ which lies in $\mathcal{O}_{X\times E}$ in the complement of a closed subset of codimension at least two lies in $\mathcal{O}_{X\times E}$. □

LEMMA 2.1.3. *Let $\mathcal{V}$ be a vector bundle over $X \times E$, where X is normal. Then $\pi_{1*}\,\mathrm{Aut}\,\mathcal{V}$ is torsion free and has the Hartogs property.*

Proof. The sheaf $\pi_{1*}\,\mathrm{Aut}\,\mathcal{V}$ is the subsheaf of invertible elements in the sheaf of algebras $\pi_{1*}Hom(\mathcal{V},\mathcal{V})$. Since $\mathcal{V}$ and hence $Hom(\mathcal{V},\mathcal{V})$ are vector bundles and X is normal, $\pi_{1*}Hom(\mathcal{V},\mathcal{V})$ is torsion free and has the Hartogs property. Hence $\pi_{1*}\,\mathrm{Aut}\,\mathcal{V}$ is torsion free. Suppose that $U \to X$ is an étale open set and $U' = U - Y$ where Y has codimension at least two in U. If $s \in \Gamma(U', \pi_{1*}\,\mathrm{Aut}\,\Xi)$, then $s^{-1} \in \Gamma(U', \pi_{1*}\,\mathrm{Aut}\,\Xi)$ as well, and both s and s^{-1} extend to elements of $\Gamma(U, \pi_{1*}Hom(\mathcal{V},\mathcal{V}))$ whose product is Id on U' and hence on U. Thus the unique extension of s lies in $\Gamma(U, \pi_{1*}\,\mathrm{Aut}\,\mathcal{V})$, so that $\pi_{1*}\,\mathrm{Aut}\,\mathcal{V}$ has the Hartogs property. □

LEMMA 2.1.4. *Let $\Xi \to X\times E$ be a principal G-bundle over $X \times E$, and suppose that X is normal.*

(i) *Let $\rho\colon G \to \mathbb{C}^N$ be a faithful representation and let $\mathcal{V} = \Xi \times_G \mathbb{C}^N$. Then $\pi_{1*}\,\mathrm{Aut}\,\Xi$ is a closed subsheaf of $\pi_{1*}\,\mathrm{Aut}\,\mathcal{V}$.*

(ii) *$\pi_{1*}\,\mathrm{Aut}\,\Xi$ is torsion free and has the Hartogs property.*

Proof. To see (i), let $f\colon U \to X$ be an étale open set and let $\Xi_U = (f \times \mathrm{Id})^*\Xi$. The scheme $\mathrm{Ad}\,\Xi_U = \Xi_U \times_G G$ is a closed subscheme of $\Xi_U \times_G GL(N,\mathbb{C})$. Thus, if $U' \to U$ is a étale open and dense subset of U and s is a section of $\Xi_U \times_G GL(N,\mathbb{C})$ whose restriction to $U' \times E$ lies in $\mathrm{Ad}\,\Xi_{U'}$, then s lies in $\mathrm{Ad}\,\Xi_U$. This proves that $\pi_{1*}\,\mathrm{Aut}\,\Xi$ is a closed subsheaf of $\pi_{1*}\,\mathrm{Aut}\,\mathcal{V}$. Since $\pi_{1*}\,\mathrm{Aut}\,\mathcal{V}$ is torsion free, $\pi_{1*}\,\mathrm{Aut}\,\Xi$ is torsion free as well, and by Lemma 1.1.2 and Lemma 2.1.3, $\pi_{1*}\,\mathrm{Aut}\,\Xi$

2.2. Statement of the main theorem. We now specialize to the case of a singular Weierstrass cubic E. Recall the notation of the introduction: let $\widetilde{\mathcal{M}} = E_{\mathrm{reg}} \otimes \Lambda$, so that $\widetilde{\mathcal{M}} = \mathfrak{h}$ if E is cuspidal and $\widetilde{\mathcal{M}} = H$ if E is nodal, and let $\mathcal{M} = \widetilde{\mathcal{M}}/W$ be the adjoint quotient.

DEFINITION 2.2.1. Fix once and for all a section of the adjoint quotient morphism, for example the Kostant or Steinberg section, and let $\Xi \to \mathcal{M} \times E$ be the corresponding G-bundle. Let $\mathcal{A} = \pi_{1*} \operatorname{Aut} \Xi$, where $\Xi \to \mathcal{M} \times E$ is a universal G-bundle as above.

Since the centralizer of a regular element is abelian, $\mathcal{A}$ is a sheaf of abelian groups over $\mathcal{M}$. It is easy to see that two different choices of a section lead to isomorphic automorphism sheaves. Note further that there is a natural inclusion of the center $Z(G)$ of G, viewed as a constant sheaf on $\mathcal{M}$, into $\mathcal{A}$.

If σ' is another section of the adjoint quotient morphism, let Ξ' be the corresponding bundle. Then, for all $x \in \mathcal{M}$, there exists a $g \in G$ such that $\mathrm{ad}(g)\sigma(x) = \sigma'(x)$, and any two such differ by right multiplication by $Z_G(\sigma(x))$. Since σ is a section of the adjoint quotient morphism, the morphism $G \times \mathfrak{h}/W \to \mathfrak{g}_{\mathrm{reg}}$ defined by $(g, x) \mapsto \mathrm{ad}\, g(\sigma(x))$ is smooth. Thus there exist local sections of this morphism in the étale topology. Similarly in the case of the adjoint quotient of G, the morphism $G \times H/W \to G_{\mathrm{reg}}$ defined by $(g, x) \mapsto \mathrm{Ad}\, g(\sigma(x))$ is smooth. A formal argument then shows:

PROPOSITION 2.2.2. *Let σ and σ' be two sections of the adjoint quotient morphism $\mathfrak{g}_{\mathrm{reg}} \to \mathfrak{h}/W = \mathcal{M}$ or $G_{\mathrm{reg}} \to H/W = \mathcal{M}$. For every $x \in \mathcal{M}$, there exists an étale neighborhood U of x and a morphism $\gamma\colon U \to G$ such that $\sigma'|U = \mathrm{ad}(\gamma)(\sigma|U)$, in the case of $\mathfrak{g}$, or $\sigma'|U = \mathrm{Ad}(\gamma)(\sigma|U)$ in the case of G. Moreover, any two choices of γ differ by right multiplication by an element of $\mathcal{A}(U)$.*

Proof. We shall just sketch the proof of the existence of the morphism γ in the case of $\mathfrak{g}$. After replacing σ' by $\mathrm{ad}(g)(\sigma')$ for some $g \in G$, we may assume that $\sigma'(x) = \sigma(x)$. Let U' be an étale neighborhood of $\sigma(x)$ over which there exists a section F of the morphism $G \times \mathfrak{h}/W \to \mathfrak{g}_{\mathrm{reg}}$ defined by $(g, x) \mapsto \mathrm{ad}\, g(\sigma(x))$, and let U be the neighborhood of x in $\mathcal{M}$ corresponding to $U' \times_{\mathfrak{g}_{\mathrm{reg}}} \sigma'(\mathcal{M})$. Then $F \circ \sigma'$, followed by the projection $G \times \mathfrak{h}/W \to G$, defines a morphism $\gamma\colon U \to G$ with the property that $\sigma'|U = \mathrm{ad}(\gamma)(\sigma|U)$, as desired. The last statement is then clear. $\square$

In the standard way, choosing a sufficiently small open cover $\{U_i\}$ of $\mathcal{M}$ and writing $\sigma'|U_i = \mathrm{ad}(\gamma_i)(\sigma|U_i)$ with $\gamma_j|U_i \times_{\mathcal{M}} U_j = \gamma_i a_{ij}|U_i \times_{\mathcal{M}} U_j$ defines a 1-cocycle $\{a_{ij}\}$ and hence an element of $H^1(\mathcal{M}; \mathcal{A})$ which is independent of the choices. This element is the identity if and only if the bundles Ξ and Ξ' are isomorphic, if and only if there exists a morphism $g\colon \mathcal{M} \to G$ such that $\sigma' = \mathrm{ad}\, g(\sigma)$, in which case we say that the sections σ and σ' are *conjugate*. This explains the relevance of the second half of the following theorem:

THEOREM 2.2.3. *Let $\mathcal{M}$ be the adjoint quotient of either $\mathfrak{g}$ or G, and let $\mathcal{A}$ denote the automorphism sheaf corresponding to the section σ with image contained in the set of regular elements. Then*

$$H^0(\mathcal{M}; \mathcal{A}) = \begin{cases} Z(G), & \text{for the case of } \mathfrak{g}; \\ Z(G) \times \mathbb{Z}, & \text{for the case of } G. \end{cases}$$

Moreover, in both cases, $H^1(\mathcal{M}; \mathcal{A}) = \{1\}$.

COROLLARY 2.2.4. *Every two sections of the adjoint quotient morphism are conjugate by a morphism from the adjoint quotient to the set of regular elements of* $\mathfrak{g}$ *or* G. □

3. Quasi-minuscule representations and spectral covers

3.1. Vector bundles attached to quasi-minuscule representations. We fix the following notation: $\rho\colon G \to \operatorname{Aut}(V)$ is a quasi-minuscule representation, with highest weight ϖ (with respect to the given fixed set of simple roots). Let $W_0 \subseteq W$ be the stabilizer of ϖ. The group W_0 is the group generated by reflections in the roots annihilated by ϖ.

With $\mathcal{M}$ as before, and Ξ the universal bundle over $\mathcal{M} \times E$ corresponding to the choice of the section σ, let $\mathcal{V} \to \mathcal{M} \times E$ be the vector bundle induced by the representation ρ. By [**10**, Theorem 2.4.3], we have:

PROPOSITION 3.1.1. *The sheaf* $\pi_{1*}\mathcal{V}$ *is a locally free sheaf over* $\mathcal{M}$ *whose rank is the multiplicity of the trivial weight in* ρ. *Moreover, the inclusion* ${\pi_1}^*\pi_{1*}\mathcal{V} \to \mathcal{V}$ *is a subbundle, so that the quotient* $\overline{\mathcal{V}}$ *is a vector bundle, and there is an exact sequence of vector bundles over* $\mathcal{M} \times E$:

$$0 \to {\pi_1}^*\pi_{1*}\mathcal{V} \to \mathcal{V} \to \overline{\mathcal{V}} \to 0. \quad \square$$

We shall need the following facts about Ξ and $\mathcal{V}$:

LEMMA 3.1.2. *There is an étale open and dense subset* $U \to \mathcal{M}$ *such that* Ξ_U *has a reduction of structure to an* H*-bundle* $\Xi_{U,H}$. *The action of the constant group scheme* $U \times H$ *over* U *as a group scheme of automorphisms of* $\Xi_{U,H}$ *defines an isomorphism from the sheaf of sections of the constant group scheme* $U \times H \to U$ *to* $\mathcal{A}|U$.

Proof. Let U be the nonempty open subset of $\widetilde{\mathcal{M}}$ where no root is the identity. Thus U is either $\mathfrak{h}_{\mathrm{reg}} = \mathfrak{h} \cap \mathfrak{g}_{\mathrm{reg}}$ or $H_{\mathrm{reg}} = H \cap G_{\mathrm{reg}}$. We just write out the Lie algebra case. The induced morphism $U \to \mathcal{M}$ is étale. For a scheme B, an H-bundle over $B \times E$ which is trivialized over $B \times \tilde{E}$ is identified with a morphism $B \to \mathfrak{h}$. Let $\Xi_{H,U}$ be the H-bundle over $U \times E$ induced by applying this remark to the inclusion $U = \mathfrak{h}_{\mathrm{reg}} \to \mathfrak{h}$ and let $\Xi'_U = \Xi_{H,U} \times_H G$. It is clear that Ξ_U and Ξ'_U have isomorphic restriction to every slice. A formal argument as in the proof of Proposition 2.2.2 shows that, possibly after replacing U by an étale open dense subset, the bundles Ξ_U and Ξ'_U become isomorphic. This gives the desired reduction of structure. Viewing $\Xi_U \cong \Xi'_U$ as the G-bundle associated to a morphism $\sigma'\colon U \to \mathfrak{h}_{\mathrm{reg}} \subseteq \mathfrak{g}$, the sections of $\mathcal{A}$ over an étale open set V of U are given by morphisms $f\colon V \to G$ such that $\operatorname{ad}(f)(\sigma'|V) = \sigma'|V$. Clearly, since the image of σ' is contained in $\mathfrak{h}_{\mathrm{reg}}$, this is the same as the group of morphisms from V to H. □

COROLLARY 3.1.3. *Let* $U \to \mathcal{M}$ *be the étale open subset of Lemma 3.1.2. Then* $\mathcal{V}|U \times E$ *is isomorphic to* $\bigoplus_\mu \mathcal{L}_\mu^{\oplus m_\mu}$, *where the* μ *are the weights of* ρ, m_μ *is the multiplicity of* μ *in* ρ, *and the* $\mathcal{L}_\mu$ *are line bundles on* $U \times E$ *with the property that* $\mathcal{L}_\mu \cong \mathcal{L}_{\mu'}$ *if and only if* $\mu = \mu'$. *Under the isomorphism of* $\mathcal{A}|U$ *with the sheaf of sections of the constant group scheme* $U \times H$, *a local section* s *of* $U \times H$ *acts on* $\mathcal{L}_\mu^{\oplus m_\mu}$ *via multiplication by the function* $\mu(s)$. □

of constant automorphisms with values in $\mathrm{Ker}(\rho) \subseteq Z(G)$. Let $\overline{\mathcal{A}}_\rho$ denote the quotient sheaf $\mathcal{A}/\mathrm{Ker}(\rho)$.

LEMMA 3.1.4. *In the above notation, there is an exact sequence*

$$\{1\} \to \mathrm{Ker}(\rho) \to H^0(\mathcal{M};\mathcal{A}) \to H^0(\mathcal{M};\overline{\mathcal{A}}_\rho) \to \{1\}$$

and the natural map $H^1(\mathcal{M};\mathcal{A}) \to H^1(\mathcal{M};\overline{\mathcal{A}}_\rho)$ *is an isomorphism.*

Proof. This is an immediate consequence of the fact that, as $\mathcal{M}$ is isomorphic to $\mathbb{A}^r$, $H^0(\mathcal{M};\mathrm{Ker}(\rho)) \cong \mathrm{Ker}(\rho)$ and $H^i(\mathcal{M};\mathrm{Ker}(\rho))$ is trivial for $i = 1, 2$. □

PROPOSITION 3.1.5. *The sheaves* $\mathcal{A}$ *and* $\overline{\mathcal{A}}_\rho$ *are torsion free and have the Hartogs property. The sheaf* $\overline{\mathcal{A}}_\rho$ *is a closed subsheaf of* $\pi_{1*}\mathrm{Aut}(\mathcal{V})$. *Finally, every local section of* $\overline{\mathcal{A}}_\rho$ *induces the identity on* $R^0\pi_{1*}\mathcal{V}$.

Proof. The first two statements follow from Lemma 2.1.4. It suffices to prove the final statement on an étale open dense of $\mathcal{M}$. By Corollary 3.1.3, it is immediate that $\overline{\mathcal{A}}_\rho$ has this property on the étale open set $U \to \mathcal{M}$ of Lemma 3.1.2. □

3.2. The spectral cover. We begin by recalling the general formalism of the Fourier-Mukai correspondence in the special case we shall need, for which a general reference is [**10**, §1]. Let B be a scheme and let $\mathcal{V}$ be a vector bundle of rank n over $B \times E$ whose restriction to every slice $\{x\} \times E$ has trivial pullback to the normalization of E. Then $\mathcal{V}$ determines functorially a relative Cartier divisor $T_\mathcal{V} \subseteq B \times E_{\mathrm{reg}}$, finite over B, and a sheaf $\mathcal{Q}_\mathcal{V}$ of $\mathcal{O}_{T_\mathcal{V}}$-modules, finite and flat over B of degree n, such that $\mathcal{V} = q_*(p_1^*(\mathcal{Q}_\mathcal{V} \otimes \pi_2^*\mathcal{O}_E(p_0)) \otimes r^*\mathcal{P})$, where $p_1 \colon B \times E \times E$ is projection onto the first two factors, $q \colon B \times E \times E$ is projection onto the first and third factors, $r \colon B \times E \times E \to E \times E$ is projection onto the last two factors, and $\pi_2 \colon B \times E \to E$ is projection onto the second factor. The divisor $T_\mathcal{V}$ is additive under exact sequences, and, viewing $\mathcal{Q}_\mathcal{V}$ as a sheaf on $B \times E$, given an exact sequence $0 \to \mathcal{V}' \to \mathcal{V} \to \mathcal{V}'' \to 0$ of vector bundles as above, there is an exact sequence of the corresponding sheaves:

$$0 \to \mathcal{Q}_{\mathcal{V}'} \to \mathcal{Q}_\mathcal{V} \to \mathcal{Q}_{\mathcal{V}''} \to 0.$$

Moreover the constructions of $T_\mathcal{V}$ and $\mathcal{Q}_\mathcal{V}$ are compatible with base change. We somewhat loosely refer to $T_\mathcal{V}$ as the *spectral cover* of B corresponding to $\mathcal{V}$. More generally, we have the following:

DEFINITION 3.2.1. Let $\nu \colon T \to B$ be a finite morphism of schemes. Suppose that $\mathcal{Q}$ is a coherent sheaf on T, flat over B, and that $\mathcal{L}$ is a line bundle on $T \times E$. The *spectral cover construction* (applied to $\mathcal{Q}$ and $\mathcal{L}$) is the vector bundle

$$\mathcal{V} = (\nu \times \mathrm{Id})_*(\pi_1^*\mathcal{Q} \otimes \mathcal{L}).$$

In the above situation, the sheaf $\nu_*\mathcal{O}_T$ acts via endomorphisms on $\mathcal{V}$, and hence $\nu_*\mathcal{O}_T^*$ acts via automorphisms of $\mathcal{V}$. Passing from a sheaf in the Zariski topology to one in the étale topology defines a homomorphism $\nu_*\mathbb{G}_{m,T} \to \pi_{1*}\mathrm{Aut}\,\mathcal{V}$.

LEMMA 3.2.2. *In the above notation, suppose that* B *is reduced irreducible and that* $\mathcal{Q}$ *is a locally free* $\mathcal{O}_T$*-module, or more generally that* T *is flat over* B *and that, for all* $x \in B$, $H^0(\nu^{-1}(x);\mathcal{Q}|\nu^{-1}(x))$ *is a faithful* $H^0(\nu^{-1}(x);\mathcal{O}_T|\nu^{-1}(x))$*-module. Then* $\nu_*\mathbb{G}_{m,T}$ *is a closed subsheaf of* $\pi_{1*}\mathrm{Aut}\,\mathcal{V}$.

Proof. The fiber of $\mathcal{V}$ over a point (x, e) of $B \times E$ is identified up to a homothety with $H^0(\nu^{-1}(x); \mathcal{Q}|\nu^{-1}(x))$ and the induced action of $\nu_*\mathcal{O}_T$ on this fiber is via the action of $H^0(\nu^{-1}(x); \mathcal{O}_T|\nu^{-1}(x))$. Since T is flat over B, $\nu_*\mathcal{O}_T$ is a vector bundle. The hypotheses imply that, for each $x \in B$, $(\nu_*\mathcal{O}_T)_x \to \pi_{1*}Hom(\mathcal{V}, \mathcal{V})_x$ is injective. Thus $\nu_*\mathcal{O}_T$ is a vector subbundle, and hence a closed subsheaf, of $\pi_{1*}Hom(\mathcal{V}, \mathcal{V})$. By (i) of Lemma 1.1.2, as sheaves of sets in the étale topology, $\pi_{1*} \operatorname{Aut} \mathcal{V} \cap \nu_*\mathcal{O}_T$ is a closed subsheaf of $\pi_{1*} \operatorname{Aut} \mathcal{V}$. It suffices to prove that $\pi_{1*} \operatorname{Aut} \mathcal{V} \cap \nu_*\mathcal{O}_T = \nu_*\mathbb{G}_{m,T}$. Replacing B by an étale open set and T by the fiber product, it is enough to show that, if $f \in \Gamma(T, \mathcal{O}_T)$ defines an automorphism of $\mathcal{V}$ and hence by functoriality of $\mathcal{Q}$, then, for all $y \in T$, f does not lie in the maximal ideal $\mathfrak{m}_y$ of y. Since f defines an automorphism of $\mathcal{Q}$, it defines an automorphism of the stalk of $\mathcal{Q}$ over y, which is a finite module over the local ring of T at y and is nonzero by our assumptions on $\mathcal{Q}$. It follows by Nakayama's lemma that $f \notin \mathfrak{m}_y$. Thus $\pi_{1*} \operatorname{Aut} \mathcal{V} \cap \nu_*\mathcal{O}_T = \nu_*\mathbb{G}_{m,T}$ is a closed subsheaf of $\pi_{1*} \operatorname{Aut} \mathcal{V}$. □

3.3. The case of a quasi-minuscule representation. We return to the case where $\mathcal{V} = \Xi \times_G \mathbb{C}^N$ is the vector bundle over $\mathcal{M} \times E$ defined by the minuscule or quasi-minuscule representation $\rho\colon G \to GL(N, \mathbb{C})$. Recall that we have the exact sequence

$$0 \to {\pi_1}^* \pi_{1*}\mathcal{V} \to \mathcal{V} \to \overline{\mathcal{V}} \to 0.$$

Let $\hat{T}_0 = \widetilde{\mathcal{M}}/W_0$ and let $\hat{\nu}\colon \hat{T}_0 \to \mathcal{M}$ be the natural morphism. The homomorphism $\varpi\colon \Lambda \to \mathbb{Z}$ defines a W_0-invariant homomorphism $\widetilde{\mathcal{M}} \to E_{\text{reg}}$ and hence a morphism $f_\varpi\colon \hat{T}_0 \to E_{\text{reg}}$. Let $T_0 \subseteq \mathcal{M} \times E_{\text{reg}}$ be the image of the product morphism $(\hat{\nu}, f_\varpi)$. By [**10**, Corollary 2.2.3], the morphism $\hat{T}_0 \to T_0$ is birational and exhibits $\hat{T}_0$ as the normalization of T_0. Moreover, by [**10**, Lemma 2.1.1], we have:

LEMMA 3.3.1. *Let $\Lambda_0 = \operatorname{Ker} \varpi$ and let $\mathcal{M}_0 = (E_{\text{reg}} \otimes \Lambda_0)/W_0$. There is a finite étale cover $C \to E_{\text{reg}}$, with C necessarily isomorphic to E_{reg}, such that $\hat{T}_0 \times_{E_{\text{reg}}} C \cong \mathcal{M}_0 \times C$ as schemes over C.* □

Let $T = T_{\mathcal{V}} \subseteq \mathcal{M} \times E_{\text{reg}}$ be the spectral cover of $\mathcal{V}$, let $\nu\colon T \to \mathcal{M}$ be the morphism induced by projection onto the first factor, and let $\mathcal{Q} = \mathcal{Q}_{\mathcal{V}}$ be the corresponding sheaf of $\mathcal{O}_T$-modules. Let $\mathcal{Q}_1$ be the sheaf corresponding to ${\pi_1}^*\pi_{1*}\mathcal{V}$ and let $\overline{\mathcal{Q}}$ be the sheaf corresponding to $\overline{\mathcal{V}}$. The general formalism of the previous subsection and [**10**, Corollary 2.2.3] imply:

PROPOSITION 3.3.2. *Let T_0 be the divisor defined above, let $\overline{T}_1$ be the reduced divisor $\mathcal{M} \times \{p_0\}$, and let m_0 be the multiplicity of the trivial weight in ρ. Then the divisor T_0 is the spectral cover of $\overline{\mathcal{V}}$, and $T = T_0 \cup T_1 \subset \mathcal{M} \times E_{\text{reg}}$ where, as a divisor, $T_1 = m_0 \cdot \overline{T}_1$. Thus $T = T_0$ if and only if ρ is minuscule. Generically along $\overline{T}_1 \cap T_0$ the varieties $\overline{T}_1$ and T_0 are smooth and meet transversely.* □

Let $\overline{T}$ be the sum $T_0 + \overline{T}_1$ considered as a (reduced) effective Cartier divisor of $\mathcal{M} \times E$. It is the reduced subscheme of T. By the above, the following is exact on the complement of a codimension two subset of $\overline{T}$:

$$0 \to \mathcal{O}_{\overline{T}} \to \mathcal{O}_{\overline{T}_1} \oplus \mathcal{O}_{T_0} \to \mathcal{O}_{\overline{T}_1 \cap T_0} \to 0.$$

In fact, since $\overline{T}$ is Cohen-Macaulay, the above sequence is everywhere exact, al-

Proof. By the formalism of spectral covers, there is an exact sequence

$$0 \to \mathcal{Q}_1 \to \mathcal{Q} \to \overline{\mathcal{Q}} \to 0.$$

Since ${\pi_1}^*\pi_{1*}\mathcal{V}$ is pulled back from a bundle on $\mathcal{M}$, it restricts to the trivial bundle on every slice and hence $\mathcal{Q}_1$ is annihilated by $I_{\overline{T}_1}$, the ideal sheaf of $\overline{T}_1$. Since $\overline{\mathcal{Q}}$ is an $\mathcal{O}_{T_0}$-module, it is annihilated by the ideal sheaf I_{T_0}. Hence $\mathcal{Q}$ is annihilated by the product ideal $I_{\overline{T}_1} \cdot I_{T_0}$, which is the ideal of $\overline{T}$. □

As for the quotient itself, by [**10**, Theorem 2.4.3] and Lemma 1.2.1, we have:

PROPOSITION 3.3.4. *The sheaf $\overline{\mathcal{Q}}$ is isomorphic to an invertible $\mathcal{O}_{\hat{T}_0}$-module, and hence is isomorphic, as a sheaf of $\mathcal{O}_{T_0}$-modules, to $\mathcal{O}_{\hat{T}_0}$.* □

COROLLARY 3.3.5. $\pi_{1*}\operatorname{Aut}\overline{\mathcal{V}} = \hat{\nu}_*\mathbb{G}_{m,\hat{T}_0}$. *In particular, if ρ is minuscule, then $\mathcal{V} = \overline{\mathcal{V}}$, $\pi_{1*}\operatorname{Aut}\mathcal{V} = \hat{\nu}_*\mathbb{G}_{m,\hat{T}_0}$, and $\overline{\mathcal{A}}_\rho$ is a closed subsheaf of $\hat{\nu}_*\mathbb{G}_{m,\hat{T}_0}$.*

Proof. By Lemma 3.2.2, $\hat{\nu}_*\mathbb{G}_{m,\hat{T}_0}$ is a closed subsheaf of $\pi_{1*}\operatorname{Aut}\overline{\mathcal{V}}$. Thus, it suffices to prove that they agree on an étale open and dense set. By Corollary 3.1.3, there is an étale open set U such that the pullback of $\mathcal{V}$ to U is of the form $\bigoplus_\mu \mathcal{L}_\mu^{\oplus m_\mu}$, where the $\mathcal{L}_\mu$ are distinct line bundles of relative degree zero on $U \times E$, indexed by the weights of ρ, and the m_μ are the multiplicities and thus $m_\mu = 1$ if $\mu \neq 0$. Hence $\overline{\mathcal{V}}|U \times E \cong \bigoplus_{\mu\neq 0}\mathcal{L}_\mu$. Thus $\pi_{1*}\operatorname{Aut}\overline{\mathcal{V}}|U$ is just a product of copies of $\mathbb{G}_{m,U}$ indexed by the nonzero weights of ρ, and this is clearly equal to $\hat{\nu}_*\mathbb{G}_{m,\hat{T}_0}|U$. □

3.4. The properly quasi-minuscule case. In the properly quasi-minuscule case, we will need a refinement of Corollary 3.3.5. Thus throughout this subsection we assume that ρ is properly quasi-minuscule. We have identified $\hat{T}_0$ with $\widetilde{\mathcal{M}}/W_0$. Let $D = \overline{T}_1 \cap T_0$ and let $\hat{D}$ be the preimage of D in $\hat{T}_0$. The preimage of $\hat{D}$ in $\widetilde{\mathcal{M}}$ is the kernel of the weight ϖ, and $\hat{D} = f_\varpi^{-1}(p_0)$ as reduced divisors. If $\Lambda_0 \subseteq \Lambda$ is the kernel of ϖ, then W_0 acts on $E_{\text{reg}} \otimes \Lambda_0$ and $\hat{D} = (E_{\text{reg}} \otimes \Lambda_0)/W_0 \subseteq \hat{T}_0$.

LEMMA 3.4.1. *Let $X = \pi_1(D) \cap \pi_1((T_0)_{\text{sing}})$. Then X has codimension at least two in $\mathcal{M}$. Moreover, the morphism $\hat{D} \to D$ is birational and identifies $\hat{D}$ with the normalization of D.*

Proof. A weight μ induces a homomorphism $\widetilde{\mathcal{M}} \to E_{\text{reg}}$, which we shall just write as μ again. By [**10**, Lemma 2.3.5], there is an open dense subset of the preimage of $(T_0)_{\text{sing}}$ in $\widetilde{\mathcal{M}}$ consisting of points x such that $\mu_1(x) = \mu_2(x)$ for two distinct weights μ_1, μ_2 of ρ and such that no root α is the identity on x. The inverse image of D in $\widetilde{\mathcal{M}}$ is the set of x such that some nonzero weight of ρ is the identity on x. Since the nonzero weights of ρ are roots, the inverse image of $\pi_1(D) \cap \pi_1((T_0)_{\text{sing}})$ in $\widetilde{\mathcal{M}}$ is thus nowhere dense in the preimage of $(T_0)_{\text{sing}}$ and so has codimension at least two in $\widetilde{\mathcal{M}}$. Since $\widetilde{\mathcal{M}} \to \mathcal{M}$ is finite and surjective, the same is true for X. The final statement is then clear. □

Let $\hat{T}$ be the scheme $\hat{T}_0 \amalg \overline{T}_1/\sim$, where $\sim$ is the identification of a point in $\hat{D}$ with its image in D. Thus the normalization of $\hat{T}$ is $\hat{T}_0 \amalg \overline{T}_1$, and the morphism $\hat{T}_0 \to T_0$ induces a morphism $\hat{T} \to \overline{T}$, and hence an algebra homomorphism $\mathcal{O}_{\overline{T}} \to \mathcal{O}_{\hat{T}}$.

PROPOSITION 3.4.2. *The sheaf $\mathcal{Q}$ is a sheaf of $\mathcal{O}_{\hat{T}}$-modules, compatibly with its structure as a sheaf of $\mathcal{O}_{\overline{T}}$-modules.*

Proof. The sheaf $\pi_{1*}Hom(\mathcal{V},\mathcal{V})$ is torsion free and has the Hartogs property. Thus, if there is an algebra homomorphism $\hat{\nu}_*\mathcal{O}_{\hat{T}} \to \pi_{1*}Hom(\mathcal{V},\mathcal{V})$ on the complement U of a closed subset of codimension at least two, then it extends to a homomorphism over $\hat{T}$, by taking the composition

$$\hat{\nu}_*\mathcal{O}_{\hat{T}} \to j_*j^*\hat{\nu}_*\mathcal{O}_{\hat{T}} \to j_*j^*\pi_{1*}Hom(\mathcal{V},\mathcal{V}) \cong \pi_{1*}Hom(\mathcal{V},\mathcal{V}),$$

where $j\colon U \to \mathcal{M}$ is the inclusion.

Let $X = \pi_1(D) \cap \pi_1((T_0)_{\mathrm{sing}})$ and let $U = \mathcal{M} - X$. By Lemma 3.4.1, X is a codimension two subset of $\mathcal{M}$. Set $U_1 = \mathcal{M} - \pi_1((T_0)_{\mathrm{sing}})$ and $U_2 = \mathcal{M} - \pi_1(D)$. Then $U = U_1 \cup U_2$, and it will suffice to check that $\mathcal{Q}$ is a sheaf of $\mathcal{O}_{\hat{T}}$-modules over both U_1 and U_2. Over U_1, $\hat{T} = \overline{T}$ and we conclude by Proposition 3.3.3. Over U_2, $\overline{T}$ is a disjoint union of $\overline{T}_1$ and T_0. Hence $\mathcal{Q}|U_2$ is a direct sum $\mathcal{Q}_1|U_2 \oplus \overline{\mathcal{Q}}|U_2$, and we conclude by Proposition 3.3.4. □

We define the sheaf $\mathcal{B}_{\hat{D}}$ to be the subsheaf of $\mathbb{G}_{m,\hat{T}_0}$ such that

$$\mathcal{B}_{\hat{D}}(U) = \{f \in \Gamma(U, \mathcal{O}_U^*) : f|U \times_{\hat{T}_0} \hat{D} = 1\}.$$

Thus $\mathcal{B}_{\hat{D}}$ is the kernel of the restriction homomorphism $\mathbb{G}_{m,\hat{T}_0} \to i_*\mathbb{G}_{m,\hat{D}}$, where i is the inclusion of $\hat{D}$ in $\hat{T}_0$ and there is a defining exact sequence

$$\{1\} \to \mathcal{B}_{\hat{D}} \to \mathbb{G}_{m,\hat{T}_0} \to i_*\mathbb{G}_{m,\hat{D}} \to \{1\}.$$

Clearly, $\hat{\nu}_*\mathcal{B}_{\hat{D}}$ has the Hartogs property.

LEMMA 3.4.3. *The sheaf $\hat{\nu}_*\mathcal{B}_{\hat{D}}$ is naturally a subsheaf of $\pi_{1*}\operatorname{Aut}(\mathcal{V})$.*

Proof. Since T_0 and $\overline{T}_1$ meet generically transversely in D, an invertible function on $\hat{T}_0$ which is 1 along $\hat{D}$ extends to a function on $\hat{T}$ by defining it to be 1 on $\overline{T}_1$. Thus the inclusion of $\mathcal{B}_{\hat{D}}$ in $\mathbb{G}_{m,\hat{T}_0}$, followed by extension by 1 over $\overline{T}_1$, induces an isomorphism from $\hat{\nu}_*\mathcal{B}_{\hat{D}}$ to the sheaf of functions on $\hat{T}$ that are identically 1 on $\overline{T}_1$. Hence there is an inclusion $\hat{\nu}_*\mathcal{B}_{\hat{D}} \to \nu_*\mathbb{G}_{m,\hat{T}}$. Since by Proposition 3.4.2 the sheaf $\mathcal{Q}$ is a sheaf of $\mathcal{O}_{\hat{T}}$-modules, there is a homomorphism $\nu_*\mathbb{G}_{m,\hat{T}} \to \pi_{1*}\operatorname{Aut}(\mathcal{V})$ and hence a homomorphism $\hat{\nu}_*\mathcal{B}_{\hat{D}} \to \pi_{1*}\operatorname{Aut}(\mathcal{V})$. It is easy to check that this homomorphism is injective. □

LEMMA 3.4.4. *Let $\operatorname{Aut}'\mathcal{V}$ be the sheaf of automorphisms of $\mathcal{V}$ which are the identity on the subbundle $\pi_1{}^*\pi_{1*}\mathcal{V}$, so that there is a natural homomorphism from $\operatorname{Aut}'\mathcal{V}$ to $\operatorname{Aut}\overline{\mathcal{V}}$. Then there is an isomorphism $\hat{\nu}_*\mathcal{B}_{\hat{D}} \to \pi_{1*}\operatorname{Aut}'\mathcal{V}$ such that the composition*

$$\hat{\nu}_*\mathcal{B}_{\hat{D}} \to \pi_{1*}\operatorname{Aut}'\mathcal{V} \to \pi_{1*}\operatorname{Aut}\overline{\mathcal{V}} = \hat{\nu}_*\mathbb{G}_{m,\hat{T}_0}$$

is the inclusion.

Proof. The previous lemma established an inclusion of $\hat{\nu}_*\mathcal{B}_{\hat{D}}$ in $\pi_{1*}\operatorname{Aut}(\mathcal{V})$. Since the image of $\hat{\nu}_*\mathcal{B}_{\hat{D}}$ is contained in the sheaf of functions on $\hat{T}$ that are identically

CLAIM 3.4.5. *The image of* $\pi_{1*}\operatorname{Aut}'\mathcal{V}$ *in* $\pi_{1*}\operatorname{Aut}\overline{\mathcal{V}} \cong \hat{\nu}_*\mathbb{G}_{m,\hat{T}_0}$ *is contained in* $\hat{\nu}_*\mathcal{B}_{\hat{D}}$.

Proof. Since $\hat{\nu}_*\mathcal{B}_{\hat{D}}$ has the Hartogs property and $\hat{\nu}_*\mathbb{G}_{m,\hat{T}_0}$ is torsion free, it suffices to establish the containment on the complement of a codimension two subset of $\mathcal{M}$. As before, let $X = \pi_1(D)\cap\pi_1((T_0)_{\text{sing}})$, let $U = \mathcal{M} - X$ let $U_1 = \mathcal{M} - \pi_1((T_0)_{\text{sing}})$ and $U_2 = \mathcal{M} - \pi_1(D)$ so that $U = U_1 \cup U_2$. By Lemma 3.4.1, X is a codimension two subset of $\mathcal{M}$ and it will suffice to prove the claim in U. On U_2, the claim is obvious since $\hat{\nu}_*\mathbb{G}_{m,\hat{T}_0}|U_2 = \hat{\nu}_*\mathcal{B}_{\hat{D}}|U_2$. So we may assume that $x \in U_1$. Let Ω be an étale open neighborhood of x in U_1, let $s \in \pi_{1*}\operatorname{Aut}'\mathcal{V}(\Omega)$, and let f_s be the corresponding function on $\Omega\times_{\mathcal{M}}\hat{T}_0$. If $\tilde{x} \in \Omega\times_{\mathcal{M}}\hat{D}$ lies over x, we must show that $f_s(\tilde{x}) = 1$.

For $x \in U_1$, let V_x be the restriction of $\mathcal{V}$ to the slice $\{x\}\times E$, and similarly for $\overline{V}_x$. By the general theory of [**11**], $\overline{V}_x$ is a regular bundle and hence, if $\nu^{-1}(x) = \{y_1,\dots,y_n\}$ where the y_i are distinct, then $\overline{V}_x \cong \bigoplus_{i=1}^n (I_{n_i}\otimes\lambda_i) = \bigoplus_{i=1}^n \overline{V}_{y_i}$, where the λ_i are line bundles on E, pairwise distinct, and indexed by the y_i, and I_n is the unique indecomposable rank n vector bundle on E with a filtration whose successive quotients are all isomorphic to $\mathcal{O}_E$. Correspondingly, the semilocal ring $H^0(\nu^{-1}(x);\mathcal{O}_{T_0}|\nu^{-1}(x))$ is a direct sum of local rings $\bigoplus_i R_i$, where $R_i \cong \mathbb{C}[t]/(t^{n_i})$ is the local ring of the fiber $\nu^{-1}(x)$ at y_i. Thus the group of units R_i^* of R_i is canonically isomorphic to $\mathbb{C}^*\times S_i$, where S_i is unipotent and the projection $R_i^*\to\mathbb{C}^*$ is the evaluation of an invertible function on $\operatorname{Spec} R_i$ at the closed point. The ring R_i is canonically identified with $\operatorname{Hom}(\overline{V}_{y_i},\overline{V}_{y_i})$. Using this identification, the filtration on $\overline{V}_{y_i}$ is defined by $t^k\overline{V}_{y_i}$ and the action of $z\in\mathbb{C}^*$ is given by multiplication by z.

By definition, $y_i \in \hat{D}$ if and only if $\lambda_i = \mathcal{O}_E$. In this case, by the local calculations of [**10**, Theorem 2.4.3], at a generic point of D, the summand of V_x corresponding to the trivial line bundle $\mathcal{O}_E$ is of the form $\mathcal{O}_E^{m_0-1}\oplus I_3$, and there is a natural exact sequence

$$0\to\mathcal{O}_E^{m_0}\to\mathcal{O}_E^{m_0-1}\oplus I_3\to I_2\to 0,$$

such that the quotient I_2 is the summand $\overline{V}_{y_i}$ of $\overline{V}_x$. Let a be the automorphism of V_x induced by the section $s\in\pi_{1*}\operatorname{Aut}'\mathcal{V}(\Omega)$ and the choice of the element $\tilde{x}\in\Omega$ lying over x. By definition, a preserves the subbundle $\mathcal{O}_E^{m_0}$ and is the identity on it, and we need to show that a acts trivially on the associated graded of the quotient I_2. But we can write $a = \operatorname{Id}+b_1+b_2$, where $b_1\colon I_3\to I_3$ is multiplication by a scalar z and b_3 is nilpotent, and the induced action on the associated graded of I_2 is multiplication by $1+z$. Restricting to $\mathcal{O}_E^{m_0}$, we see that $z=0$, and hence a induces the constant function 1, as claimed. Thus, f_s is identically 1 on $\Omega\times_{\mathcal{M}}\hat{D}$. □

Returning to the proof of Lemma 3.4.4, we have constructed homomorphisms in both directions between $\pi_{1*}\operatorname{Aut}'\mathcal{V}$ and $\hat{\nu}_*\mathcal{B}_{\hat{D}}$. On U_2, $\pi_{1*}\operatorname{Aut}'\mathcal{V}|U_2$ and $\hat{\nu}_*\mathcal{B}_{\hat{D}}|U_2$ are both naturally identified with $\hat{\nu}_*\mathbb{G}_{m,\hat{T}_0}|U_2$ and the homomorphisms we have constructed are compatible with these identifications. Hence the two homomorphisms are inverse to each other on U_2. Since both $\pi_{1*}\operatorname{Aut}'\mathcal{V}|U_2$ and $\hat{\nu}_*\mathcal{B}_{\hat{D}}|U_2$ are torsion free and U_2 is open and dense, these homomorphisms induce isomorphisms between $\pi_{1*}\operatorname{Aut}'\mathcal{V}$ and $\hat{\nu}_*\mathcal{B}_{\hat{D}}$. □

COROLLARY 3.4.6. *The composition* $\overline{\mathcal{A}}_\rho \to \pi_{1*}\operatorname{Aut}'\mathcal{V} \to \pi_{1*}\operatorname{Aut}\overline{\mathcal{V}} \cong \hat{\nu}_*\mathbb{G}_{m,\hat{T}_0}$ *defines an embedding of* $\overline{\mathcal{A}}_\rho$ *as a closed subsheaf of* $\hat{\nu}_*\mathcal{B}_{\hat{D}}$.

Proof. This follows from Proposition 3.1.5 and the previous result, using the fact that $\overline{\mathcal{A}}_\rho$ is closed in $\pi_{1*}\operatorname{Aut}\mathcal{V}$ and hence in $\pi_{1*}\operatorname{Aut}'\mathcal{V}$. □

3.5. Some auxiliary cohomology computations. We collect some of the general cohomological results needed to compute $H^0(\mathcal{M};\mathcal{A})$ and $H^1(\mathcal{M};\mathcal{A})$. These computations will ultimately reduce to calculations for $H^i(\mathcal{M};\hat{\nu}_*\mathbb{G}_{m,\hat{T}_0})$ as well as $H^i(\mathcal{M};\hat{\nu}_*\mathcal{B}_{\hat{D}})$, $i=0,1$, or related sheaves, and so we analyze these cohomology groups here.

LEMMA 3.5.1. *Suppose that* ρ *is minuscule. Then*

$$H^0(\mathcal{M};\hat{\nu}_*\mathbb{G}_{m,\hat{T}_0}) = \begin{cases} \mathbb{C}^*, & \text{if } E \text{ is cuspidal;} \\ \mathbb{C}^*\times\mathbb{Z}, & \text{if } E \text{ is nodal.}\end{cases}$$

In both cases, $H^1(\mathcal{M};\hat{\nu}_*\mathbb{G}_{m,\hat{T}_0}) = \{1\}$.

Proof. Since $\hat{\nu}$ is finite, $R^i\hat{\nu}_*\mathcal{S}=0$ for every sheaf $\mathcal{S}$ on $\hat{T}_0$, by e.g. [**3**, Prop. (3.6), p. 24] or [**6**, Corollary 3.4, p. 32]. Thus, by the Leray spectral sequence, $H^i(\mathcal{M};\hat{\nu}_*\mathbb{G}_{m,\hat{T}_0}) = H^i(\hat{T}_0;\mathbb{G}_{m,\hat{T}_0})$ for all i. For $i=0$, $H^0(\hat{T}_0;\mathbb{G}_{m,\hat{T}_0})$ is the group of morphisms from $\hat{T}_0$ to $\mathbb{G}_m$, and for $i=1$, $H^1(\hat{T}_0;\mathbb{G}_{m,\hat{T}_0}) = \operatorname{Pic}\hat{T}_0$.

If E is cuspidal, $\hat{T}_0 = \mathfrak{h}/W_0$. Since W_0 is generated by the reflections in the coroots annihilated by ϖ, $\hat{T}_0 \cong \mathbb{A}^r$. The result is immediate in this case.

Now suppose that E is nodal. Since ρ is minuscule, ϖ is a fundamental weight and hence annihilates all the simple roots except for one, say α. Thus, W_0 is the Weyl group of a subroot system R_0 of R which is generated by all the simple roots of R except α. Clearly Λ_0, the kernel of ϖ on Λ, is exactly the coroot lattice for R_0. It follows that $\mathcal{M}_0 = (E_{\mathrm{reg}}\otimes\Lambda_0)/W_0 \cong \mathbb{A}^{r-1}$. By Lemma 3.3.1, after an étale base change, the morphism $\hat{T}_0 \to E_{\mathrm{reg}}$ becomes a product $\mathcal{M}_0\times C$, where $C\cong E_{\mathrm{reg}}$ is a finite étale cover of E_{reg}. By Lemma 1.2.1, $H^0(\hat{T}_0;\mathbb{G}_{m,\hat{T}_0}) = \mathbb{C}^*\times \mathsf{X}(\mathbb{G}_m) = \mathbb{C}^*\times\mathbb{Z}$ and $H^1(\hat{T}_0;\mathbb{G}_{m,\hat{T}_0}) = \operatorname{Pic}\hat{T}_0 = \{1\}$. □

LEMMA 3.5.2. *Suppose that* ρ *is properly quasi-minuscule. Then*

$$H^0(\mathcal{M};\hat{\nu}_*\mathcal{B}_{\hat{D}}) = \begin{cases} \{1\}, & \text{if } E \text{ is cuspidal;} \\ \mathbb{Z} = \langle f_\varpi\rangle, & \text{if } E \text{ is nodal.}\end{cases}$$

In both cases $H^1(\mathcal{M};\hat{\nu}_*\mathcal{B}_{\hat{D}}) = \{1\}$.

Proof. As in the proof of Lemma 3.5.1, an argument using the Leray spectral sequence shows that it suffices to compute $H^i(\hat{T}_0;\mathcal{B}_{\hat{D}})$. We have a short exact sequence of sheaves

$$\{1\} \to \mathcal{B}_{\hat{D}} \to \mathbb{G}_{m,\hat{T}_0} \to i_*\mathbb{G}_{m,\hat{D}} \to \{1\}.$$

Applying the long exact cohomology sequence to the defining exact sequence for $\mathcal{B}_{\hat{D}}$ then establishes the lemma in this case.

Now suppose that E is nodal. Write $\hat{T}_0 = H/W_0$ and $\hat{D} = H_0/W_0$, where $H_0 = \mathbb{G}_m \otimes \Lambda_0 \subseteq H = \mathbb{G}_m \otimes \Lambda$, and the embedding of $\hat{D}$ in $\hat{T}_0$ is induced by the inclusion $H_0 \subseteq H$. Clearly,

$$H^0(H/W_0; \mathbb{G}_m) = H^0(H; \mathbb{G}_m)^{W_0} = (\mathbb{C}^* \times \mathsf{X}(H))^{W_0} = \mathbb{C}^* \times \mathsf{X}(H)^{W_0}.$$

Similarly, $H^0(H_0/W_0; \mathbb{G}_m) \cong \mathbb{C}^* \times \mathsf{X}(H_0)^{W_0}$, and the restriction map is the identity on the $\mathbb{C}^*$ factors and is induced by restriction of characters on the first factor. We claim that the homomorphism $\mathsf{X}(H)^{W_0} \to \mathsf{X}(H_0)^{W_0}$ is surjective with kernel $\langle f_\varpi \rangle$. The exact sequence of tori

$$\{1\} \to H_0 \to H \xrightarrow{\varpi} \mathbb{G}_m \to \{1\}$$

induces an exact sequence on character groups

$$\{1\} \to \mathsf{X}(\mathbb{G}_m) = \mathbb{Z} \to \mathsf{X}(H) \to \mathsf{X}(H_0) \to \{1\},$$

where the first inclusion is via $\langle \varpi \rangle$. Taking invariants under W_0, we obtain the sequence

$$\{1\} \to \langle f_\varpi \rangle \to \mathsf{X}(H)^{W_0} \to \mathsf{X}(H_0)^{W_0} \to H^1(W_0; \mathbb{Z}),$$

where $H^1(W_0; \mathbb{Z})$ is the group cohomology of W_0 acting trivially on $\mathbb{Z}$. Since every homomorphism from W_0 to $\mathbb{Z}$ is zero, $\mathsf{X}(H)^{W_0} \to \mathsf{X}(H_0)^{W_0}$ is surjective with kernel $\langle f_\varpi \rangle$ as claimed. The result now follows by considering the long exact cohomology sequence associated to the defining exact sequence for $\mathcal{B}_{\hat{D}}$ as before. □

We will need to consider one more computation related to the existence of an involution on $\hat{T}_0$. Suppose that the weights of ρ are invariant under multiplication by -1. Equivalently, ρ is isomorphic to its dual, and it follows that there is a G-invariant nondegenerate bilinear form on V, necessarily either symmetric or skew-symmetric. In this case, multiplication by -1 on E_{reg} induces an involution of $\mathcal{M} \times E_{\text{reg}}$ which leaves T_0 invariant and lifts to the normalization $\hat{T}_0$. We denote the corresponding involution of $\hat{T}_0$ by τ. In terms of the Weyl group, let W_1 be the stabilizer of the pair $\{\pm\varpi\}$ and let $w \in W_1$ be an element such that $w(\varpi) = -\varpi$. Then w acts on $\hat{T}_0$ as the involution τ. Let $\hat{S} = \hat{T}_0/\langle \tau \rangle = \widetilde{\mathcal{M}}/W_1$ and let $\varphi \colon \hat{S} \to \mathcal{M}$ be the induced morphism. Finally, we set

$$(\hat{\nu}_* \mathbb{G}_{m,\hat{T}_0})^- = \{f \in \hat{\nu}_* \mathbb{G}_{m,\hat{T}_0} : \tau^* f = f^{-1}\}$$

in the appropriate sense. Equivalently, we have the norm homomorphism

$$N \colon \hat{\nu}_* \mathbb{G}_{m,\hat{T}_0} \to \varphi_* \mathbb{G}_{m,\hat{S}},$$

and by definition $(\hat{\nu}_* \mathbb{G}_{m,\hat{T}_0})^- = \operatorname{Ker} N$. If f is the pullback of a function g on $\hat{S}$, $N(f) = g^2$ so that the image of N contains the image of $\varphi_* \mathbb{G}_{m,\hat{S}}$ under the squaring homomorphism. Thus N is surjective in the étale topology, and there is an exact sequence

$$\{1\} \to (\hat{\nu}_* \mathbb{G}_{m,\hat{T}_0})^- \to \hat{\nu}_* \mathbb{G}_{m,\hat{T}_0} \to \varphi_* \mathbb{G}_{m,\hat{S}} \to \{1\}.$$

Clearly, $(\hat{\nu}_* \mathbb{G}_{m,\hat{T}_0})^-$ is a closed subsheaf of $\hat{\nu}_* \mathbb{G}_{m,\hat{T}_0}$.

If ρ is properly quasi-minuscule, then the weights of ρ are always invariant under -1. Clearly $w(f_\varpi) = f_\varpi^{-1}$, and so $\hat{D} = f_\varpi^{-1}(1)$ is invariant under τ. In fact, since ϖ is a root, the reflection r_ϖ in ϖ is an element of W sending ϖ to $-\varpi$ and fixing $\hat{D} = (E_{\text{reg}} \otimes (\operatorname{Ker} \varpi))/W_0$ pointwise. Let $(\hat{\nu}_*\mathcal{B}_{\hat{D}})^- = \hat{\nu}_*\mathcal{B}_{\hat{D}} \cap (\hat{\nu}_*\mathbb{G}_{m,\hat{T}_0})^-$. Since $\tau|\hat{D}$ is trivial, a function f on $\hat{D}$ such that $\tau^*f = f^{-1}$ is necessarily ± 1. Since there is also an inclusion of $\{\pm 1\}$ in $(\hat{\nu}_*\mathbb{G}_{m,\hat{T}_0})^-$, it follows that

$$(\hat{\nu}_*\mathbb{G}_{m,\hat{T}_0})^- \cong (\hat{\nu}_*\mathcal{B}_{\hat{D}})^- \times \{\pm 1\}$$

Since $(\hat{\nu}_*\mathbb{G}_{m,\hat{T}_0})^-$ is a closed subsheaf of $\hat{\nu}_*\mathbb{G}_{m,\hat{T}_0}$, $(\hat{\nu}_*\mathcal{B}_{\hat{D}})^-$ is a closed subsheaf of $\hat{\nu}_*\mathcal{B}_{\hat{D}}$.

LEMMA 3.5.3. *In the above notation,*

$$H^0(\mathcal{M}; (\hat{\nu}_*\mathbb{G}_{m,\hat{T}_0})^-) = \begin{cases} \{\pm 1\}, & \text{if } E \text{ is cuspidal;} \\ \{\pm 1\} \times \mathbb{Z}, & \text{if } E \text{ is nodal.} \end{cases}$$

Moreover, in both cases $H^1(\mathcal{M}; (\hat{\nu}_*\mathbb{G}_{m,\hat{T}_0})^-) = \{1\}$.

Proof. Using the defining exact sequence for $(\hat{\nu}_*\mathbb{G}_{m,\hat{T}_0})^-$ and Lemma 3.5.1, it suffices to compute the kernel and cokernel of the norm homomorphism

$$H^0(\mathcal{M}; \hat{\nu}_*\mathbb{G}_{m,\hat{T}_0}) \to H^0(\mathcal{M}; \varphi_*\mathbb{G}_{m,\hat{S}}).$$

Clearly $H^0(\mathcal{M}; \varphi_*\mathbb{G}_{m,\hat{S}}) = H^0(\hat{S}; \mathbb{G}_{m,\hat{S}})$ is the group of τ-invariant morphisms from $\hat{T}_0$ to $\mathbb{G}_m$. Thus, if E is cuspidal, the norm map on H^0 is just $\mathbb{C}^* \to \mathbb{C}^*$ given by squaring, so that its kernel is $\{\pm 1\}$ and its cokernel is trivial. In the nodal case, $H^0(\hat{T}_0; \mathbb{G}_m) \cong \mathbb{C}^* \times \mathbb{Z}$, where we can take f_ϖ to be a generator for $\mathbb{Z}$. The involution τ acts trivially on the $\mathbb{C}^*$ factor and via -1 on the $\mathbb{Z}$ factor, so that $H^0(\hat{S}; \mathbb{G}_{m,\hat{S}}) \cong \mathbb{C}^*$ and the norm map $\mathbb{C}^* \times \mathbb{Z}$ is squaring on the first factor and zero on the second. Thus, in this case, the kernel is $\{\pm 1\} \times \mathbb{Z}$ and the cokernel is trivial. □

Using the isomorphism $(\hat{\nu}_*\mathbb{G}_{m,\hat{T}_0})^- \cong (\hat{\nu}_*\mathcal{B}_{\hat{D}})^- \times \{\pm 1\}$ gives:

LEMMA 3.5.4. *Suppose that ρ is properly quasi-minuscule. Then the weights of ρ are invariant under -1 and $\hat{D} \subset \hat{T}_0$ is non-empty and is invariant under -1. Moreover*

$$H^0(\mathcal{M}; (\hat{\nu}_*\mathcal{B}_{\hat{D}})^-) = \begin{cases} \{1\}, & \text{if } E \text{ is cuspidal;} \\ \mathbb{Z} = \langle f_\varpi \rangle, & \text{if } E \text{ is nodal.} \end{cases}$$

In both cases $H^1(\mathcal{M}; (\hat{\nu}_*\mathcal{B}_{\hat{D}})^-) = \{1\}$. □

4. Explicit examples

In this section, we give a concrete description of the sheaf $\overline{\mathcal{A}}_\rho$ in case ρ is the standard representation of $SL(n+1, \mathbb{C})$, $Sp(2n, \mathbb{C})$, or $Spin(n, \mathbb{C})$ or one of the minuscule representations of a group of type E_6 or E_7. These cases cover every group which has a minuscule representation, and all of the minuscule representations with the exception of the higher exterior powers of the standard representation of $SL(n+1, \mathbb{C})$ and the spin or half spin representations for $Spin(n, \mathbb{C})$. The remaining groups of type E_8, F_4, G_2 do not have any minuscule representations. In the next section, we will give a general procedure for dealing with all of the prop-

4.1. Type A_n: $G = SL(n+1,\mathbb{C})$. Let $\rho\colon SL(n+1,\mathbb{C}) \to GL(n+1,\mathbb{C})$ be the defining (minuscule) representation and let $\mathcal{V}$ be the corresponding vector bundle over $\mathcal{M} \times E$. The restriction $\mathcal{V}|(\{x\} \times E)$ is regular for every $x \in \mathcal{M}$. It follows that $T = T_0 = \hat{T}_0$. Since ρ is faithful, $\overline{\mathcal{A}}_\rho = \mathcal{A}$.

PROPOSITION 4.1.1. *$\mathcal{A} \subseteq \nu_*\mathbb{G}_{m,T}$ is the kernel of the norm map $N\colon \nu_*\mathbb{G}_{m,T} \to \mathbb{G}_{m,\mathcal{M}}$.*

Proof. This is immediate from the identification of N with the determinant homomorphism. □

COROLLARY 4.1.2. *There is an exact sequence*

$$\{1\} \to \mathcal{A} \to \nu_*\mathbb{G}_{m,T} \xrightarrow{N} \mathbb{G}_{m,\mathcal{M}} \to \{1\}.$$

Proof. It suffices to observe that the image of N contains the $(n+1)^{\text{st}}$ powers of functions in $\mathbb{G}_{m,\mathcal{M}}$ and hence N is surjective in the étale topology. □

Thus, there is an exact sequence

$$0 \to H^0(\mathcal{M};\mathcal{A}) \to H^0(T;\mathbb{G}_{m,T}) \to H^0(\mathcal{M};\mathbb{G}_{m,\mathcal{M}}) \to$$
$$\to H^1(\mathcal{M};\mathcal{A}) \to \operatorname{Pic} T \to \operatorname{Pic}\mathcal{M} \to 0.$$

Since $\mathcal{M} \cong \mathbb{A}^n$, $H^0(\mathcal{M};\mathbb{G}_{m,\mathcal{M}}) = \mathbb{C}^*$. By Lemma 3.5.1 (or directly), $\operatorname{Pic} T = 0$, and $H^0(T;\mathbb{G}_{m,T}) \cong \mathbb{C}^*$, in the cuspidal case, and $H^0(T;\mathbb{G}_{m,T}) \cong \mathbb{C}^* \times \mathbb{Z}$ in the nodal case. The norm map N is raising to the $(n+1)^{\text{st}}$ power on the $\mathbb{C}^*$ factors and is trivial on the $\mathbb{Z}$ factor, if it is present (because $N(f_\varpi) = \det$). Hence N is surjective, so that $H^1(\mathcal{M};\mathcal{A}) = \{1\}$, and $H^0(\mathcal{M};\mathcal{A}) = \operatorname{Ker} N$ is either $\mathbb{Z}/(n+1)\mathbb{Z}$ or $(\mathbb{Z}/(n+1)\mathbb{Z}) \times \mathbb{Z}$, depending on whether E is cuspidal or nodal.

REMARK 4.1.3. More generally, let E be any Weierstrass cubic, let B be an arbitrary scheme, and let $\mathcal{V} \to B \times E$ be a family of regular semistable bundles of rank $n+1$ with trivial determinant. In case E is singular, we do not require that $\mathcal{V}|\{x\} \times E$ pulls back to the trivial vector bundle on $\widetilde{E}$ for every $x \in B$. Let $\nu\colon T_{\mathcal{V}} \to B$ be the spectral cover; it is compatible with base change. In this case, if $\Xi_{\mathcal{V}}$ is the associated $SL(n+1,\mathbb{C})$-bundle, then arguments as above show that there is an exact sequence

$$\{1\} \to \pi_{1*}\operatorname{Aut}\Xi_{\mathcal{V}} \to \nu_*\mathbb{G}_{m,T_{\mathcal{V}}} \xrightarrow{N} \mathbb{G}_{m,B} \to \{1\}.$$

A similar remark holds for elliptic fibrations.

4.2. Type C_n: $G = Sp(2n,\mathbb{C})$. Let $\rho\colon G \to GL(2n,\mathbb{C})$ be the standard representation. Thus ρ is faithful and minuscule. In particular $\overline{\mathcal{A}}_\rho = \mathcal{A}$. The corresponding vector bundle $\mathcal{V}$ has regular restriction to each slice $\{x\} \times E$ since the difference of any two distinct weights is a root. Hence $T = T_0 = \hat{T}_0$ in this case.

There is a nondegenerate symplectic form B on $\mathcal{V}$, and hence an isomorphism $\varphi\colon \mathcal{V} \to \mathcal{V}^*$ via $\varphi(v)(w) = B(v,w)$. Note that A is symplectic if and only if $\varphi \circ A^{-1} = A^* \circ \varphi$. As we have seen, the isomorphism φ induces an involution τ on $T = \hat{T}_0$ covering the identity on $\mathcal{M}$. We can see τ quite concretely as follows: we have $T = \mathbf{Spec}\,\pi_{1*}\mathcal{H}om(\mathcal{V},\mathcal{V})$, viewed as a sheaf of $\mathcal{O}_{\mathcal{M}}$-algebras, and the involution is the one defined by

$$\tau(A) = \varphi^{-1} \circ A^* \circ \varphi.$$

It is easy to check that this agrees with the involution τ on T as defined previously. In particular, A is symplectic if and only if $\tau(A) = A^{-1}$. By definition, we have:

PROPOSITION 4.2.1. $\mathcal{A} = (\nu_*\mathbb{G}_{m,T})^-$. □

Applying Lemma 3.5.3 then gives:

COROLLARY 4.2.2. *$H^1(\mathcal{M};\mathcal{A}) = \{1\}$, and $H^0(\mathcal{M};\mathcal{A})$ is isomorphic to $\{\pm 1\}$ if E is cuspidal and $\{\pm 1\} \times \mathbb{Z}$ if E is nodal.*

As in the case of $SL(n+1, \mathbb{C})$, there is a relative version of the sheaf $\mathcal{A}$, including the case where E is smooth or certain cases where the G-bundles $\Xi|\{x\} \times E$ need not pull back to the trivial bundle on the normalization.

4.3. Type D_n: $G = Spin(2n, \mathbb{C})$. In this case, let ρ be the $2n$-dimensional orthogonal representation, i.e. ρ is the composition $Spin(2n, \mathbb{C}) \to SO(2n, \mathbb{C}) \to GL(2n, \mathbb{C})$. In this case the kernel of ρ is a subgroup of order 2. The representation ρ is minuscule, so that $T = T_0$. However, since there are pairs of weights whose difference is not a multiple of a root, there exist $x \in \mathcal{M}$ such that the restriction $\mathcal{V}|(\{x\} \times E)$ is not regular. Thus T_0 is not normal, and hence $\hat{T}_0 \neq T_0$.

As in the symplectic case, the quadratic form on $\mathcal{V}$ defines an isomorphism $\varphi\colon \mathcal{V} \to \mathcal{V}^*$. The involution $\tau\colon \hat{T}_0 \to \hat{T}_0$ is clearly induced by the involution of $\pi_{1*}Hom(\mathcal{V}, \mathcal{V})$ defined by $A \mapsto \varphi^{-1} \circ A^* \circ \varphi$. Moreover, A is orthogonal if and only if $\tau(A) = A^{-1}$. Thus, via Corollary 3.3.5,

PROPOSITION 4.3.1. $\overline{\mathcal{A}}_\rho = (\hat{\nu}_*\mathbb{G}_{m,\hat{T}_0})^-$. □

COROLLARY 4.3.2. *$H^1(\mathcal{M};\mathcal{A}) = \{1\}$, and $H^0(\mathcal{M};\mathcal{A})$ is isomorphic to $Z(G)$ if E is cuspidal and $Z(G) \times \mathbb{Z}$ if E is nodal, where $Z(G) \cong \mathbb{Z}/2\mathbb{Z} \times \mathbb{Z}/2\mathbb{Z}$ if n is even and $Z(G) \cong \mathbb{Z}/4\mathbb{Z}$ if n is odd.*

Proof. By Lemma 3.5.3, $H^1(\mathcal{M};\overline{\mathcal{A}}_\rho) = \{1\}$, and $H^0(\mathcal{M};\overline{\mathcal{A}}_\rho)$ is either $\{\pm 1\}$ if E is cuspidal or $\mathbb{Z} \times \{\pm 1\}$ if E is nodal. By Lemma 3.1.4, $H^1(\mathcal{M};\mathcal{A}) = \{1\}$ and there is an exact sequence

$$0 \to \mathbb{Z}/2\mathbb{Z} \to H^0(\mathcal{M};\mathcal{A}) \to H^0(\mathcal{M};\overline{\mathcal{A}}_\rho) \to \{1\}.$$

It follows that the torsion subgroup of $H^0(\mathcal{M};\mathcal{A})$ has order 4 and its rank is zero if E is cuspidal and one if E is nodal. Since $H^0(\mathcal{M};\mathcal{A})$ contains a subgroup isomorphic to $Z(G)$, it follows that $H^0(\mathcal{M};\mathcal{A})$ is as claimed. □

4.4. Type B_n: $G = Spin(2n+1, \mathbb{C})$. In this case, let ρ be the standard representation induced by $Spin(2n+1) \to SO(2n+1, \mathbb{C}) \subseteq GL(2n+1, \mathbb{C})$. The kernel of ρ has order two. The representation ρ is properly quasi-minuscule and the trivial weight has multiplicity one. The spectral cover is $T = T_0 \cup T_0$, where T_0 is reduced. It is straightforward to check that, if E is cuspidal, then T_0 is normal, whereas T_0 is not normal in case E is nodal. The quadratic form on $\mathcal{V}$ induces an involution on T_0 and $\hat{T}_0$. With this said, we have:

PROPOSITION 4.4.1. $\overline{\mathcal{A}}_\rho = (\hat{\nu}_*\mathcal{B}_{\hat{D}})^-$.

Proof. By Corollary 3.4.6, $\overline{\mathcal{A}}_\rho$ is a closed subsheaf of $\hat{\nu}_*\mathcal{B}_{\hat{D}}$. On the other hand, $(\hat{\nu}_*\mathcal{B}_{\hat{D}})^-$ is also a closed subsheaf of $\hat{\nu}_*\mathcal{B}_{\hat{D}}$. Thus, it suffices to prove that via the character μ, and hence on $\mathcal{L}_{-\mu}$ via the character μ^{-1}. Clearly, this identifies $\overline{\mathcal{A}}_\rho|U$ with $(\hat{\nu}_*\mathcal{B}_{\hat{D}})^-|U$. □

Standard arguments, using Lemma 3.5.4, then show:

COROLLARY 4.4.2. *$H^1(\mathcal{M};\mathcal{A}) = \{1\}$, and $H^0(\mathcal{M};\mathcal{A})$ is isomorphic to $\mathbb{Z}/2\mathbb{Z}$ if E is cuspidal and $(\mathbb{Z}/2\mathbb{Z})\times\mathbb{Z}$ if E is nodal.* □

4.5. Type E_6. Let G be of type E_6 and let $\rho\colon G\to GL(V)$ be one of the 27-dimensional minuscule representations of G. (The two minuscule representations are dual to each other, and the argument below works equally well for either.) The representation ρ is faithful and minuscule. Thus $T = T_0$. However, T_0 is never normal, since there are pairs of weights of ρ whose difference is not a multiple of a root. The group G is the stabilizer of a symmetric cubic form C in the automorphism group of V. If we decompose V into a direct sum of weight spaces $V = \bigoplus_\mu L_\mu$, then the form C preserves the decomposition of V into weight spaces. It is non-trivial on $L_{\mu_1}\otimes L_{\mu_2}\otimes L_{\mu_3}$ if and only if $\mu_1+\mu_2+\mu_3 = 0$.

CLAIM 4.5.1. *Let $\widetilde{H}$ be the set of $A\in GL(V)$ such that $A|L_\mu$ is multiplication by a nonzero scalar a_μ and such that the a_μ satisfy: for every triple of weights (μ_1,μ_2,μ_3) such $\mu_1+\mu_2+\mu_3 = 0$, $a_{\mu_1}a_{\mu_2}a_{\mu_3} = 1$. Then $\rho(H) = \widetilde{H}$.*

Proof. Clearly, if $h\in H$, then $\rho(h)|L_\mu$ is multiplication by $a_\mu = \mu(h)$ and hence the a_μ satisfy the multiplicative property of the claim. Conversely, if $A\in\widetilde{H}$, then A preserves C, so that $A = \rho(g)$ for some $g\in G$. As A commutes with $\rho(H)$ and ρ is faithful, g commutes with H and hence $g\in H$. □

We turn now to a description of $\mathcal{A} = \overline{\mathcal{A}}_\rho$. By Corollary 3.3.5, $\mathcal{A}$ is a closed subsheaf of $\hat{\nu}_*\mathbb{G}_{m,\hat{T}_0}$. We will describe this subsheaf by means of a correspondence on $\hat{T}_0$. Recall that W_0 is the stabilizer of the highest weight ϖ of ρ.

LEMMA 4.5.2. *If μ_1 is a weight of ρ, then there exist weights μ_2,μ_3 of ρ such that $\mu_1+\mu_2+\mu_3 = 0$. The Weyl group W acts transitively on the set of ordered pairs $(\{\mu_1,\mu_2,\mu_3\},\mu_1)$, where $\{\mu_1,\mu_2,\mu_3\}$ is an unordered triple of weights of ρ whose sum is zero.*

Proof. Since the Weyl group acts transitively on the weights of ρ, it suffices to show that the stabilizer W_0 of ϖ acts transitively on the set of weights μ with the property that $\varpi+\mu$ is the negative of a weight of ρ. The kernel of ϖ in the coroot lattice of E_6 is the coroot lattice of D_5. Thus W_0 contains a subgroup isomorphic to the Weyl group of D_5. As a representation of $Spin(10)$, V is a direct sum $\mathbb{C}\oplus V^{10}\oplus S^{16}$, where V^{10} is the orthogonal representation and S^{16} is one of the half spinor representations. The weight spaces L_μ, where μ has the property that $\varpi+\mu$ is the negative of a weight of ρ, are all weight spaces in V^{10}. Since the Weyl group of D_5 acts transitively on these weights, the result follows. □

In terms of cubic surfaces, the weights of ρ may be identified with lines on a smooth cubic surface, and the lemma says that the stabilizer W_0 in W of a line ℓ acts transitively on the 10 lines which meet ℓ.

Given the weight ϖ, fix weights μ_2,μ_3 such that $\varpi+\mu_2+\mu_3 = 0$. Let W' be the stabilizer of the unordered triple $\{\varpi,\mu_2,\mu_3\}$ and let W'' be the stabilizer of the ordered pair $(\{\varpi, \mu_2, \mu_3\}, \varpi)$. Thus W'' is a subgroup both of W_0 and W'. Set $S' = \widetilde{\mathcal{M}}/W'$ and $S'' = \widetilde{\mathcal{M}}/W''$. Thus there is a diagram of schemes

$$\begin{array}{ccccc} & & S'' & & \\ & \swarrow & & \searrow & \\ \hat{T}_0 & & & & S' \\ & \searrow & & \swarrow & \\ & & \mathcal{M} & & \end{array} .$$

Let $\nu' \colon S' \to \mathcal{M}$ and $\nu'' \colon S'' \to \mathcal{M}$ be the induced morphisms. Then there is a homomorphism $c \colon \hat{\nu}_*\mathbb{G}_{m,\hat{T}_0} \to \nu'_*\mathbb{G}_{m,S'}$ defined by pulling a function on $\hat{T}_0$ back to S'', followed by the norm homomorphism $\nu''_*\mathbb{G}_{m,S''} \to \nu'_*\mathbb{G}_{m,S'}$. By Lemma 1.1.2, $\operatorname{Ker} c$ is a closed subsheaf of $\hat{\nu}_*\mathbb{G}_{m,\hat{T}_0}$. If $x \in \mathcal{M}$ is a point such that the morphism $\widetilde{\mathcal{M}} \to \mathcal{M}$ is étale and $\tilde{x} \in \widetilde{\mathcal{M}}$ lies in the fiber over x, the fiber $\nu^{-1}(x)$ is identified with the set of weights of ρ and the fiber $(\nu')^{-1}(x)$ is identified with the set of unordered triples $\{\mu_1, \mu_2, \mu_3\}$ such that $\mu_1 + \mu_2 + \mu_3 = 0$. Clearly,

$$c(f)(\{\mu_1, \mu_2, \mu_3\}) = f(\mu_1)f(\mu_2)f(\mu_3)$$

and hence $c(f) = 1$ if and only if $f(\mu_1)f(\mu_2)f(\mu_3) = 1$ for all such triples.

PROPOSITION 4.5.3. $\mathcal{A} = \operatorname{Ker} c$.

Proof. Since both $\mathcal{A}$ and $\operatorname{Ker} c$ and closed subsheaves of the torsion free sheaf $\hat{\nu}_*\mathbb{G}_{m,\hat{T}_0}$, it suffices to check the equality on an étale open dense set. Let U be the étale open set of Lemma 3.1.2 and Corollary 3.1.3. The result over U follows immediately from the above description of $\operatorname{Ker} c$ and Claim 4.5.1. □

PROPOSITION 4.5.4. *$H^1(\mathcal{M}; \mathcal{A}) = \{1\}$, and $H^0(\mathcal{M}; \mathcal{A})$ is isomorphic to $\mathbb{Z}/3\mathbb{Z}$ if E is cuspidal and $(\mathbb{Z}/3\mathbb{Z}) \times \mathbb{Z}$ if E is nodal.*

Proof. Let $\mathcal{I} = \operatorname{Im} c$. Then $\mathcal{I}$ is a subsheaf of $\nu'_*\mathbb{G}_{m,S'}$ and there is an exact sequence

$$\{1\} \to \mathcal{A} \to \hat{\nu}_*\mathbb{G}_{m,\hat{T}_0} \to \mathcal{I} \to \{1\}.$$

First suppose that E is cuspidal. Since there is a finite morphism from $\widetilde{\mathcal{M}} = \mathfrak{h}$ to S', $\Gamma(S', \mathbb{G}_{m,S'}) \subseteq \Gamma(\mathfrak{h}, \mathbb{G}_{m,\mathfrak{h}^*}) = \mathbb{C}^*$ as $\mathfrak{h}$ is affine. Thus $\Gamma(S', \mathbb{G}_{m,S'}) = \mathbb{C}^*$ consists of the nonzero constant functions. It follows that the homomorphism $c \colon H^0(\mathcal{M}; \hat{\nu}_*\mathbb{G}_{m,\hat{T}_0}) \to H^0(\mathcal{M}; \nu'_*\mathbb{G}_{m,S'})$ is the cubing homomorphism from $\mathbb{C}^*$ to $\mathbb{C}^*$ and factors through the subgroup $H^0(\mathcal{M}; \mathcal{I})$ of $H^0(\mathcal{M}; \nu'_*\mathbb{G}_{m,S'})$. This forces $H^0(\mathcal{M}; \mathcal{I}) = H^0(\mathcal{M}; \nu'_*\mathbb{G}_{m,S'}) = \mathbb{C}^*$, and the induced homomorphism

$$H^0(\mathcal{M}; \hat{\nu}_*\mathbb{G}_{m,\hat{T}_0}) \to H^0(\mathcal{M}; \mathcal{I})$$

is again the cubing homomorphism $\mathbb{C}^* \to \mathbb{C}^*$. Thus it is surjective with kernel $\mathbb{Z}/3\mathbb{Z}$. It follows that $H^0(\mathcal{M}; \mathcal{A}) = \mathbb{Z}/3\mathbb{Z}$ and that $H^1(\mathcal{M}; \mathcal{A}) \subseteq H^1(\mathcal{M}; \hat{\nu}_*\mathbb{G}_{m,\hat{T}_0})$, which is trivial by Lemma 3.5.1. This concludes the proof in the cuspidal case.

Now assume that E is nodal. By Lemma 3.5.1, $H^0(\mathcal{M}; \hat{\nu}_*\mathbb{G}_{m,\hat{T}_0}) \cong \mathbb{C}^* \times \mathbb{Z}$.

$c(f_\varpi) = 1$, so that it suffices to show that $H^0(\mathcal{M}; \nu'_*\mathbb{G}_{m,S'}) = H^0(S'; \mathbb{G}_{m,S'}) = \mathbb{C}^*$. As in the proof of Lemma 3.5.2,

$$H^0(S'; \mathbb{G}_{m,S'}) = H^0(H; \mathbb{G}_{m,H})^{W'} = (\mathbb{C}^* \times \mathsf{X}(H))^{W'} = \mathbb{C}^* \times \mathsf{X}(H)^{W'}.$$

Since $\mathsf{X}(H) = \Lambda^*$, it suffices to prove that $(\mathfrak{h}^*)^{W'} = 0$, or equivalently that $\mathfrak{h}^{W'} = 0$. It is easy to check that, if ϖ, μ_2, μ_3 are three weights of ρ whose sum is zero, then $\{\varpi, \mu_2, \mu_3\}^\perp$ is a root lattice of type D_4 and hence W' contains the Weyl group of D_4. Also, by Lemma 4.5.2, W' acts transitively on the set $\{\varpi, \mu_2, \mu_3\}$. From this, it follows that there are no W'-invariant vectors in $\mathfrak{h} = \mathrm{span}\{\varpi, \mu_2, \mu_3\} \oplus \{\varpi, \mu_2, \mu_3\}^\perp$.

Thus $H^0(S'; \mathbb{G}_{m,S'}) = \mathbb{C}^*$, and the rest of the proof proceeds as in the cuspidal case. □

4.6. Type E_7. Let G be of type E_7, and let $\rho\colon G \to GL(V)$ be the 56-dimensional irreducible representation of E_7. Since ρ is faithful and minuscule, $\overline{\mathcal{A}}_\rho = \mathcal{A}$ and $T = T_0$. Since there are pairs of weights of ρ whose difference is not a multiple of a root, T_0 is never normal. The group G is the stabilizer of a non-degenerate skew form and a symmetric quartic form on V. Decomposing V as a direct sum of weight spaces, $V = \bigoplus_\mu L_\mu$, the skew form is non-zero exactly on tensor products of the form $L_\mu \otimes L_{-\mu}$ and the quartic form is nontrivial exactly on tensor products of the form $L_{\mu_1} \otimes L_{\mu_2} \otimes L_{\mu_3} \otimes L_{\mu_4}$ when $\mu_1 + \mu_2 + \mu_3 + \mu_4 = 0$ and $\mu_1 + \mu_i \neq 0$ for $i = 2, 3, 4$. We say that an unordered quadruple $\{\mu_1, \mu_2, \mu_3, \mu_4\}$ of weights *properly sums to zero* $\mu_1 + \mu_2 + \mu_3 + \mu_4 = 0$ and $\mu_i + \mu_j \neq 0$ for all i, j, $1 \le i, j \le 4$.

An argument as for the case of E_6 then shows:

CLAIM 4.6.1. *Let $\widetilde{H}$ be the set of $A \in GL(V)$ such that $A|L_\mu$ is multiplication by a scalar a_μ and such that the a_μ satisfy: $a_\mu a_{-\mu} = 1$ for all weights μ and $a_{\mu_1} a_{\mu_2} a_{\mu_3} a_{\mu_4} = 1$ whenever $\{\mu_1, \mu_2, \mu_3, \mu_4\}$ properly sums to zero. Then $\rho(H) = \widetilde{H}$.* □

We turn now to the description of $\mathcal{A}$. Since the skew form identifies V and V^*, there is an involution τ on $\hat{T}_0$. By Claim 4.6.1, $\mathcal{A}$ is a closed subsheaf of $(\hat{\nu}_*\mathbb{G}_{m,\hat{T}_0})^-$. As in the case of E_6, we will describe this subsheaf by a correspondence. We begin with a lemma on the Weyl group:

LEMMA 4.6.2. *The group W acts transitively on the set of all pairs*

$$(\{\mu_1, \mu_2, \mu_3, \mu_4\}, \mu_1)$$

consisting of a quadruple which properly sums to zero and one of the weights in the quadruple. Moreover, if $\{\varpi, \mu_2, \mu_3, \mu_4\}$ is such a quadruple, the stabilizer W_0 of ϖ contains an element which cyclically permutes the set $\{\mu_2, \mu_3, \mu_4\}$.

Proof. Since W acts transitively on the weights of ρ, to prove the first statement, it suffices to show that W_0 acts transitively on the set of all quadruples $\{\varpi, \mu_2, \mu_3, \mu_4\}$ which properly sum to zero. The kernel of ϖ is the coroot lattice Λ_1 of E_6, and hence W_0 contains a subgroup W_1 isomorphic to the Weyl group of E_6. Let G_1 be the corresponding subgroup of G of type E_6. The representation V viewed as a representation of G_1, is isomorphic to a direct sum $\mathbb{C} \oplus \mathbb{C} \oplus A \oplus \overline{A}$ where A and $\overline{A}$ are the two non-isomorphic minuscule representations of G_1. The weight spaces L_μ with μ as above form one of the minuscule factors A or $\overline{A}$. In fact, possibly after relabeling, the quadruples containing ϖ which properly sum to zero consist of ϖ and a triple of weights which become weights for the maximal torus of G_1 acting on A and which sum to zero as weights acting on Λ_1. Since W_1 acts transitively on these weights, the first statement follows. The second is a consequence of Lemma 4.5.2. $\square$

There is an interpretation of the proof of Lemma 4.6.2 in terms of exceptional curves on a del Pezzo surface X of degree two. Viewing X as the blowup of a smooth cubic surface X_1 at a general point p_7, let E_7 be the proper transform of p_7. The weights of V are identified with the set of all exceptional curves on X. Two exceptional curves on X can meet in 0, 1 or 2 points, and they meet in two points if and only if their sum as weights is zero. If we let ϖ correspond to $3H - \sum_{i=1}^{6} E_i - 2E_7$, the proper transform of a nodal cubic on X_1 whose singular point is p_7, then the remaining weights μ_2, μ_3, μ_4 correspond to exceptional curves on X, not equal to E_7, whose sum is $3H - \sum_{i=1}^{6} E_i$, and hence to lines on X_1 whose sum is a hyperplane section.

Returning to our study of $\mathcal{A}$, fix a quadruple $\{\varpi, \mu_2, \mu_3, \mu_4\}$ which properly sums to zero. Let W' be the stabilizer of $\{\varpi, \mu_2, \mu_3, \mu_4\}$ and let W'' be the stabilizer of the pair $(\{\varpi, \mu_2, \mu_3, \mu_4\}, \varpi)$. Thus W'' is a subgroup both of W_0 and of W'. As in the case of E_6, we form the spaces $S' = \widetilde{\mathcal{M}}/W'$ and $S'' = \widetilde{\mathcal{M}}/W''$. Clearly, the degree of the morphism $S'' \to S'$ is 4. If $\nu'\colon S' \to \mathcal{M}$ is the induced morphism, then we can again define the homomorphism $c\colon \hat{\nu}_*\mathbb{G}_{m,\hat{T}_0} \to \hat{\nu}'_*\mathbb{G}_{m,S'}$. Clearly, the involution τ extends to involutions on S'' and S', also denoted τ, and c is equivariant. Thus the restriction of c to $(\hat{\nu}_*\mathbb{G}_{m,\hat{T}_0})^-$ takes values in $(\hat{\nu}'_*\mathbb{G}_{m,S'})^-$. An argument as in the proof of Proposition 4.5.3 gives:

PROPOSITION 4.6.3. $\mathcal{A} = (\hat{\nu}_*\mathbb{G}_{m,\hat{T}_0})^- \cap \operatorname{Ker} c.$ $\square$

PROPOSITION 4.6.4. *$H^1(\mathcal{M};\mathcal{A}) = \{1\}$, and $H^0(\mathcal{M};\mathcal{A})$ is isomorphic to $\mathbb{Z}/2\mathbb{Z}$ if E is cuspidal and $(\mathbb{Z}/2\mathbb{Z}) \times \mathbb{Z}$ if E is nodal.*

Proof. Let $\mathcal{I} = \operatorname{Im}(c|(\hat{\nu}_*\mathbb{G}_{m,\hat{T}_0})^-)$. Then $\mathcal{I}$ is a subsheaf of $(\nu'_*\mathbb{G}_{m,S'})^-$ and there is an exact sequence

$$\{1\} \to \mathcal{A} \to (\hat{\nu}_*\mathbb{G}_{m,\hat{T}_0})^- \to \mathcal{I} \to \{1\}.$$

The proof is completed, using Lemma 3.5.3, by showing that $H^0(\mathcal{M};\mathcal{I}) = 0$.

We begin by showing that $H^0(\mathcal{M};(\nu'_*\mathbb{G}_{m,S'})^-) = \{\pm 1\}$. In the case when E is cuspidal, S' is a quotient of $\mathfrak{h}$ by a finite group and hence $H^0(\mathcal{M};\nu'_*\mathbb{G}_{m,S'}) = \mathbb{C}^*$. Since τ acts trivially on this $\mathbb{C}^*$, $H^0(\mathcal{M};(\nu'_*\mathbb{G}_{m,S'})^-) = \{\pm 1\}$. When E is nodal, $H^0(\mathcal{M};\nu'_*\mathbb{G}_{m,S'})$ is equal to the product $\mathbb{C}^* \times \mathsf{X}(H)^{W'}$ and we must compute the second factor. Given the quadruple $\{\varpi, \mu_2, \mu_3, \mu_4\}$ which properly sums to zero, let Λ' be the intersection of the kernels of the weights $\varpi, \mu_2, \mu_3, \mu_4$ and let $H' \subseteq H$ be the corresponding closed connected subgroup. Note that Λ' is a coroot lattice of type D_4. Let H'' be the torus whose Lie algebra is spanned by the weights, which we view as elements of $\mathfrak{h}$ via the inner product. Note that W' acts on H' and H'', and there is a W'-equivariant isogeny $H' \times H'' \to H$. Correspondingly, there is a W'-equivariant embedding $\mathsf{X}(H) \to \mathsf{X}(H') \oplus \mathsf{X}(H'')$. Since W' contains a subgroup generated by the reflections in the coroots contained in Λ', which is the Weyl group which sends ϖ to μ_2. These two elements generate a subgroup of the symmetric group S_4 which is of index at most two and hence contains the alternating group. It follows that $\mathsf{X}(H'')^{W'} = 0$ as well. Thus again $H^0(\mathcal{M}; \nu'_*\mathbb{G}_{m,S'}) = \mathbb{C}^*$ and $H^0(\mathcal{M}; (\nu'_*\mathbb{G}_{m,S'})^-) = \{\pm 1\}$.

Thus $H^0(\mathcal{M}; (\nu'_*\mathbb{G}_{m,S'})^-) = \{\pm 1\}$, where the functions in question are constant functions on S'. The final step is to show that the image of $H^0(\mathcal{M}; \mathcal{I})$ in $H^0(\mathcal{M}; (\nu'_*\mathbb{G}_{m,S'})^-) = \{\pm 1\}$ is trivial. Let $x_0 \in \mathcal{M}$ be the image of the identity in $\widetilde{\mathcal{M}}$. Since the identity is fixed by W, there is a unique preimage of x_0 in $\hat{T}_0$, as well as in S' and S'', and all of these points are τ-invariant. Let $U \to \mathcal{M}$ be an étale neighborhood of x_0 and u a point in the preimage of x_0. Then there is a unique point $\hat{u} \in U \times_{\mathcal{M}} \hat{T}_0$ lying above u, and similarly there are unique points $u' \in U \times_{\mathcal{M}} S'$ and $u'' \in U \times_{\mathcal{M}} S''$ above u. In particular, $\tau(u'') = u''$, where we continue to denote by τ the induced involution on $U \times_{\mathcal{M}} S''$. Hence, if $f \in (\hat{\nu}_*\mathbb{G}_{m,\hat{T}_0})^-(U)$, then $f(\hat{u}) = \pm 1$, and the value of the pullback of f to $U \times_{\mathcal{M}} S''$ at u'' is again ± 1. Finally, since u'' is a point of total ramification for the morphism $S'' \to S'$, and the degree of the morphism is 4, the value of $c(f)$ at u' is $(\pm 1)^4 = 1$. It follows that every function in $H^0(\mathcal{M}; \mathcal{I})$ takes the value 1 at u', and since these functions are constant, it must be identically 1. Thus $H^0(\mathcal{M}; \mathcal{I}) = \{1\}$ as claimed. □

5. The general case

Our goal in this section is to give a unified proof of Theorem 2.2.3. Throughout this section, let $\rho: G \to GL(V)$ denote the unique properly quasi-minuscule representation of G. Hence $\operatorname{Ker} \rho = Z(G)$ and $\overline{\mathcal{A}}_\rho = \mathcal{A}/Z(G)$.

5.1. Identification of the image torus. Let R_s be the set of short roots of G. It is the set of nonzero weights of ρ. Let $m_0 \geq 1$ be the multiplicity of the trivial weight in ρ. For $\alpha \in R_s$, let L_α denote the corresponding one-dimensional weight space of V. Let L_0 be the subspace of V where H acts trivially.

LEMMA 5.1.1. *$\rho(H)$ is the subgroup $\widetilde{H}$ of $GL(V)$ consisting of the A such that $A|L_0 = \mathrm{Id}$, $A|L_\alpha$ is multiplication by a nonzero scalar a_α and the a_α satisfy: $a_\alpha a_{-\alpha} = 1$, and, for every triple $(\alpha, \beta, \gamma) \in R_s^3$ such that $\alpha+\beta+\gamma = 0$, $a_\alpha a_\beta a_\gamma = 1$.*

Proof. The element $\rho(h)$ acts on L_0 as the identity and on L_α via $\alpha(h)$. Thus clearly $\rho(H) \subseteq \widetilde{H}$. To see the opposite inclusion, we begin with the following general discussion. Under the identification of W/W_0 with R_s, there is a homomorphism $\varphi: \mathbb{Z}[W/W_0] \to \Lambda^*$. Here, if e_α is the basis vector corresponding to α, $\varphi(e_\alpha) = \alpha$. The image of φ is the span of the short roots, which is just the root lattice $Q(R)$. The kernel K of φ is the set of all relations among the elements of R_s.

LEMMA 5.1.2. *The kernel K is generated by the following two types of relations:*

(i) *For $\alpha \in R_s$, $e_\alpha + e_{-\alpha}$;*

(ii) *For $\alpha, \beta, \gamma \in R_s$ such that $\alpha + \beta = \gamma$, $e_\alpha + e_\beta - e_\gamma$.*

Proof. If G is not simply laced, the set R_s forms a root system whose span is that of R. Thus it suffices to consider the case $R = R_s$, i.e. G is simply laced (but not necessarily irreducible). Let $v = \sum_{\alpha \in R_s} n_\alpha e_\alpha \in K$, so that $\sum_{\alpha \in R_s} n_\alpha \alpha = 0$. Using relations of type (i), we may assume that all of the α are positive roots. If $\alpha_1, \dots, \alpha_r$ are the simple roots, and $\alpha = \sum_i n_i \alpha_i$ is a positive root, let $\ell(\alpha) = \sum_i n_i$. The proof is by induction on the largest value of $\ell(\alpha)$ where $n_\alpha \neq 0$. If this number is greater than one, then $\alpha = \alpha_i + \beta$ for some i, where $\ell(\beta) = \ell(\alpha) - 1$. Applying the relation $n_\alpha(e_\alpha - e_{\alpha_i} - e_\beta)$ then eliminates e_α. Eventually we reach the stage where all α are simple. Since the simple roots are linearly independent, it follows that at this stage $v = 0$. $\square$

Let F be the free $\mathbb{Z}$-module with basis $f_\alpha, \alpha \in R_s^+$ and $f_{\alpha,\beta}$, where α, β, and $\alpha + \beta \in R_s$. Let $\psi\colon F \to \mathbb{Z}[W/W_0]$ be defined by $\psi(f_\alpha) = e_\alpha + e_{-\alpha}$ and $\psi(f_{\alpha,\beta}) = e_\alpha + e_\beta - e_{\alpha+\beta}$. The preceding lemma implies that

$$F \xrightarrow{\psi} \mathbb{Z}[W/W_0] \xrightarrow{\varphi} Q(R) \to 0$$

is exact, where $Q(R)$ is the root lattice.

Applying $\mathrm{Hom}(\cdot, \mathbb{C}^*)$ to this sequence gives an exact sequence of tori:

$$\{1\} \to \mathrm{Hom}(Q(R), \mathbb{C}^*) \to \mathrm{Hom}(\mathbb{Z}[W/W_0], \mathbb{C}^*) \to \mathrm{Hom}(F, \mathbb{C}^*).$$

Moreover, $\mathrm{Hom}(Q(R), \mathbb{C}^*) = \overline{H}$ is the maximal torus in the adjoint form $\rho(G)$ of G, and the induced homomorphism $H \to \overline{H} \to \mathrm{Hom}(\mathbb{Z}[W/W_0], \mathbb{C}^*)$ is just the homomorphism which sends $h \in H$ to the homomorphism $\mathbb{Z}[W/W_0] \to \mathbb{C}^*$ defined by $\alpha \mapsto \alpha(h)$. It follows that, given $a_\alpha \in \mathbb{C}^*$ for $\alpha \in R_s$ satisfying the conditions of the lemma, there is an $h \in H$ such that $\alpha(h) = a_\alpha$. Thus $\rho(h) = A$ as claimed. $\square$

5.2. The spectral cover and the basic correspondence. Since ρ is not minuscule, the spectral cover T is of the form $m_0 \cdot \overline{T}_1 + T_0$ for a positive integer m_0. The divisor $D = \overline{T}_1 \cap T_0$ is nonempty, as is its preimage $\hat{D}$ in the normalization $\hat{T}_0$ of T_0. As the weights of ρ are invariant under -1, there is an involution τ of $\hat{T}_0$.

We shall need the following lemma on roots:

LEMMA 5.2.1. *The Weyl group W acts transitively on the set of unordered triples $\{\alpha, \beta, \gamma\}$ with $\alpha, \beta, \gamma \in R_s$ and $\alpha + \beta + \gamma = 0$. The stabilizer in W of such a triple contains a cyclic permutation of the set $\{\alpha, \beta, \gamma\}$.*

Proof. To prove the first statement, it suffices to prove that W acts transitively on the set of unordered pairs $\{\alpha, \beta\}$ such that α, β, and $\alpha + \beta$ are short roots, i.e. with the set of embeddings of a root system of type A_2 into the subroot system of short roots G up to the action of the outer automorphism group of A_2. Given two such pairs $\{\alpha_1, \beta_1\}$ and $\{\alpha_2, \beta_2\}$, we may assume that $\alpha_1 = \alpha_2$ since W acts transitively on the set of short roots. Then $\{\alpha_1, \beta_1, \beta_2\}$ spans a root system of type A_2 (if $\beta_1 = \beta_2$ or if the Cartan integer $n(\beta_1, \beta_2) = -1$) or of type A_3 (if the Cartan integer $n(\beta_1, \beta_2)$ is 0 or 1), and the result can be checked directly in either case.

The second statement is an immediate consequence of the fact that the Weyl group of a root system of type A_2 is the symmetric group S_3 and contains an index two subgroup as claimed. $\square$

We now define a correspondence on $\hat{T}_0$ which commutes with τ in an appropriate sense. As we have seen in Lemma 5.2.1, W acts transitively on the set of unordered triples $\{\alpha, \beta, \gamma\}$ with $\alpha, \beta, \gamma \in R_s$ and $\alpha + \beta + \gamma = 0$. Let $\beta \in R_s$ be such that $\varpi + \beta = -\gamma \in R_s$. (Such a β exists except in case G is of type B_n or A_1.) Let $W' \subset W$ be the stabilizer of the set $\{-\beta, \gamma\}$ and set $S' = \widetilde{M}/W'$. Let $T_?$ be irreducible, i.e. that W acts transitively on the set of ordered pairs (α, β) of short roots whose sum is a short root, as long as G is not of type A_n. However, we shall not need this fact. Viewing $\hat{T}_0 = \widetilde{\mathcal{M}}/W_0$ as the quotient $(\widetilde{\mathcal{M}} \times R_s)/W$ and similarly for S', we see that there are morphisms $S'' \to \hat{T}_0$ and $S'' \to S'$. Thus as in the case of E_6 and E_7 there is a diagram

$$\begin{array}{ccccc} & & S'' & & \\ & \swarrow & & \searrow & \\ \hat{T}_0 & & & & S' \\ & \searrow & & \swarrow & \\ & & \mathcal{M} & & \end{array} .$$

There are obvious analogues of the involution τ on S' and S'', for which the above morphisms are equivariant. For example, the involution on S'' is induced by the function $(x, (\alpha, \beta, \gamma)) \mapsto (x, (-\alpha, -\beta, -\gamma))$. We denote all of these involutions by τ. Let $\nu' \colon S' \to \mathcal{M}$ and $\nu'' \colon S'' \to \mathcal{M}$ be the induced morphisms, so that $\nu' \circ \tau = \nu'$, and similarly for ν''. We define the correspondence homomorphism $c \colon \hat{\nu}_* \mathbb{G}_{m,\hat{T}_0} \to \nu'_* \mathbb{G}_{m,S'}$ in the usual way, by pulling a function up to S'' and taking the image of the norm homomorphism from $\nu''_* \mathbb{G}_{m,S''}$ to $\nu'_* \mathbb{G}_{m,S'}$. Clearly, c commutes with τ, and so there is an induced homomorphism, which we also denote by c, from $(\hat{\nu}_* \mathbb{G}_{m,\hat{T}_0})^-$ to $(\nu'_* \mathbb{G}_{m,S'})^-$.

For G of type A_1 or B_n, there are no triples $\{\alpha, \beta, \gamma\}$ with $\alpha, \beta, \gamma \in R_s$ such that $\alpha + \beta + \gamma = 0$. In this case, the constructions above are vacuous and there is no correspondence to consider. In this case, we define c to be the trivial homomorphism.

For future reference, we shall need the following lemma:

LEMMA 5.2.2. *Assume that G is not of type A_1 or B_n. There is a nonempty codimension two subscheme X of S' such that, for every étale open set $U \to \mathcal{M}$, if $f \in \hat{\nu}_* \mathcal{B}_{\hat{D}}(U)$, then $c(f)|U \times_{\mathcal{M}} X = 1$. If $\beta \in R_s$ is such that $\varpi + \beta = -\gamma \in R_s$ and W' is the stabilizer of the set $\{\varpi, \beta, \gamma\}$, then X is the subset of $S' = \widetilde{\mathcal{M}}/W'$ where the functions from S' to E_{reg} defined by ϖ, β, γ are all equal to the identity.*

Proof. Let $\mathcal{R}_1$ be the subset of ordered triples $(\alpha, \beta, \gamma) \in R_s^3$ such that $\alpha + \beta + \gamma = 0$, and let $\mathcal{R}_2$ be the set of unordered triples $\{\alpha, \beta, \gamma\}$ such that $\alpha, \beta, \gamma \in R_s$ and $\alpha + \beta + \gamma = 0$. There is the natural correspondence $\tilde{c}$ from $\widetilde{\mathcal{M}} \times R_s$ to $\widetilde{\mathcal{M}} \times \mathcal{R}_2$ defined as follows. Let $\varphi_1 \colon \widetilde{\mathcal{M}} \times \mathcal{R}_1 \to \widetilde{\mathcal{M}} \times R_s$ and $\varphi_2 \colon \widetilde{\mathcal{M}} \times \mathcal{R}_1 \to \widetilde{\mathcal{M}} \times \mathcal{R}_2$ be the obvious morphisms: $\varphi_1(x, (\alpha, \beta, \gamma)) = (x, \alpha)$ and $\varphi_2(x, (\alpha, \beta, \gamma)) = (x, \{\alpha, \beta, \gamma\})$. Then $\tilde{c}(g)$ is the pullback of g under φ_1, followed by the norm map induced by φ_2. Explicitly,

$$\tilde{c}(g)(x, \{\alpha, \beta, \gamma\}) = g(x, \alpha) g(x, \beta) g(x, \gamma).$$

The function g is the pullback of a function f on $\hat{T}_0 = (\widetilde{\mathcal{M}} \times R_s)/W$, if and only if it is W-invariant. In this case, $\tilde{c}(g)$ is also W-invariant, and hence is the pullback of a function on S', which is clearly $c(f)$.

Let f be a function on $\hat{T}_0$ which is 1 on $\hat{D}$. Then its pullback g to $\widetilde{\mathcal{M}} \times R_s$ is identically 1 on $\coprod_{\alpha \in R_s} D_\alpha$, where

$$D_\alpha = \{(x, \alpha) \in \widetilde{\mathcal{M}} \times R_s : \alpha(x) \text{ is the identity } p_0\}.$$

Thus $\tilde{c}(g)$ is identically 1 on the subscheme $\coprod_{\{\alpha,\beta,\gamma\}\in\mathcal{R}_2} X_{\alpha,\beta,\gamma}$, where

$$X_{\alpha,\beta,\gamma} = \{(x, \{\alpha,\beta,\gamma\}) \in \widetilde{\mathcal{M}} \times \mathcal{R}_2 : \alpha(x) = \beta(x) = \gamma(x) \text{ is the identity } p_0\}.$$

Clearly $X_{\alpha,\beta,\gamma}$ is nonempty and W-invariant, and it is of codimension two since $\alpha+\beta+\gamma = 0$. Hence $\coprod_{\{\alpha,\beta,\gamma\}\in\mathcal{R}_2} X_{\alpha,\beta,\gamma}$ is the pullback of a nonempty codimension two subset X of S' which has the required properties. □

We now claim:

PROPOSITION 5.2.3. *The sheaf $\overline{\mathcal{A}}_\rho$ is isomorphic to the closed subsheaf* $\operatorname{Ker} c \cap (\hat{\nu}_*\mathcal{B}_{\hat{D}})^-$ *of* $(\hat{\nu}_*\mathcal{B}_{\hat{D}})^-$.

Proof. By Corollary 3.4.6, $\overline{\mathcal{A}}_\rho$ is isomorphic to a closed subsheaf of $\hat{\nu}_*\mathcal{B}_{\hat{D}}$. Thus it suffices to prove that, via this isomorphism, $\overline{\mathcal{A}}_\rho \cong \hat{\nu}_*\mathcal{B}_{\hat{D}} \cap (\hat{\nu}_*\mathbb{G}_{m,\hat{T}_0})^- \cap \operatorname{Ker} c$. Since both the image of $\overline{\mathcal{A}}_\rho$ and $\hat{\nu}_*\mathcal{B}_{\hat{D}} \cap (\hat{\nu}_*\mathbb{G}_{m,\hat{T}_0})^- \cap \operatorname{Ker} c$ are closed in $\hat{\nu}_*\mathcal{B}_{\hat{D}}$, it suffices to prove that their restrictions to an open dense set agree. As usual, we use the open set $U \to \mathcal{M}$ of Lemma 3.1.2 and Corollary 3.1.3, and we shall prove that their restrictions to U agree. Note that $\hat{\nu}(\hat{D})$ does not lie in the image of U. Over U, $\mathcal{A}$ is the sheaf of sections of the constant group scheme $U \times H$ and hence $\overline{\mathcal{A}}_\rho$ is the sheaf of sections of the constant group scheme $U \times \overline{H}$. The fiber of $\hat{T}_0$ over a point x in the image of U is indexed by the weights of ρ, i.e. by R_s. Thus if $f \in \hat{\nu}_*\mathbb{G}_{m,\hat{T}_0}(U)$, its value at x is given by elements $a_\alpha \in \mathbb{C}^*$. To say that $f \in (\hat{\nu}_*\mathbb{G}_{m,\hat{T}_0})^- \cap \operatorname{Ker} c$ means exactly that $a_\alpha a_{-\alpha} = 1$ and that, for every triple $(\alpha,\beta,\gamma) \in R_s^3$ such that $\alpha+\beta+\gamma = 0$, $a_\alpha a_\beta a_\gamma = 1$. By Lemma 5.1.1, this is exactly the condition that f lies in the image of $\overline{\mathcal{A}}_\rho$. □

Proof of Theorem 2.2.3. By Lemma 3.1.4, to prove Theorem 2.2.3, it will suffice to prove that $H^1(\mathcal{M};\overline{\mathcal{A}}_\rho) = \{1\}$ and that $H^0(\mathcal{M};\overline{\mathcal{A}}_\rho) = \{1\}$ or $\mathbb{Z}$ depending on whether E is cuspidal or nodal. If c is trivial, then $\overline{\mathcal{A}}_\rho = (\hat{\nu}_*\mathcal{B}_{\hat{D}})^-$ and the result is just Lemma 3.5.4. So we may assume that c is not trivial. Restricting c to $(\hat{\nu}_*\mathcal{B}_{\hat{D}})^-$ and applying Proposition 5.2.3 gives an exact sequence

$$\{1\} \to \overline{\mathcal{A}}_\rho \to (\hat{\nu}_*\mathcal{B}_{\hat{D}})^- \to \mathcal{I} \to \{1\},$$

where $\mathcal{I}$ is the image of $(\hat{\nu}_*\mathcal{B}_{\hat{D}})^-$ in $\nu'_*\mathbb{G}_{m,S'}$.

As usual, assume first that E is cuspidal. Since S' is the quotient of the affine space $\mathfrak{h}$ by a finite group, $H^0(\mathcal{M};\nu'_*\mathbb{G}_{m,S'}) \cong \mathbb{C}^*$ is identified with the group of nonzero constant functions on S'. Since $H^0(\mathcal{M};(\hat{\nu}_*\mathcal{B}_{\hat{D}})^-) = H^1(\mathcal{M};(\hat{\nu}_*\mathcal{B}_{\hat{D}})^-) = \{1\}$ in this case, by Lemma 3.5.4, it suffices to show that $H^0(\mathcal{M};\mathcal{I}) = \{1\}$. In any case, $H^0(\mathcal{M};\mathcal{I})$ is a subsheaf of $H^0(\mathcal{M};\nu'_*\mathbb{G}_{m,S'}) \cong \mathbb{C}^*$ and hence consists of constant functions on S'. To determine the value of these constant functions, it suffices to restrict to the nonempty subset X of Lemma 5.2.2, where we have seen that the image of any local section of $\hat{\nu}_*\mathcal{B}_{\hat{D}}$ takes the value 1. Hence $H^0(\mathcal{M};\mathcal{I}) = \{1\}$, concluding the proof in the cuspidal case.

Finally assume that E is nodal. In this case, $H^0(\mathcal{M};(\hat{\nu}_*\mathcal{B}_{\hat{D}})^-) \cong \mathbb{Z}$ and $H^1(\mathcal{M};(\hat{\nu}_*\mathcal{B}_{\hat{D}})^-) = \{1\}$ by Lemma 3.5.4. We must again show that $H^0(\mathcal{M};\mathcal{I}) =$

We have seen that $S' = H/W'$, where W' is the stabilizer of an unordered triple of short roots $\{\varpi, \beta, \gamma\}$ whose sum is zero. Thus

$$H^0(S'; \mathbb{G}_{m,S'}) \subseteq (H^0(H; \mathbb{G}_{m,H}))^{W'} = (\mathbb{C}^* \times \mathsf{X}(H))^{W'} = \mathbb{C}^* \times \mathsf{X}(H)^{W'}.$$

Suppose that $g \in H^0(S'; \mathbb{G}_{m,S'})$ and $g|X = 1$. Then, since the origin 1 of H is contained in the pullback of X, the pullback of g is the identity at $1 \in H$. It follows that the pullback of g lies in $\mathsf{X}(H)^{W'}$.

By Lemma 5.2.2, the preimage of X in H contains the identity component of $\operatorname{Ker}\varpi \cap \operatorname{Ker}\beta \cap \operatorname{Ker}\gamma$, which is a subtorus H' of codimension two. In particular, the pullback of g is equal to 1 on H'. Let H'' be the subtorus of H whose Lie algebra is spanned by the coroots $\varpi^\vee$, $\beta^\vee$, and $\gamma^\vee$. The induced homomorphism $H' \times H'' \to H$ is a W'-equivariant isogeny, so that

$$\mathsf{X}(H)^{W'} \subseteq \mathsf{X}(H')^{W'} \oplus \mathsf{X}(H'')^{W'}.$$

Since g has trivial restriction to H', the image of g lies in $\mathsf{X}(H'')^{W'}$. Here $\mathsf{X}(H'') \cong \mathbb{Z}^2$ is the weight lattice of a root system of type A_2, and W' contains a subgroup of order 3 acting on $\mathbb{Z}^2$ in the standard way. It follows that $\mathsf{X}(H'')^{W'} = 0$, so that $H^0(\mathcal{M}; \mathcal{I}) = \{1\}$ as claimed.

This completes the proof of Theorem 2.2.3. □

References

[1] J. F. Adams, *Lectures on Exceptional Lie Groups*, University of Chicago Press, Chicago, 1996.

[2] N. Bourbaki, *Groupes et Algèbres de Lie*, Chap. 4, 5, et 6, Masson, Paris, 1981.

[3] P. Deligne, *Cohomologie Etale: Séminaire de Géométrie Algébrique du Bois-Marie SGA $4\frac{1}{2}$*, Lecture Notes in Mathematics **569**, Springer, Berlin/Heidelberg/New York, 1977.

[4] R. Donagi, *Spectral covers*, in *Current Topics in Complex Algebraic Geometry* (Berkeley 1992/93), MSRI Publications **28**, Cambridge University Press, Cambridge , 1995.

[5] R. Donagi and D. Gaitsgory, *The gerbe of Higgs bundles*, Transform. Groups **7** (2002), 109–153.

[6] E. Freitag and R. Kiehl, *Etale Cohomology and the Weil Conjecture*, Ergebnisse der Mathematik und ihrer Grenzgebiete 3. Folge **13**, Springer, Berlin Heidelberg, 1988.

[7] R. Friedman and J.W. Morgan, *Holomorphic principal bundles over elliptic curves*, math.AG/9811130.

[8] R. Friedman and J.W. Morgan, *Holomorphic principal bundles over elliptic curves II: The parabolic construction*, J. Differential Geometry **56** (2001), 301–379, math.AG/0006174.

[9] R. Friedman and J.W. Morgan, *Holomorphic principal bundles over elliptic curves III: Singular curves and fibrations*, math.AG/0108104.

[10] R. Friedman and J.W. Morgan, *Minuscule representations, invariant polynomials, and spectral covers*, math.AG/0011082.

[11] R. Friedman, J.W. Morgan, and E. Witten, *Vector bundles over elliptic fibrations*, J. Algebraic Geometry **8** (1999), 279–401.

[12] V. Kanev, *Spectral covers and Prym-Tjurin varieties I*, in *Abelian Varieties*, Proceedings of the International Conference (Egloffstein 1993), W. Barth, K. Hulek, and H. Lange eds., de Gruyter, Berlin-New York, 1995.

[13] F. Knop, *Automorphisms, root systems, and compactifications of homogeneous varieties*, Jour. Amer. Math. Soc. **9** (1996), 153–174.

[14] B. Kostant, *Lie group representations on polynomial rings*, Am. J. Math. **85** (1963), 327–404.

[15] R. Steinberg, *Regular elements of semisimple algebraic groups*, Publ. Math. Inst. Hautes Études Sci. **25** (1965), 49–80.

Department of Mathematics, Columbia University, New York, NY 10027

E-mail address: rf@math.columbia.edu, jm@math.columbia.edu

Titles in This Series

326 **Kenneth D. T-R McLaughlin and Xin Zhou, Editors,** Recent developments in integrable systems and Riemann-Hilbert problems, 2003

325 **Seok-Jin Kang and Kyu-Hwan Lee, Editors,** Combinatorial and geometric representation theory, 2003

324 **Jean-Paul Brasselet, Gary Kennedy, Kristin Lauter, Lee McEwan, and Caroline Grant Melles, Editors,** Topics in algebraic and noncommutative geometry, 2003

323 **Vadim Olshevsky, Editor,** Fast algorithms for structured matrices; theory and applications, 2003

322 **S. Dale Cutkosky, Dan Edidin, Zhenbo Qin, and Qi Zhang, Editors,** Vector bundles and representation theory, 2003

321 **Anna Kamińska, Editor,** Trends in Banach spaces and operator theory, 2003

320 **William Beckner, Alexander Nagel, Andreas Seeger, and Hart F. Smith, Editors,** Harmonic analysis at Mount Holyoke, 2003

319 **W. H. Schikhof, C. Perez-Garcia, and A. Escassut, Editors,** Ultrametric functional analysis, 2003

318 **David E. Radford, Fernando J. O. Souza, and David N. Yetter, Editors,** Diagrammatic morphisms and applications, 2003

317 **Hui-Hsiung Kuo and Ambar N. Sengupta, Editors,** Finite and infinite dimensional analysis in honor of Leonard Gross, 2003

316 **O. Cornea, G. Lupton, J. Oprea, and D. Tanré, Editors,** Lusternik-Schnirelmann category and related topics, 2002

315 **Theodore Voronov, Editor,** Quantization, Poisson brackets and beyond, 2002

314 **A. J. Berrick, Man Chun Leung, and Xingwang Xu, Editors,** Topology and Geometry: Commemorating SISTAG, 2002

313 **M. Zuhair Nashed and Otmar Scherzer, Editors,** Inverse problems, image analysis, and medical imaging, 2002

312 **Aaron Bertram, James A. Carlson, and Holger Kley, Editors,** Symposium in honor of C. H. Clemens, 2002

311 **Clifford J. Earle, William J. Harvey, and Sevín Recillas-Pishmish, Editors,** Complex manifolds and hyperbolic geometry, 2002

310 **Alejandro Adem, Jack Morava, and Yongbin Ruan, Editors,** Orbifolds in mathematics and physics, 2002

309 **Martin Guest, Reiko Miyaoka, and Yoshihiro Ohnita, Editors,** Integrable systems, topology, and physics, 2002

308 **Martin Guest, Reiko Miyaoka, and Yoshihiro Ohnita, Editors,** Differentiable geometry and integrable systems, 2002

307 **Ricardo Weder, Pavel Exner, and Benoit Grébert, Editors,** Mathematical results in quantum mechanics, 2002

306 **Xiaobing Feng and Tim P. Schulze, Editors,** Recent advances in numerical methods for partial differential equations and applications, 2002

305 **Samuel J. Lomonaco, Jr. and Howard E. Brandt, Editors,** Quantum computation and information, 2002

304 **Jorge Alberto Calvo, Kenneth C. Millett, and Eric J. Rawdon, Editors,** Physical knots: Knotting, linking, and folding geometric objects in $\mathbb{R}^3$, 2002

303 **William Cherry and Chung-Chun Yang, Editors,** Value distribution theory and complex dynamics, 2002

302 **Yi Zhang, Editor,** Logic and algebra, 2002

301 **Jerry Bona, Roy Choudhury, and David Kaup, Editors,** The legacy of the inverse

300 **Sergei Vostokov and Yuri Zarhin, Editors,** Algebraic number theory and algebraic geometry: Papers dedicated to A. N. Parshin on the occasion of his sixtieth birthday, 2002

299 **George Kamberov, Peter Norman, Franz Pedit, and Ulrich Pinkall,** Quaternions, spinors, and surfaces, 2002

298 **Robert Gilman, Alexei G. Myasnikov, and Vladimir Shpilrain, Editors,** Computational and statistical group theory, 2002

297 **Stephen Berman, Paul Fendley, Yi-Zhi Huang, Kailash Misra, and Brian Parshall, Editors,** Recent developments in infinite-dimensional Lie algebras and conformal field theory, 2002

296 **Sean Cleary, Robert Gilman, Alexei G. Myasnikov, and Vladimir Shpilrain, Editors,** Combinatorial and geometric group theory, 2002

295 **Zhangxin Chen and Richard E. Ewing, Editors,** Fluid flow and transport in porous media: Mathematical and numerical treatment, 2002

294 **Robert Coquereaux, Ariel García, and Roberto Trinchero, Editors,** Quantum symmetries in theoretical physics and mathematics, 2002

293 **Donald M. Davis, Jack Morava, Goro Nishida, W. Stephen Wilson, and Nobuaki Yagita, Editors,** Recent progress in homotopy theory, 2002

292 **A. Chenciner, R. Cushman, C. Robinson, and Z. Xia, Editors,** Celestial Mechanics, 2002

291 **Bruce C. Berndt and Ken Ono, Editors,** q-series with applications to combinatorics, number theory, and physics, 2001

290 **Michel L. Lapidus and Machiel van Frankenhuysen, Editors,** Dynamical, spectral, and arithmetic zeta functions, 2001

289 **Salvador Pérez-Esteva and Carlos Villegas-Blas, Editors,** Second summer school in analysis and mathematical physics: Topics in analysis: Harmonic, complex, nonlinear and quantization, 2001

288 **Marisa Fernández and Joseph A. Wolf, Editors,** Global differential geometry: The mathematical legacy of Alfred Gray, 2001

287 **Marlos A. G. Viana and Donald St. P. Richards, Editors,** Algebraic methods in statistics and probability, 2001

286 **Edward L. Green, Serkan Hoşten, Reinhard C. Laubenbacher, and Victoria Ann Powers, Editors,** Symbolic computation: Solving equations in algebra, geometry, and engineering, 2001

285 **Joshua A. Leslie and Thierry P. Robart, Editors,** The geometrical study of differential equations, 2001

284 **Gaston M. N'Guérékata and Asamoah Nkwanta, Editors,** Council for African American researchers in the mathematical sciences: Volume IV, 2001

283 **Paul A. Milewski, Leslie M. Smith, Fabian Waleffe, and Esteban G. Tabak, Editors,** Advances in wave interaction and turbulence, 2001

282 **Arlan Ramsay and Jean Renault, Editors,** Groupoids in analysis, geometry, and physics, 2001

281 **Vadim Olshevsky, Editor,** Structured matrices in mathematics, computer science, and engineering II, 2001

280 **Vadim Olshevsky, Editor,** Structured matrices in mathematics, computer science, and engineering I, 2001

279 **Alejandro Adem, Gunnar Carlsson, and Ralph Cohen, Editors,** Topology, geometry, and algebra: Interactions and new directions, 2001

278 **Eric Todd Quinto, Leon Ehrenpreis, Adel Faridani, Fulton Gonzalez, and Eric Grinberg, Editors,** Radon transforms and tomography, 2001

TITLES IN THIS SERIES

277 **Luca Capogna and Loredana Lanzani, Editors,** Harmonic analysis and boundary value problems, 2001

276 **Emma Previato, Editor,** Advances in algebraic geometry motivated by physics, 2001

275 **Alfred G. Noël, Earl Barnes, and Sonya A. F. Stephens, Editors,** Council for African American researchers in the mathematical sciences: Volume III, 2001

274 **Ken-ichi Maruyama and John W. Rutter, Editors,** Groups of homotopy self-equivalences and related topics, 2001

273 **A. V. Kelarev, R. Göbel, K. M. Rangaswamy, P. Schultz, and C. Vinsonhaler, Editors,** Abelian groups, rings and modules, 2001

272 **Eva Bayer-Fluckiger, David Lewis, and Andrew Ranicki, Editors,** Quadratic forms and their applications, 2000

271 **J. P. C. Greenlees, Robert R. Bruner, and Nicholas Kuhn, Editors,** Homotopy methods in algebraic topology, 2001

270 **Jan Denef, Leonard Lipschitz, Thanases Pheidas, and Jan Van Geel, Editors,** Hilbert's tenth problem: Relations with arithmetic and algebraic geometry, 2000

269 **Mikhail Lyubich, John W. Milnor, and Yair N. Minsky, Editors,** Laminations and foliations in dynamics, geometry and topology, 2001

268 **Robert Gulliver, Walter Littman, and Roberto Triggiani, Editors,** Differential geometric methods in the control of partial differential equations, 2000

267 **Nicolás Andruskiewitsch, Walter Ricardo Ferrer Santos, and Hans-Jürgen Schneider, Editors,** New trends in Hopf algebra theory, 2000

266 **Caroline Grant Melles and Ruth I. Michler, Editors,** Singularities in algebraic and analytic geometry, 2000

265 **Dominique Arlettaz and Kathryn Hess, Editors,** Une dégustation topologique: Homotopy theory in the Swiss Alps, 2000

264 **Kai Yuen Chan, Alexander A. Mikhalev, Man-Keung Siu, Jie-Tai Yu, and Efim I. Zelmanov, Editors,** Combinatorial and computational algebra, 2000

263 **Yan Guo, Editor,** Nonlinear wave equations, 2000

262 **Paul Igodt, Herbert Abels, Yves Félix, and Fritz Grunewald, Editors,** Crystallographic groups and their generalizations, 2000

261 **Gregory Budzban, Philip Feinsilver, and Arun Mukherjea, Editors,** Probability on algebraic structures, 2000

260 **Salvador Pérez-Esteva and Carlos Villegas-Blas, Editors,** First summer school in analysis and mathematical physics: Quantization, the Segal-Bargmann transform and semiclassical analysis, 2000

259 **D. V. Huynh, S. K. Jain, and S. R. López-Permouth, Editors,** Algebra and its applications, 2000

258 **Karsten Grove, Ib Henning Madsen, and Erik Kjær Pedersen, Editors,** Geometry and topology: Aarhus, 2000

257 **Peter A. Cholak, Steffen Lempp, Manuel Lerman, and Richard A. Shore, Editors,** Computability theory and its applications: Current trends and open problems, 2000

256 **Irwin Kra and Bernard Maskit, Editors,** In the tradition of Ahlfors and Bers: Proceedings of the first Ahlfors-Bers colloquium, 2000

255 **Jerry Bona, Katarzyna Saxton, and Ralph Saxton, Editors,** Nonlinear PDE's, dynamics and continuum physics, 2000

For a complete list of titles in this series, visit the